Practical Preimplantation Genetic Diagnosis

Yury Verlinsky and Anver Kuliev

Practical Preimplantation Genetic Diagnosis

With 125 Figures

Springer

Yury Verlinsky, PhD
Anver Kuliev, MD, PhD
Reproductive Genetics Institute, Chicago, IL, USA

British Library Cataloguing in Publication Data
A catalogue record for this book is available from the British Library

Library of Congress Control Number: 2005923550

ISBN-10: 1-85233-920-9 Printed on acid-free paper
ISBN-13: 978-1-85233-920-3

Printed in Singapore (TB/KYO)

9 8 7 6 5 4 3 2 1

Springer Science+Business Media
Springeronline.com

Contents

Foreword

The casual reader may not fully appreciate the significance of the well-chosen title for this unique and welcome volume, *Practical Preimplantation Genetic Diagnosis*. The title is not only apropos but provided by the team that truly made it possible.

Prenatal genetic diagnosis began in the 1950s, with detection of X-chromatin in amniotic fluid cells. By the late 1960s amniocentesis allowed diagnosis of chromosomal abnormalities and selected inborn errors of metabolism. Labs sprung up worldwide, and in the 1970s antenatal testing became standard in certain patients. The 1980s brought two key advances. First was noninvasive screening, with an invasive procedure like amniocentesis performed only if the risk was high enough to be justified. Concurrently, there was movement toward prenatal diagnosis earlier in pregnancy, as witnessed by chorionic villus sampling (CVS). Still, diagnosis was not possible before 10 to 12 weeks, and, hence, clinical pregnancy termination was still necessary.

Although proof of the principle for preimplantation genetic diagnosis (PGD) had been shown in 1967 by Professor Robert Edwards through X-chromatin analysis in rabbit blastocysts, application in the human awaited advances in human assisted reproductive technologies (ARTs) and in molecular techniques. The first human PGD finally came only in 1990. During the next few years, there was great interest in PGD in the genetic community; however, relatively few cases were performed, mostly for mendelian disorders. With introduction of fluorescent in situ hybridization (FISH), however, PGD was burgeoning. By 2000, novel indications were added, including aneuploidy testing to identify chromosomally normal embryos. Indeed, this is now the most common indication for PGD.

Practical Preimplantation Genetic Diagnosis systematically covers indications and technology. One is taken through PGD from start (embryo) biopsy to end (consequences), and we learn from the group that has performed half the world's cases. The technique for obtaining embryonic DNA is first described (polar body biopsy or blastomere biopsy), followed by single-cell genetic analysis (FISH or polymerase chain reaction–based approaches). The spectrum of mendelian disorders and their detection follows. Once controversial topics like late-onset disorders are covered, with especially lucid coverage on PGD to obtain human leukocyte antigen (HLA)-compatible embryos. Preimplantation genetic diagnosis not only avoids another affected offspring, but also provides umbilical cord–derived stem cells for transplantation into an older moribund sib. Both numerical and structural chromosomal abnormalities are discussed. We are given novel information on the unexpectedly high *sequential* error rate in meiosis I followed by meiosis II. Clinical outcome and the safety of PGD are discussed (and reassurance is given). Throughout, the discussion flows seamlessly between indication and counseling (reflecting the impeccable genetic credentials of the authors) and laboratory performance (likewise). Illustrations are highly informative and lucid. This is invaluable in understanding PGD for mendelian disorders, the diagnosis of which requires linkage analysis. The illustrations make a potentially arcane topic quite clear.

Overall, one comes away with a keen appreciation of how PGD has moved from "boutique" medicine, practiced only rarely by the cognoscenti, to a technique rapidly becoming de rigueur in all in vitro fertilization (IVF) centers. Within 5 to 10 years, surely most, if not all, ART will require genetic testing prior to embryo transfer.

Considerable information is new and not heretofore available even in peer review format. The result is a pleasing combination of reading a textbook peppered with peer review information. The sense of movement in PGD is thus palpable.

Practical Preimplantation Genetic Diagnosis is thus a gem. In this age of edited works, a text written by only one or two authors is rare. That our authors practically invented the field, and certainly have made it practical, makes this volume truly extraordinary. This volume should thus be at the reach of every reproductive geneticist, ART lab, and reproductive biologist. The avid student will come away fully informed and up to date, and also ready for tomorrow's advances.

Joe Leigh Simpson, M.D.
Ernst W. Bertner Chairman and Professor
Department of Obstetrics and Gynecology
Houston, Texas
June 2005

Preface

Although treatment is the major goal in the control of genetic disease, as in other fields of medicine, this is not yet a reality for most inherited conditions. Even with dramatic advancement in the field of gene therapy, there are not, unfortunately, many success stories to allow predicting its sound impact in the near future. Therefore, in the absence of radical treatment, prevention of genetic disorders is still one of the available options for genetically disadvantaged people. Among the measures for prevention, the avoidance of the birth of an affected child to couples at genetic risk has become a quite acceptable option in many populations. This is based on population screening and prenatal diagnosis, and has been quite successful, resulting in almost eradicating new cases of some genetic diseases from several Mediterranean populations, such as those in Sardinia and Cyprus. However, this has generated an increasing number of abortions following prenatal diagnosis, leading to a growing concern and negative reaction in society to the preventive genetics programs. For example, some ethnic groups cannot accept any control measures regarding congenital diseases because pregnancy termination is not allowed due to social or religion reasons. Even in those communities where abortion is allowed, some couples have to experience two or more terminations of pregnancies before they can have a normal child. No doubt these families might require an alternative option to achieve their desired family size without the need for termination of pregnancy.

The present book is devoted to a principally new approach in prevention of genetic disorders, which avoids the need for prenatal diagnosis and termination of pregnancy. Accordingly, a novel concept of prepregnancy diagnosis is introduced, which is based on the control of the processes of the oocyte maturation, fertilization, and implantation, so to select and transfer back to patients only normal embryos, and achieve an unaffected pregnancy resulting in the birth of a healthy child. In this way, the couples at high risk of having an offspring with genetic disease have an option to control the outcome of their pregnancy from the onset. So the place of this approach in the context of other available approaches for prevention of genetic disorders is discussed, together with other primary preventive measures, which may allow presently avoiding up to a half of congenital malformations presented at birth (Chapter 1). Although the option of prepregnancy diagnosis involves ovarian stimulation and in vitro fertilization (IVF), the description of the available experience demonstrates that this has appeared to be an acceptable procedure in many ethnic groups all over the world.

As described in the methodological section (Chapter 2), preimplantation genetic diagnosis (PGD) may be achieved by testing of the female, and perhaps in the future also male, gametes or single cells from the preimplantation embryo. Both of these methods have been used in clinical practice and have shown to be useful in prediction of the genotype of the resulting pregnancy, although many problems still remain to be resolved, requiring careful consideration. The experience in the application of both of these approaches and the progress in overcoming the existing problems are described, based on the original practice of more than 3500 PGD cycles, presently comparable to the overall available world experience, as well as the experiences from the other active PGD centres. Because basic clinical methods for performing PGD are similar to those in IVF, which have been described previously in detail, the emphasis in this book is mainly on the laboratory aspects. However, the clinical impact of PGD, and some of the ethical, social, and legal aspects are also briefly explored (Chapters 6 and 8). The present progress of PGD makes clear that the at-risk couples have to be aware of all available preventive options, including PGD, which is a complement rather than an alternative to the available preventive genetics services, allowing the couples to make their own choices.

Because PGD of single-gene disorders involves extensive preparatory work for each particular condition, a special section is devoted to the description of the PGD strategy for each group of these disorders. This includes the description of primer sequences, PCR conditions, polymorphic markers, the experience of the application of conventional and fluorescence PCR, allele drop out (ADO) rates, and other expected problems leading to misdiagnosis, with the detailed presentation of relevant case reports (Chapter 3). A special section is also devoted to preimplantation nongenetic testing, involving HLA typing, which has recently become one of the most attractive PGD applications allowing improved access to HLA-compatible stem cell transplantation (Chapter 4).

The application of PGD is of particular interest for assisted reproduction practices, so this is described in detail in a special section, providing strong evidence of PGD impact on pregnancy outcome (Chapters 5 and 6). This is not surprising, as the introduction of methods for preselection of cytogenetically normal and avoiding the transfer of cytogenetically abnormal embryos should considerably improve the efficiency of IVF, based on the fact that aneuploidies are major contributors to spontaneous abortions and may also explain some of the implantation failures. Although more data are still needed to confirm positive impact on birth outcome, an extensive original experience of FISH analysis of over 15,000 oocytes and embryos is presented (Chapter 5), together with the analysis of previous reproductive histories of patients, demonstrating strong evidence of clinical impact (Chapter 6). These chapters also include the follow-up studies of detected cases with chromosomal abnormalities in the different stages of preimplantation development to explore the problems of mosaicism and its biological significance at the cleavage stage.

It is presently being further confirmed that PGD is the most attractive option for couples carrying balanced translocations (Chapter 5). In light of these data, a pioneering work on conversion of interphase nuclei of single biopsied blastomeres into metaphase chromosomes, based on the use of the modern nuclear transfer techniques, is described with presentation of the original experience of the application of these methods for PGD of translocations.

For the first time, the original experience of the latest applications of PGD to embryonic stem cells is described, providing the possibility of obtaining preimplantation embryos with known genotypes as a source for the establishment of custom-made embryonic stem cells (Chapter 7).

Finally, the current status of PGD is summarized, with prospects of the application of PGD to medical practice. This information may help not only in planning and organization of such services, but also in providing a working manual for the establishment and performance of PGD in the framework of IVF and genetic practices.

Acknowledgements

We are indebted to our colleagues in DNA, FISH, and embryology laboratories, O. Verlinsky, T. Sharapova, S. Ozen, K. Lazyuk, G. Wolf, Y. Illkevitch, N. Strelchenko, V. Galat, V. Kuznyetsov, Z. Zlatoposlky, M. Chmura, V. Koukharenko, and I. Kirillova, and also to our genetic counsellors, C. Lavine, R. Beck, R. Genoveze, and D. Pauling for their participation in acquisition of the data and technical assistance.

Contributors

Svetlana Rechitsky, PhD, Single-Gene Disorders and HLA Typing

Jeanine Cieslak, MLT, Micromanipulation and Chromosomal Disorders

1

Primary Prevention of Genetic Disorders and Place of Preimplantation Genetic Diagnosis

Despite recent progress in gene therapy it is still not a realistic option for genetic disorders, prevention remaining the main approach. Preventive measures may be applied at the community level by avoiding new mutations, protecting from all possible environmental hazards, predictive testing for genetic and complex disorders, and prospective screening for common genetic disorders specific for each ethnic group. The optimal time for offering these preventive measures is preconception or preimplantation stage, as any detection afterwards will involve the decision either to keep the pregnancy with the long-term social, familial, and financial consequences of a seriously affected child, or to terminate the planned and wanted pregnancy.

So the strategies for prevention range from preventing environmental hazards and providing vitamin supplementation programs to the prepregnancy or prenatal diagnosis. One example of the highly effective population-based preventive measures realized at the preconception stage has been prevention of neural tube defects and some other congenital abnormalities by folic acid or folic acid–containing multivitamins. On the other hand, preconception and preimplantation genetic diagnosis (PGD) has been established as a realistic option for primary prevention of genetic disorders, which is described in the detail in this book.

The list of most relevant approaches for prevention of congenital disorders presently includes (1) the avoidance of new mutations through environmental programs, (2) discouragement of pregnancy at advanced ages through community education and family planning, (3) periconception folic acid supplementation or multivitamin fortification of basic foodstuffs, (4) rubella vaccination, (5) avoidance of alcohol consumption and smoking during pregnancy, and (6) prenatal and (7) prepregnancy (preimplantation) diagnosis. The decision to adopt any of the available preventive programmes depends on the differences in health services development, ethnic distribution of congenital disease, and the local attitudes to genetic screening and termination of pregnancy. For example, induced abortions are still not permissible in many countries on eugenic grounds. On the other hand, an increasing number of countries are gradually permitting prenatal diagnosis and termination of pregnancies for medical indications even in some strict religious settings.

The impact of the community-based preventive approaches is obvious from the Down syndrome prevention program in the majority of industrialised countries of Europe and the United States. The programs are based on the introduction of prenatal maternal serum screening for all pregnant women to detect pregnancies at increased risk, followed by prenatal diagnosis and selective pregnancy termination, combined with prenatal diagnosis offered to all women of advanced maternal age, which have resulted in the reduction of the birth prevalence of Down syndrome at least by half. However, this led to an increasing number

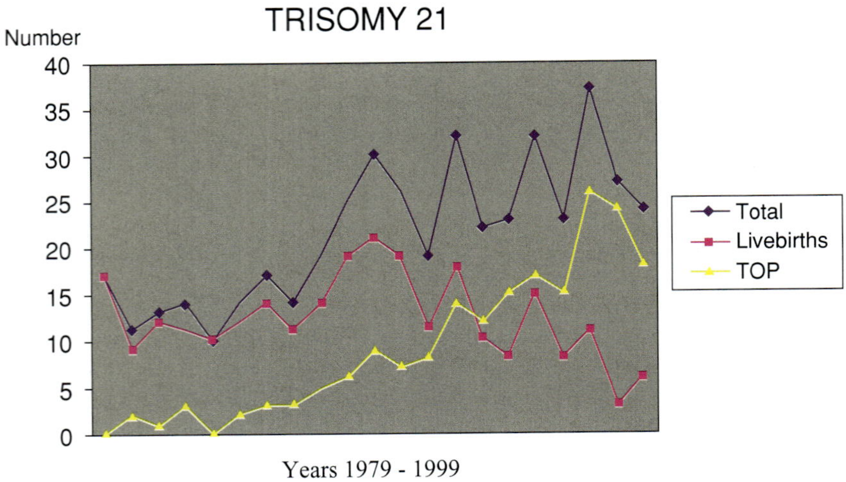

Figure 1.1. Prevention of Down syndrome, 1979–1999, in Strasbourg, France, by prenatal diagnosis and termination of pregnancy [1]. The data are representative for all industrialized countries where population screening and prenatal diagnosis are operational for more than two decades, resulting in tremendous reduction of Down syndrome prevalence (orange curve) compared with the expected prevalence in the absence of preventive measures (black curve), which is achieved by the increase of the number of pregnancy terminations (yellow curve).

of pregnancy terminations (Figure 1.1), creating an increasing concern even in those populations where abortions are allowed [1–2]. This is particularly relevant for those countries where the proportion of mothers over age 35 is above 10%. One of the approaches to overcome this problem is the application and improvement of family planning in these countries, where the reduction of Down syndrome prevalence by one third may be achieved solely by initiating community education to discourage childbearing in women over age 35.

The other current approach for avoiding congenital disorders at the community level involves the expanded use of fetal ultrasound, which has improved the detection rate of affected pregnancies, the efficiency of which may be evaluated by community-based birth defect monitoring systems available in some countries. As termination is usually requested only for the most severe disorders, fetal anomaly scanning selectively reduces the proportion of children born with lethal or incurable conditions, including neural tube defects. According to EUROCAT, approximately 50–80% of affected pregnancies are terminated, representing a growing concern (2). So the most attaractive strategy might be to concentrate on the primary preventive measures, such as the prepregnancy vitamin supplementation, which has been shown to be one of the most efficient approaches for primary prevention of congenital disorders [3–12]. In some populations, the application of this approach resulted in the overall reduction of the prevalence of congenital disorders by as much as by half (from 40.6 to 20.6 per 1000). This included the reduction of the prevalence of neural tube defects and some other congenital abnormalities.

The fact that the diet supplementation with folic acid or folic acid–containing multivitamins may substantially reduce the population prevalence of four groups of congenital disorders, i.e., neural tube defects, cardiovascular, urinary tract, and limb deficiencies, represents an important breakthrough in prevention of congenital disorders. Although more data are needed to further confirm this and investigate the possibility of reduction of other birth defects, such as pyloric stenosis, the impact of folic acid on the prevalence of congenital disorders is in agreement with the fact that (1) mothers who give birth to a child with neural tube defects have mildly elevated blood and amniotic fluid levels of homocysteine, (2) hyperhomocysteinaemia and/or lack of methionine can induce neural tube defects in animal experiments, (3) low maternal folate status appears to be a risk factor for neural tube defects, (4) vitamins of B group including folate/folic acid are important in homocysteine metabolism, and (5) vitamins B_6, B_{11}, and B_{12} are able to reduce hyperhomocysteinaemia. It is known that homocysteine accumulates if conversion to methionine is slowed because of shortage of folate or vitamin B_{12}, or both, and a raised plasma homocysteine suggests suboptimal nucleic acid and amino acid metabolism. It also has direct harmful effects, e.g., it increases the risk of cardiovascular disease through thickening the lining of blood vessels, and may also increase the risk of certain cancers and dementia [13–16].

It is also known that reactions catalysed by tetrahydrofolate are crucial for cell growth and multiplication, making rapidly dividing cells particularly vulnerable to deficiency of either folate or vitamin B_{12}. This may affect the embryo morphogenetic movements, which depend on focal rapid cell multiplication, increasing the risk of congenital malformation. Folate deficiency may be caused also by genetic factors, leading to rare variants of several enzymes involved in one-carbon transfer, which cause problems ranging from greatly increased plasma homocysteine levels with very-early-onset cardiovascular disease, to developmental delay and neurological problems, with or without megaloblastic anaemia. Variants of lesser effect in the same enzymes may contribute to genetic predisposition to cardiovascular disease and neural tube defects. For example, about 10% of many populations are homozygous for a common polymorphism of the enzyme methyl tetrahydrofolate (MTHFR) (valine replaces alanine at codon 677). In homozygotes this reduces the enzyme activity by 50–70%, slows the regeneration of methionine leading to the raised plasma homocysteine, and increases risk of cardiovascular disease. So fetuses homozygous for the variant are at increased risk of neural tube defects. Variants of other enzymes, and other vitamins, may also influence homocysteine levels and the risk of neural tube defects. Accordingly, folic acid supplementation increases the supply of tetrahydrofolate, accelerates most folate-dependent metabolic reactions, and reduces plasma homocysteine levels [17–20].

The available data suggest that the cause of neural tube defects is not a primary lack of folate in the diet but an inborn error of vitamin B_{11} and/or homocysteine metabolism. An interaction between genetic predisposition and nutrition, therefore, may have a causal role in the development of neural tube defects, i.e., a dietary deficiency may trigger the genetic predisposition. The genetic–nutrient interaction through genetic predisposition and low folate status is associated with a greater risk for neural tube defects than either variable alone.

There are different options for ensuring appropriate multivitamin/vitamin B_{11} consumption in women of childbearing age, with each of them having its disadvantages and social feasibilities. The optimal daily intake of folate in the periconception period is 0.66 mg, whereas the usual intake per day is only about 0.18 mg. It seems also impossible to achieve a 3.7-fold increase in consumption through food intake alone, since this would require about 15 daily servings of broccoli or brussel sprouts. In addition, a large increase in the consumption of extra folate from natural foods is relatively ineffective at increasing folate levels. So the consumption of folate-rich foods may not be the most appropriate way to prevent the development of neural tube defects and other congenital abnormalities, requiring the appropriate strategy for folic acid supplementation.

Over 90% of pregnancies where the fetus has a neural tube defect occur among women without any previous indication of increased risk. Identifiable risk groups include women with a prior affected pregnancy, who have a 3–4% recurrence risk, and women who are heterozygous for the MTHFR mutation. However, these groups account for only a small proportion of affected pregnancies. Trials of the effect of folic acid supplementation on the prevalence of neural tube defects provide conclusive scientific evidence for its preventive effect [3–12], suggesting that (1) dietary supplementation with folic acid or with multivitamin preparations containing folic acid before and during early pregnancy (periconceptional supplementation) markedly reduces both the first occurrence of neural tube defects and recurrence among women with a previous affected pregnancy, who have an increased (3–4%) risk; (2) the effect is greatest in areas with a high baseline prevalence of neural tube defects, but also applies in lower prevalence areas; (3) no harmful effects have ever been observed, with levels of supplementation ranging from 360 μg to 5 mg of folic acid daily.

Because of the expected strong impact on prevalence of congenital malformations, the current dietary folate intake for adults in the United States is recommended at 400 μg, representing a daily intake of 200 μg folic acid equivalents, the recommended intake for pregnant women being 400 μg dietary folate plus 400 μg folic acid representing a daily intake of 600 μg folic acid equivalents [21].

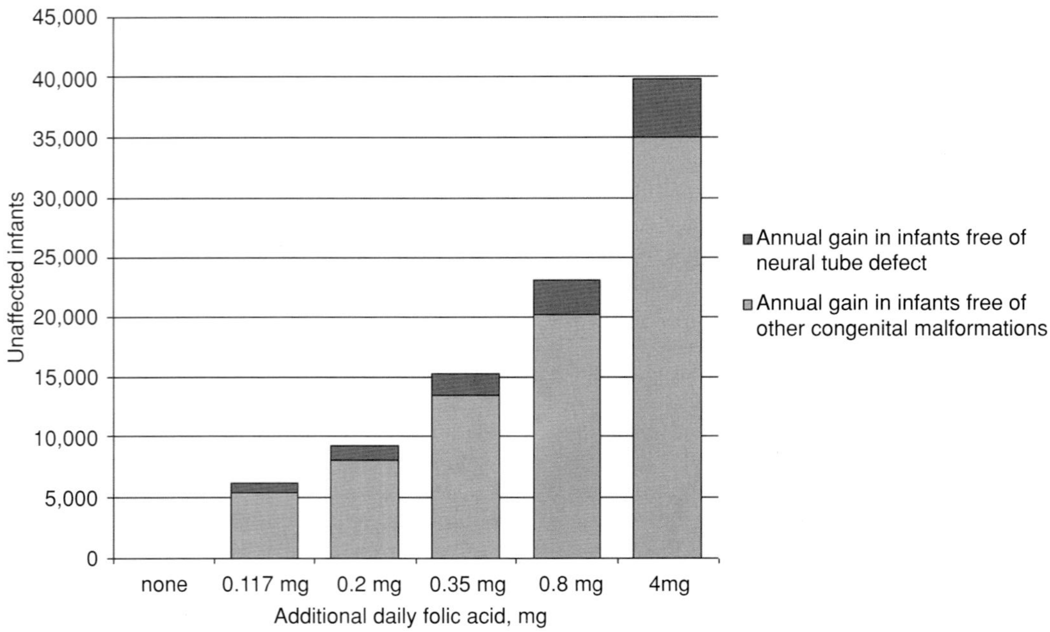

Figure 1.2. Projected effect of folic acid fortification on annual gain of infants free of congenital malformations in North America. The estimates were prepared by Bernadette Modell for WHO/EURO meetings on Prevention of Congenital Disorders, Copenhagen 2001 and Rome 2002 [25, 26], providing the expected reduction of congenital malformations overall depending on the the additional folic acid intake (lighter portion of the bars), and the proportion of avoided neural tube defects (darker portion of the bars).

The UK Committee on Medical Aspects of Food and Nutrition Policy recommendation of 240 μg of folic acid per 100 g flour [22] is approximately equivalent to an additional folic acid intake of 200 μg a day. To obtain adequate protection against risk of neural tube defects, the mean plasma folate should be about 10 ng/ml, while the mean plasma folate level in most populations is around 5 ng/ml [21]. There is considerable variation within and between populations, but few have a mean plasma folate in the recommended range, and very few individuals could meet the above recommendations without food fortification or the use of folic acid supplements. The groups of greatest concern are those with folate intakes and plasma folate at the lower end of the range, for whom only food fortification is capable of bringing them into the recommended range.

Folic acid fortification of all cereal grain products at a level of 140 μg/100 g flour is mandatory in the United States and Canada, where the birth prevalence of neural tube defects has since fallen by about 20%, with no adverse effects reported [16, 23]. In Hungary, folic acid, vitamin B_{12}, and vitamin B_6 are being added to bread, with average daily intake of folic acid, vitamin B_{12}, and B_6 from this source approximately 200, 1, and 1080 μg, respectively [24]. The prevalence of neural tube defects has fallen by 41%. Based on this study and other relevant data on the potential reduction of congenital disorders through primary preventive measures, the expected overall reduction of the prevalence of congenital disorders in North America may be projected to be approximately 40,000 births of children with major congenital disorders [25] (Figure 1.2).

It is of importance to mention that folic acid fortification costs only about $1 per metric ton of flour. It is so minor that the extra cost is not sufficient even to change the price of a loaf of bread. However, the multivitamin food fortification does not substitute completely periconception supplementation, so prospective pregnant women should be informed about the need for folic acid and multivitamin supplementation in addition to the intake of the multivitamin fortified food. Taking into consideration the estimated lifetime cost for a single patient with spina bifida, which is approximately $250,000, a complementary periconception supplementation would be highly cost effective, although the major benefit is the birth of an unaffected child. This approach may be best realized by advising oral contraceptive users to start taking folic acid–containing multivitamins as soon as they stop using contraception, using a multivitamin supplement containing a physiological dose (0.4–0.8 mg) of folic acid, which contributes to preventing in addition to more efficient reduction of neural tube defects (about 90% versus 70%), prevention of some other congenital disorders of public health importance.

However, prevention of congenital malformations is not actually the major objective of the above programs, as replacing an affected live birth by abortion, as shown in Figure 1.1, does not solve the problem, as the abortion of a wanted pregnancy is also an unfavourable pregnancy outcome. This makes primary prevention by food fortification, preconception vitamin supplementation, or PGD much more attractive interventions not associated with distress, because, on the contrary, they increase the number of healthy wanted babies born and have a highly positive effect on the quality of life of parents and children. In evaluating the effects of folic acid fortification, the appropriate evaluation of the program is therefore gain in number of unaffected pregnancies (see Figure 1.2), rather than reduction in the number of affected live births [26]. The same consideration applies to PGD, the main objective of which is to assist couples to have an unaffected child of their own.

Thus, one of the most recent possibilities to avoid the genetic disease before pregnancy is preconception and preimplantation genetic diagnosis, which has recently become available worldwide. Instead of prenatal screening and termination of affected pregnancies, which is not tolerated in many communities and ethnic groups, the prepregnancy diagnosis provides an option for couples at risk to plan unaffected pregnancies from the onset (Figure 1.3). This is the reason why the preconception and prepregnancy diagnosis has already become an integral part of preventive services for congenital disorders, providing a choice for those couples who are unable to accept prenatal screening and termination of pregnancy. Together with other approaches for primary prevention, PGD may soon represent an important component of preconception clinics, which may soon shift services from secondary preventive measures based on prospective carrier screening and prenatal diagnosis to preconception prevention and PGD, to ensure only unaffected pregnancy from the onset.

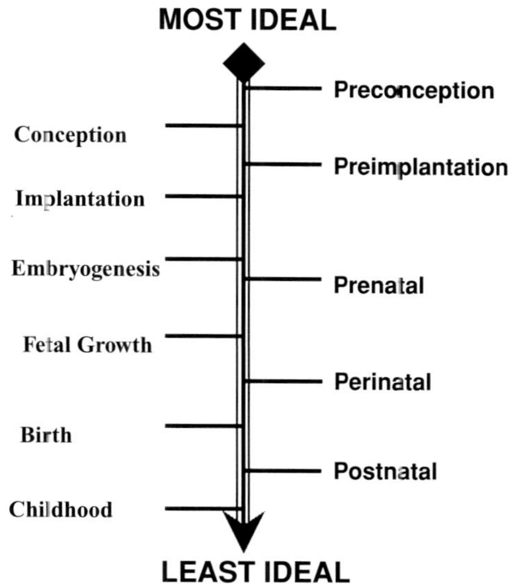

Figure 1.3. Options for diagnosis and prevention offered for couples at genetic risk. Stages of development are shown on the left, and the points for application of preventive measures on the right.

References

1. Stoll C, Alembik Y, Dott B, Roth MP. Impact of prenatal diagnosis on livebirth prevalence of children with congenital anomalies. Annales de Genetique 2002;45:115–121.
2. EUROCAT Report No. 8. Surveillance of congenital anomalies in Europe 1980–1999. Edited by EUROCAT Working Group. EUROCAT Central Registry, Room 1F08, University of Ulster Nowtownabbey, County Antrim, Northern Ireland BT37 0QB. Email eurocat@ulster.ac.uk.
3. MRC Vitamin Study Research Group. Prevention of neural tube defects: Results of the Medical Research Council Vitamin Study. Lancet 1991;338:131–137.
4. Centers for Disease Control and Prevention. Use of folic acid for prevention of spina bifida and other neural tube defects 1983–1991. MMWR 1991;40:513–516.
5. Czeizel AE, Dudás I. Prevention of the first occurrence of neural tube defects by periconceptional vitamin supplementation. N Engl J Med 1992;327:1832–1835.
6. US Department of Health and Human Services, Food and Drug Administration. Food standards: Amendment of standards of identity for enriched grain products to require addition of folic acid. Federal Register 1996;61:8781–8807.
7. Czeizel AE. Primary prevention of neural-tube defects and some other major congenital abnormalities. Pediatr Drugs 2000;2:437–449.
8. Wald NJ, Noble J. Primary prevention of neural tube defects. In: Rodeck CH, Whittle MJ (eds.). Fetal

Medicine: Basic Science and Clinical Practice. Churchill Livingstone, London, 1999, pp. 283–290.
9. Berry CJ, Li Z, Erickson JD. Prevention of neural-tube defects with folic acid in China. N Engl J Med 1999;341: 1485–1490.
10. Selhub J, Jacques PF, Rosenberg IH, et al. Serum total homocysteine concentrations in the Third National Health and Nutrition Examination Survey (1991–1994): Population reference ranges and contribution of vitamin status to high serum concentrations. Ann Intern Med 1999;131:331–339.
11. Ruddell LJ, Chisholm A, Williams S, Mann JI. Dietary strategies for lowering homocysteine concentrations. Am J Clin Nutr 2000;71:1448–1454.
12. Olney RS, Mulinare J. Trends in neural tube defect prevalence, folic acid fortification, and vitamin supplement use. Semin Perinatol 2002;26:277–285.
13. Wald DS, Law M, Morris JK. Homocysteine and cardiovascular disease: Evidence on causality from a meta-analysis. Br Med J 2002; 325:1202.
14. Schnyder G, Roffi M, Pin R, Flammer Y, Lange H, Eberli FR, Meier B, Turi ZG, Hess OM. Decreased rate of coronary restenosis after lowering of plasma homocysteine levels. N Engl J Med 2001;345:1593–1600.
15. La Vecchia C, Negri E, Pelucchi C, Franceschi S. Dietary folate and colorectal cancer. Int J Cancer 2002;102:545–547.
16. McIlroy SP, Dynan KB, Lawson JT, Patterson CC, Passmore AP. Moderately elevated plasma homocysteine, methylenetetrahydrofolate reductase genotype, and risk for stroke, vascular dementia, and Alzheimer disease in Northern Ireland. Stroke 2002;33:2351–2356.
17. Klerk M, Verhoef P, Clarke R, Blom HJ, Kok FJ, Schouten EG; MTHFR Studies Collaboration Group. MTHFR 677C→T polymorphism and risk of coronary heart disease: A meta-analysis. JAMA 2002;288:2023–2031.
18. Shields DC, Kirke PN, Mills JL, Ramsbottom D, Molloy AM, Burke H, Weir DG, Scott JM, Whitehead AS. The "thermolabile" variant of methylenetetrahydrofolate reductase and neural tube defects: An evaluation of genetic risk and the relative importance of the genotypes of the embryo and the mother. Am J Hum Genet 1999;64:1045–1055.
19. Brody LC, Conley M, Cox C, Kirke PN, McKeever MP, Mills JL, Molloy AM, O'Leary VB, Parle-McDermott A, Scott JM, Swanson DA. A polymorphism, R653Q, in the trifunctional enzyme methylenetetrahydrofolate dehydrogenase/methenyltetrahydrofolate cyclohydrolase/formyltetrahydrofolate synthetase is a maternal genetic risk factor for neural tube defects: Report of the Birth Defects Research Group. Am J Hum Genet 2002;71:1207–1215.
20. Moat SJ, Ashfield-Watt PA, Powers HJ, Newcombe RG, McDowell IF. Effect of riboflavin status on the homocysteine-lowering effect of folate in relation to the MTHFR (C677T) genotype. Clin Chem 2003;49:295–302.
21. Wald NJ. Folic acid and neural tube defects. In: Walter P, Hornig D, Moser U (eds.). Functions of vitamins beyond recommended dietary allowances. Bibl Nutr Diet Karger, Basel, 2001, No. 55, pp. 22–33.
22. Department of Health. Report on Health and Social Subjects. 50. Folic Acid and the Prevention of Disease: Report of Committee on Medical Aspects of Food and Nutrition Policy. The Stationery Office, London, 2000.
23. Moore LL, Bradlee ML, Singer MR, Rothman KJ, Milunsky A. Folate intake and the risk of neural tube defects: An estimation of dose-response. Epidemiology 2003;14: 200–205.
24. Raats M, Thorpe L, Hurren C, Elliott K. Changing preconceptions: The HEA folic acid campaign 1995–98. Health Education Authority 2, London, 1998.
25. Report of WHO/EURO Meeting on Development of EURO Strategy on Congenital Disorders. Minsk, November 29–30, 2001. Unpublished WHO/EURO Document (#51203630). WHO, Copenhagen, Denmark.
26. Report of the WHO/EURO Meeting on the Regional Policy for Prevention of Congenital Disorders. Folic acid: From research to public health practice. Rome, November 11–12, 2002. Instituto Superiore di Sanita, Rome, Italy.

2

Approaches to Preimplantation Genetic Diagnosis

Introduced in 1990 as an experimental procedure, preimplantation genetic diagnosis (PGD) is now becoming an established clinical option in reproductive medicine. The number of apparently healthy children born after PGD is already close to 2000, validating that there is no ostensible evidence of any incurred adverse effect. Approximately 7000 PGD cases have been performed in more than 50 centres around the world, allowing hundreds of at-risk couples not only to avoid producing offspring with genetic disorders, but more importantly, to have unaffected healthy babies of their own without facing the risk of pregnancy termination after traditional prenatal diagnosis. Without PGD it is likely that few of these children would have been born.

Applied first for preexisting Mendelian diseases [1, 2], such as cystic fibrosis (CF) and X-linked disorders, PGD initially did not seem to be practical. Only a few babies were born during the first three years of work, and several misdiagnoses were reported [3, 4]. After the introduction of fluorescent in situ hybridization (FISH) analysis in 1993–1994 for PGD of chromosomal disorders [5–10] (see Chapter 5), however, the number of PGD cycles began to double annually, yielding more than 100 unaffected children by the year 1996 (11, 12) (Figure 2.1).

Application of PGD increased further when the ability to detect translocations became possible in 1996, first using locus-specific FISH probes and then more widely available subtelomeric probes [13, 14] (see Section 5.2). Because many carriers of balanced translocations have a poor chance of having an unaffected pregnancy, PGD has a clear advantage over the traditional prenatal diagnosis in assisting these couples to establish an unaffected pregnancy and deliver a child free from unbalanced translocation [15, 16]. The experience of over 500 PGD cycles for translocations accumulated to date demonstrates at least a fivefold reduction of spontaneous abortions in these couples compared with their experience before PGD [17–19] (see Chapter 6).

The natural extension of PGD's ability to allow transfer of euploid embryos is its positive impact on the liveborn pregnancy outcome. This is especially applicable to poor prognosis IVF patients (prior IVF failures, maternal age over 37). Introduction of commercially available five-color probes in 1998–1999 led to the accumulated experience of more than 5000 clinical cycles worldwide for the aneuploidy testing [17, 19, 20–23] (see Section 5.1). This has resulted in birth of over 1000 children including a few with misdiagnosis, suggesting the continued need for the improvement of accuracy of FISH analysis. The overall pregnancy rate per transfer is 28.1%, much higher than in IVF patients of comparable age group (average age over 39 years). Widespread confirmation of these results would indicate that the current IVF practice of transferring embryos based solely on morphological criteria is inefficient and in need of revision, given that half of these embryos are chromosomally abnormal and would compromise outcome (see Chapter 6).

Further expansion of PGD occurred in 1999, when it was applied for late-onset diseases with

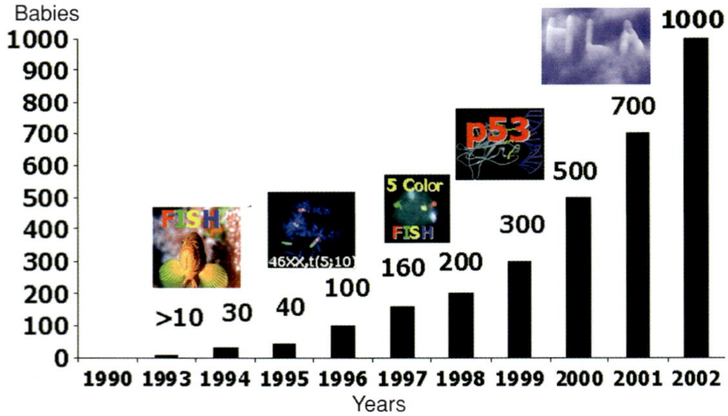

Figure 2.1. Babies born after PGD: 1990–2002. Bar graphs showing the number of children born each year after application of PGD. Application of each novel technique for PGD is shown above the bar graphs, leading to the expanding indications, such as chromosomal aneuploidies in 1993–1994 (shown as FISH), translocations in 1996 (shown by example of translocation t(5;10), five-color FISH in 1998, application to late-onset disease with genetic predisposition in 1999 (shown as an example of cancer predisposition caused by mutation in p53 gene mutations), and application to non-disease testing in 2000 (shown as an example of preimplantation HLA typing).

genetic predisposition [24] (see Section 3.6), a novel indication never previously considered for the traditional prenatal diagnosis. For patients with inherited pathological predisposition, this option provided the realistic reason for undertaking pregnancy with a reasonable chance of unaffected offspring. Prospective parents at such risk should be aware of this emerging option.

Another unique option that can be considered now, albeit involving ethical debate (see Chapter 8), concerns HLA typing as part of PGD, which has never been considered in traditional prenatal diagnosis [25] (see Chapter 4). In this application, PGD offers not only preventative technology to avoid affected offspring, but also a new method for treating (older) siblings with congenital or acquired bone marrow diseases. Preimplantation HLA typing was first applied to couples desiring an unaffected (younger) child free from the genetic disorder in the older sibling. In addition to diagnosis to assure a genetically normal embryo, HLA matched, unaffected embryos were replaced. At delivery, cord blood (otherwise to be discarded) was gathered for stem cell transplantation. This approach can also be used without testing of the causative gene, with the sole purpose of finding a matching HLA progeny for a source of stem cell transplantation for affected siblings with congenital or acquired bone marrow disease or cancer [26] (see Chapter 4).

The 15-year PGD experience clearly demonstrates considerable progress. As mentioned, approximately 7000 PGD attempts worldwide have resulted in the birth of close to 2000 apparently unaffected children, with overall congenital malformation rate not different from population prevalence [27, 28]. With the highly improved accuracy of genetic analysis and indications expanding well beyond those for prenatal diagnosis, more than 1000 PGD cycles are now performed annually. As seen in Figure 2.1, the experience during the years 2001–2002 shown has resulted in the birth of nearly the same number of children as during the entire preceding decade since the introduction of PGD. PGD offers special attractions not possible with traditional prenatal diagnosis. One is to avoid clinical pregnancy termination. This is especially attractive for couples carrying translocations, couples at risk for producing offspring with common diseases of autosomal dominant or recessive etiology, and, finally, for couples wishing to have not only an unaffected child, but an HLA-compatible cord blood donor for treatment of an older sibling with a congenital disorder. Yet the greatest numerical impact of PGD is in standard assisted reproduction practices (Chapter 6), where improved IVF efficiency through aneuploidy testing is surely evolving to become standard.

2.1 Obtaining Biopsy Material

At present biopsy material for performing PGD may be obtained from three major sources:

1. Matured and fertilized oocytes from which the first and second polar body (PB1 and PB2) are removed.
2. Eight-cell cleavage-stage embryos, from which one or two blastomeres are biopsied.
3. Blastocyst-stage embryos, from which more than a dozen cells are removed.

The biopsied material is tested for single-gene disorders using PCR analysis, or used for PGD for chromosomal abnormalities by fluorescent in situ hybridization (FISH) analysis (see below). Each of these PGD methods has advantages and disadvantages, and their choice depends on circumstances; however, in some cases, combination of two or three methods might be required. Despite a possible embryo cell number reduction, which might have a potential influence on the embryo viability, blastomere biopsy allows detecting paternally derived abnormalities. Removal of PB1 and PB2, on the other hand, should not have any effect on the embryo viability as they are naturally extruded from oocytes as a result of maturation and fertilization, but they provide no information on the paternally derived anomalies, even if this constitutes less than 10% of chromosomal errors in preimplantation embryos. Approaches for preconception testing of the paternally derived mutations are presently being developed and will be described in Section 2.1.2.

To perform PGD for chromosomal aneuploidies, both PB1 and PB2 are removed simultaneously, on the next day after insemination of the matured oocytes or introcytoplasmic sperm injection (ICSI), and analysed by FISH as described in Section 2.2.2. PB1 is the by-product of the first meiotic division and normally contains a double signal for each chromosome, each representing a single chromatid (see below). Accordingly, in case of meiosis I error, instead of a double signal, four different patterns might be observed, ranging from no or one signal to three or four signals, suggesting either chromosomal nondisjunction, evidenced by no or four signals, or chromatid missegregation, represented by one or three signals [29]. The genotype of the oocytes will, accordingly, be opposite to PB1 genotype, i.e., missing signals will suggest extra chromosome material in the corresponding oocyte, while an extra signal (or signals) will indicate monosomy or nullisomy status of the tested chromosome. In contrast to PB1, the normal FISH pattern of PB2 is represented by one signal for each chromosome (chromatid), so any deviation from this, such as no or two signals instead of one, will suggest a meiosis II error.

The method of blastomere biopsy was extensively used in PGD, despite its limitation due to high mosaicism rate in cleaving embryos (see Section 5.1). The FISH pattern of blastomeres is represented by two signals for each chromosome tested, so any deviation from this will suggest the chromosomal abnormality. The same pattern applies to blastocyst analysis, which has an advantage of analysing not one but a group of cells, obviating the problem of mosaicism at least to some extent.

Although more data have to be collected to exclude completely the short-term and long-term side effects, the data available show no evidence for any detrimental effect of the PB, blastomere, or blastocyst biopsy. Overall, these methods were used in thousands of clinical cycles, and resulted in birth of close to two thousand unaffected children, showing a comparable prevalence of congenital abnormalities to that in general population, which suggests no detrimental effect of any of the above biopsy procedures mentioned [23, 27, 28].

2.1.1 Polar Body Sampling

Introduced 14 years ago [2], PB biopsy has become one of the established approaches for PGD. The idea of performing PB PGD is based on the fact that polar bodies are the by-products of female meiosis, which allows predicting the resulting genotype of the maternal contribution to the embryos. Neither PB1, which is extruded as a result of the first meiotic division, nor PB2, extruded following the second meiotic division, have any known biological value for preimplantation and postimplantation development of the embryo. Initially, only PB1 was tested, based on the fact that in the absence of crossing-over, PB1 will be homozygous for the allele not contained in the oocyte and PB2 [30]. However, the PB1 approach was not applicable for predicting the eventual genotype of the oocytes if crossing-over occurred, because the primary oocyte in this case will be heterozygous for the abnormal gene. As the frequency of crossing-over varies with the distance between the locus and the centromere, approaching as much as 50% for telomeric genes, the PB1 approach appeared to be of limited value, unless the oocytes can be tested further on. So, the analysis of PB2 has been introduced to detect hemizygous normal oocytes resulting after the second meiotic division. As will be described below, this PGD technique presently involves a two-step oocyte analysis, which requires a sequential testing of PB1 and PB2 [31].

PB1 and PB2 are removed following stimulation and oocyte retrieval using a standard IVF protocol. Following extrusion of PB1, the zona pellucida (ZP) is opened mechanically using a microneedle, and PB1 aspirated into a blunt micropipette (see micromanipulation setup and procedure steps in Figures 2.2–2.4). The oocytes are then inseminated with motile sperm or using ICSI (Figure 2.4, Step 9), and examined for the presence of pronuclei and extrusion of PB2s, which are removed in the same manner as PB1 (Figure 2.5). To avoid additional invasive procedures, both PB1 and PB2 are removed simultaneously for FISH analysis (Figure 2.6), and are fixed and analysed on the same slide. The biopsied

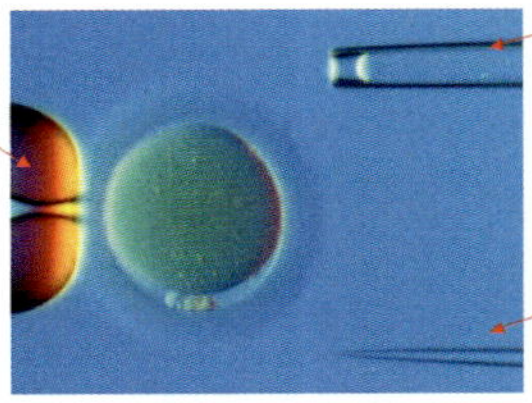

Figure 2.2. Required microtools for first polar body removal. Image of a metaphase II oocyte along with the three microtools that are required to perform first polar body (PB1) removal. A holding pipette is used (*left*) to hold the oocyte in position by gentle suction created by a hydraulic microsyringe system. On the right, a microneedle for partial zona dissection (*bottom*) and a micropipette (inner diameter 15 μm) for polar body removal are placed into a double tool holder. The micropipette is also attached to a hydraulic microsyringe (100 μl) system for fine control during the procedure. The same tools are required for embryo biopsy with the exception of the micropipette which has a larger inner diameter of 30 –35 μm.

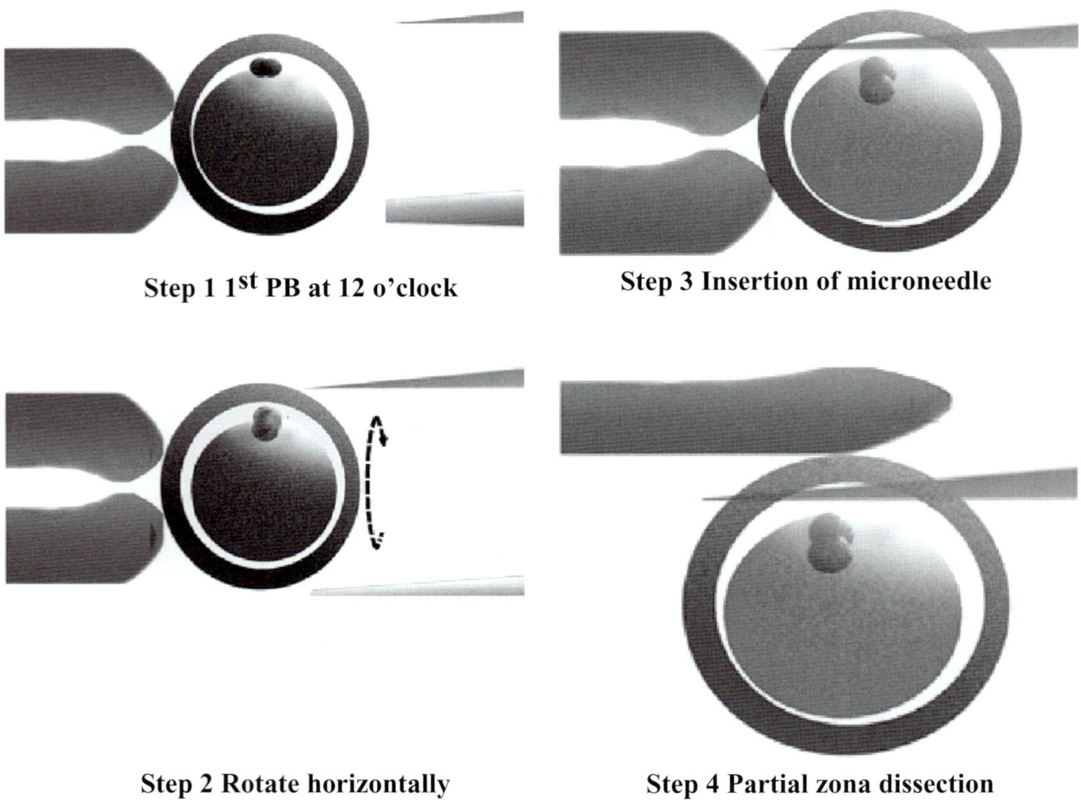

Figure 2.3. First polar body removal prior to ICSI. Step 1: The oocyte is secured by gentle suction from the holding pipette and the oocyte is rotated so that the PB1 is visualized at 12 o'clock. Step 2: The oocyte is rotated horizontally, slightly forward, so that it faces the operator. Step 3: The microneedle is passed through the zona pellucida at the 1–2 o'clock position and passed tangentially through the perivitelline space and out the 10–11 o'clock position. Step 4: The oocyte is released from the holding pipette and held by the microneedle. The microneedle is brought to the bottom of the holding pipette and pressed to it, pinching a portion of the zona pellucida. By gently rubbing against the holding pipette with a sawing motion, the cut is accomplished and the oocyte is released.

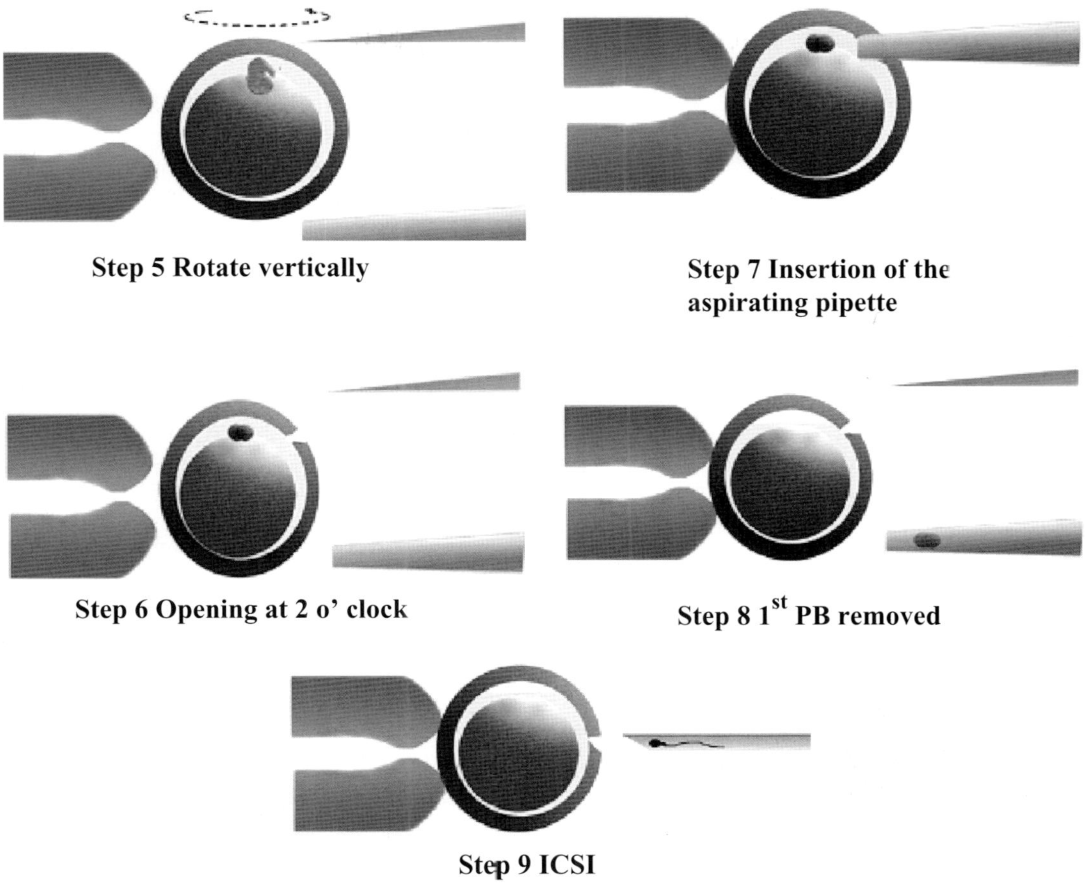

Figure 2.4. First polar body removal (*continued*) (see Figure 2.3). Step 5: The oocyte is brought to the holding pipette by moving the microscope stage and is rotated vertically using the microneedle until the slit opening is visualized at the 2 o'clock position. Step 6: Once again it is held in place by gentle suction from the holding pipette making sure the slit opening is at the 2 o'clock position and is in focus. Step 7: The micropipette is brought into the same focal plane as the slit opening and PB1. The micropipette is passed under the zona pellucida to PB1. Step 8: Using gentle suction created by the hydraulic microsyringe system, PB1 is aspirated into the micropipette and then deposited in a separate microdrop of medium. Step 9: The oocyte is transferred to another micromanipulation dish for intracytoplasmic sperm injection (ICSI), which is performed by passing the microtool through the slit opening.

oocytes are then returned to culture, checked for cleavage, and transferred, depending on the genotype of the corresponding PB1 and PB2 [29].

Although PB1 and PB2 have no any known biological significance in preimplantation and postimplantation development, follow-up studies have been carried out to investigate possible detrimental effects. Following PGD and ICSI, zygotes with two pronuclei were observed in 1192 (81.8%) of 1458 oocytes, compared with 30,972 (77.3%) of 40,092 in routine non-PGD cycles, which suggests no difference in fertilization rate observed after PB1 in comparison to nonbiopsied oocytes. Also there was no difference in blastocyst formation of the embryos resulting from the biopsied oocytes. We compared the blastocyst formation of embryos resulting from biopsied oocytes observed in 1653 (50.2%) of 3293 embryos, which was not different from 49.8% (9726 of 19,529 nonbiopsied embryos) observed in routine IVF. Similarly, no detrimental effect was noted after PB2 removal, which was evident from cleavage rate, blastocyst formation, and the number of cells in the respective blastocysts [32]. As will be seen below, there were no differences either after sequential PB1, PB2, and blastomere sampling (see Section 2.1.4).

PB approach will have an increasing impact in those ethnic groups in which PGD may be done only before fertilization, such as in Germany and

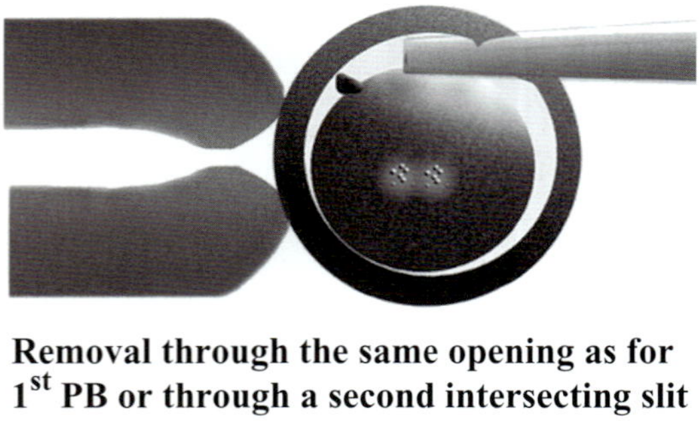

Figure 2.5. Second polar body removal on day 1 after fertilization assessment. The second polar body (PB2) is removed following fertilization assessment. The same microtools are required as with PB1 removal. The zygote is held in place by gentle suction from the holding pipette and is rotated using the microneedle. The same slit opening can be used or if PB2 has been extruded away from the opening, a second intersecting slit (3D-PZD) (see Figure 2.9) can be made in order to have easier access to the second polar body.

Malta. In these countries, the testing may be limited to PB1, which, as mentioned above, might not be sufficient to predict embryo genotype, unless PB2 is tested before fusion of pronuclei, which may be combined with freezing of the oocytes at the pronuclear stage. After the analysis, in a subsequent menstrual cycle, only the oocytes predicted of having the normal maternal allele may be thawed and cultured to allow fusion of the pronuclei, embryo development, and the embryo transfer. Since zygotes are not considered to be embryos until pronuclear fusion, and no abnormal oocytes may be thawed and cultured, obviating the establishment of the affected embyo, this technique may be ethically more acceptable to many couples. Therefore, the technique may allow creating a new class of genetic diagnosis, which may be called preembryonic genetic diagnosis, pushing the frontier of genotyping to an even earlier stage, as shown in the preembryonic diagnosis of sickle cell disease presented below.

A 33-year old woman and spouse at risk for producing a child with sickle cell disease was referred for PGD to avoid a possible termination of a pregnancy following prenatal diagnosis. A standard IVF protocol was initiated but the patient

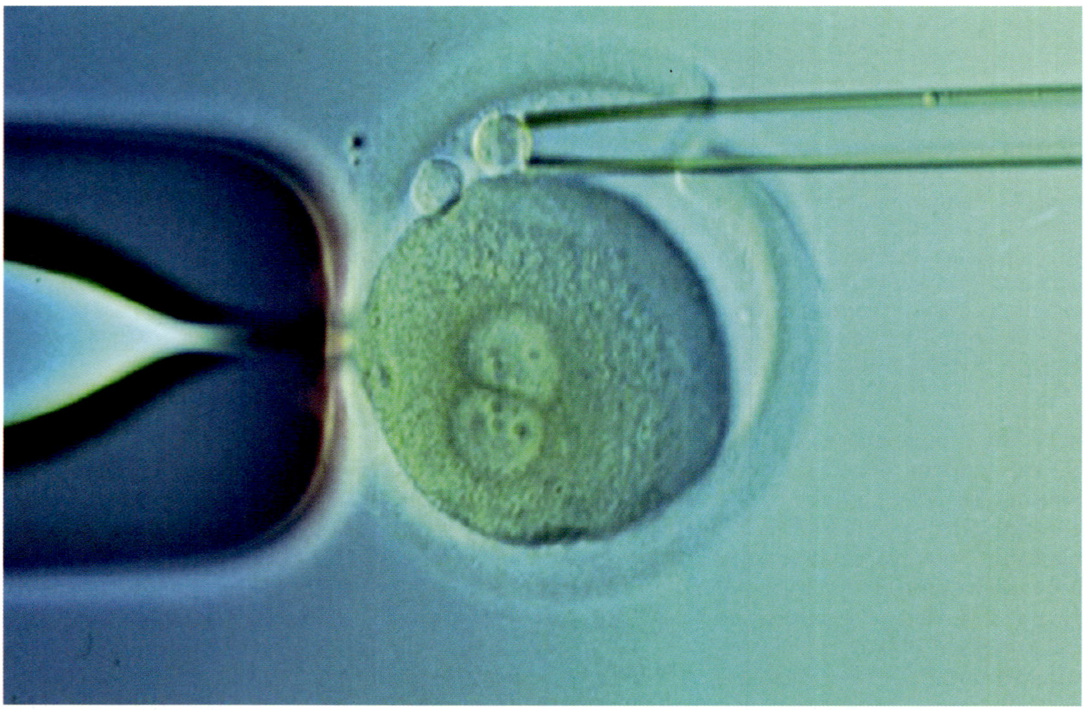

Figure 2.6. Simultaneous removal of the first and second polar body. Both PB1 and PB2 are removed following fertilization assessment, which is used for aneuploidy testing. The same microtools are required as with PB1 removal. The zygote is held in place by gentle suction from the holding pipette and is rotated using the microneedle. The same slit opening can be used as shown in Figure 2.3 allowing to have sufficient opening to remove both PB1 and PB2.

suffered hyperstimulation syndrome, which precluded transfer of embryos in that cycle. Twenty-eight mature oocytes were aspirated and placed in culture medium. Of the 28 aspirated oocytes, 14 extruded PB1 that were removed. The oocytes were then fertilized by intracytoplasmic sperm injection. As soon as the PB2s were removed, and prior to the fusion of the male and female pronuclei, all embryos were frozen. The PB1 and PB2 were analyzed by multiplex nested PCR to avoid misdiagnosis due to allele dropout (ADO), which occurs in approximately 5–10% of PB analyses. This involved nested, multiplex PCR with primer sets for the sickle cell mutation and two linked short tandem repeat (STR) markers; one located at the 5′ end of the beta-globin gene (5′-STR) and the other in the human thyrosin hydroxylase gene (THO-STR), for both of which the mother was heterozygous.

To detect potential contamination with extraneous DNA and identify the embryo that implanted and established pregnancy, three additional nonlinked STRs were amplified, including STR at the 5′ untranslated region of human coagulation factor A subunit gene (HUMF13A01), STR for von Willebrand's Disease (vWF), and an STR for chromosome 21 (D21S11). The list of primer sequences, reaction conditions, and details of the nested PCR were described earlier [29, 33].

The pronuclear-stage oocytes predicted to be normal were thawed, cultured to develop into the cleaving embryos of the 6–8-cell stage and transferred back to the patient in the two subsequent clinical cycles. The oocytes predicted to contain the mutant maternal gene were not thawed, but analyzed directly at the pronuclear stage for the confirmation of PB diagnosis. Of 28 aspirated oocytes, 14 extruded their PB1 and were studied for the presence of sickle cell mutation. Following intracytoplasmic sperm injection, PB2s were extruded from 13 of them and studied. Results of both PB1 and PB2 were available in 12 of these 13 oocytes. Overall, 6 oocytes were predicted to contain a normal allele, based on heterozygous status of PB1 and hemizygous mutant status of PB2.

In one of the oocytes, although the sickle cell analysis of PB1 showed only the normal allele in agreement with the 5′-STR, it was heterozygous for the THO-STR, suggesting that this is a case of ADO. Therefore, without simultaneous amplification of linked STRs, a misdiagnosis of the heterozygous oocyte as homozygous due to ADO would have occurred leading to a misdiagnosis of the maternal contribution to the zygote. In this particular instance, the error would have caused an unaffected zygote to be misdaignosed as affected, but the reverse could also occur.

Subsequently, the patient was prepared for a frozen embryo transfer. In the first frozen cycle, 4 zygotes determined to have the maternal unaffected allele were thawed and cultured. Three developed into cleaving embryos of acceptable quality, and were transferred, resulting in a singleton pregnancy, which was spontaneously aborted [31]. In the second frozen cycle, 2 unaffected embryos were transferred and resulted in a singleton pregnancy and birth of an unaffected child, following confirmation of PB diagnosis by chorionic villus sampling (CVS). The results of the application of nonlinked markers allowed not only the exclusion of possible DNA contamination but also the identification of the embryo that implanted yielding a clinical pregnancy. All the remaining oocytes predicted to contain an abnormal gene were not further cultured, but exposed directly to PCR analysis at the pronuclear stage for confirmation of diagnosis as frozen sample, and shown to be abnormal as predicted by PB analysis.

Presented data demonstrate feasibility of performing preembryonic diagnosis for single-gene disorders, which resulted in obtaining two unaffected pregnancies and birth of a healthy child. Of course, PGD for single-gene disorders may be performed by the use of PB1 analysis alone, as described in the first case of PGD by PB1 [2]. Although this allowed preselection of a few mutation-free oocytes inferred from the homozygous abnormal status of PB1, the majority of oocytes were heterozygous after the first meiotic division, so the genotype of the resulting embryos could not be predicted, thus limiting the number of normal embryos for transfer.

The data also show that to avoid discard of the preimplantation embryos reaching the cleavage stage by the time the PB genotyping results were obtained, freezing of oocytes may be used immediately after fertilization and extrusion of PB2 and prior to fusion of the male and female pronuclei, the actual point which is considered to be the beginning of the embryonic period of development [34]. In fact, freezing also may be omitted, as recent developments in PCR analysis allow completing the genetic diagnosis before pronuclei fusion. This opens a possibility for application of PGD for couples unable to accept any intervention and discard of the human embryos.

2.1.2 Preconception Testing for Paternally Derived Mutations by Sperm Duplication

As seen above, the genetic composition of oocytes may reliably be tested through removal and testing of PB1 and PB2. On the other hand, no method has yet been available for testing the outcome of male meiosis, as genetic analysis destroys the sperm, making it useless for fertilization. To overcome this problem, a new technique has been introduced, allowing duplicating a sperm before genetic analysis, so one of the duplicated sperm can be used for testing and the other for fertilization and consequent transfer of the resulting embryos, provided that the genetic analysis of the corresponding duplicate shows normal genotype [35]. To demonstrate the reliability of the technique, over 100 human sperm from chromosomally normal donors as well as from translocation carriers, were injected into enucleated mouse oocytes, and the duplicated cells resulting from an overnight culture were tested by FISH to compare the chromosomal status of both daughter cells. All but 3% of the haploid cell pairs derived from the normal donors were identical for the chromosomes tested, while, as expected, a high proportion of the paired nuclei derived from sperm of translocation carriers were chromosomally unbalanced, suggesting that ooplasm from mature mouse eggs can support the faithful replication of any human sperm genome, irrespective of the genotype.

A similar technique was developed to duplicate human sperm using human oocytes (Figure 2.7). However, it showed that the duplication of sperm may be done faithfully in approximately two thirds of cases, in contrast to the use of murine oocytes, so the technique has still to be tested further before applying it clinically, with the expected important practical implications for PGD of paternally derived conditions, such as translocations, known to produce as much as 70% abnormal sperm on the average.

The technique also has potential for research purposes, as shown in preliminary work devoted to study of mosaicism nature [36]. Following duplication of human sperm in cow oocytes, 31 resulting embryos were cultured up to the 8-cell stage, and tested by probes specific to chromosomes 13, 16, 18, 21, and 22. As much as 16% of the resulting sperm duplicates appeared not to be identical, which may further be related to the genetic differences between the donors involved. In fact, one of the three sperm donors for the above experiment produced mostly mosaic embryos in two PGD cycles. However, the rate of mosaicism in sperm duplicates of the three donors involved in this small series was similar indicating that the generation of mosaic embryos, at least in the patients previously tested by PGD, may not be related to a sperm genotype, but to the sperm centrosome [36].

Preconception diagnosis may be realized also in the future development of human gametes, using the techniques of haploidisation, which is described below.

2.1.3 Development of Artificial Human Gametes in Vitro

Attempts were undertaken for creating both female and male gametes, both demonstrating a strong morphological evidence for haploidisation [37, 38]. The technique is based on inducing nuclei of mitotic somatic cells to skip the S-phase of the cell cycle and undergo haploidisation when introduced into oocytes, which allows obtaining artificial gametes from somatic cells through the process of haploidisation. We showed that the efficiency of haploidisation of donor cumulus cell nuclei differs depending of the stage of development of the enucleated recipient oocyte [39, 40]. This may be tested using the extruded PBs or generated pronulei, which also allow investigating the correctness of chromosomal segregation. As seen from the flow chart in Figure 2.8, the first step involves enucleation of in vitro matured MII oocytes under control of ultraviolet (UV) luminescence, which is important to ensure the accuracy of chromosome analysis of the resulting haploid nuclei. Then the cumulus cell nuclei, which are at the G0 of the cell cycle, are introduced into ooplasts by injection and the oocytes are activated by electrostimulation delivered by the electrofusion device (XRONOS, RGI-4, Chicago). Following oocyte activation, the chromosomes of transferred nuclei segregate with the extrusion of polar bodies or form two pronuclei, both evidencing the formation of artificial gametes through somatic cell haploidization [39, 40]. The FISH analysis and DNA fingerprinting of PB1 and pronuclei resulting from the haploidisation procedure showed the haploid chromosomal set, with the resulting DNA originating from the donor nuclei, so the extruded PB1

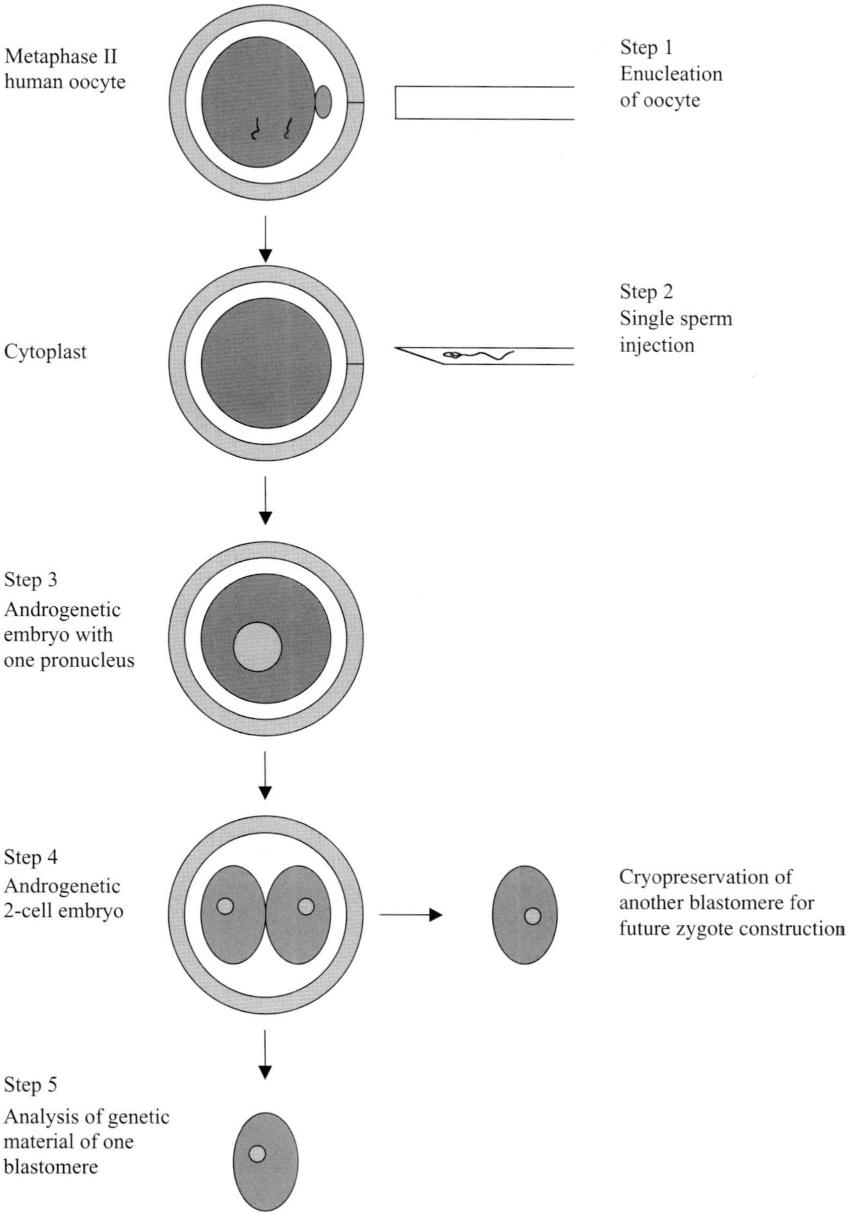

Figure 2.7. Flow chart of sperm duplication in human MII cytoplast. Step1: Enucleation of human metaphase II oocyte; Step 2: Injection of single human sperm into the cytoplast; Step 3: Reconstructed androgenic embryo with one pronucleus; Step 4: Development of this pronucleus into 2-cell embryo; Step 5: Testing of one of the cells, with the other one available for further zygote construction of known male contribution.

may be used to investigate the genetic contents of the corresponding pronucleus. However, preliminary data showed that as much as 90% of these haploid nuclei appeared to be with chromosomal aneuploidies. This suggests that the use of the resulting haploid nuclei in gamete reconstruction procedures may not be acceptable at the present time.

To determine if an incubation time of nuclei in ooplast improves chromosomal segregation, two groups of a total of 122 reconstructed MII oocytes were studied, one activated 5–7 hr after the nuclear

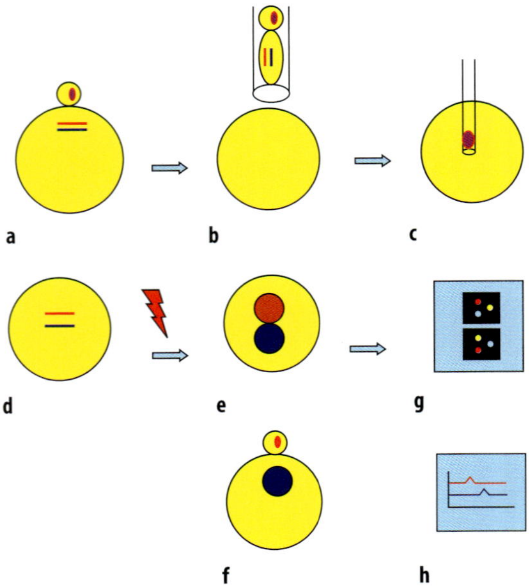

Figure 2.8. Flow chart of haploidization procedure for obtaining gametes from somatic cells (*a*) In vitro matured MII oocyte with extruded first polar body (PB1); (*b*) Both PB1 and metaphase II chromosomes are removed (shown in pipette) to prepare a recipient enucleated ooplast; (*c*) Single diploid cumulus cells were prepared as a nuclei donor, from which a nucleus was mechanically isolated, for the introduction into ooplast; (*d*) Reconstructed oocyte converts somatic cell nucleus into metaphase chromosomes; (*e*) Two pronuclei or (*f*) pronucleus and PB1 obtained after electrostimulation using electrofusion device RGI-4; (*g*) Fluorescent in situ hybridization analysis or (*h*) PCR-based chromosomal analysis confirming the formation of two haploid sets of chromosomes in both e and f scenarios.

transfer, and the other after 12–21 hr. However, as high as approximately 90% aneuploidy rate was observed irrespective of incubation time, with the majority being of the complex nature, suggesting no improvement of prolonged incubation on the accuracy of chromosomal segregation [39, 40]. So although haploidisation of somatic cells may be achieved using MII oocyte cytoplasm, the aneuploidy rate is much higher than in normal meiosis, which currently makes the techniques not acceptable for clinical practice.

2.1.4 Embryo Biopsy

Embryo biopsy is performed as soon as the embryo reaches a minimum of 6 cells or more so as not to cause a considerable decrease in cell number at later stages of development. A mechanical opening of zona pellucida has been developed, called 3D-PZD, allowing the creation of a V-shaped triangular flap or square flap opening, sufficient in size for a micropipette to pass through in order to remove blastomere(s) [29]. Micromanipulation dishes are prepared the same way as for PB1 removal, except for sucrose use, which is eliminated. The micromanipulation setup is the same as for PB1 removal with one exception – the micropipette has a larger diameter of 25–30 μm. The embryo biopsy procedure is shown and described in Figure 2.9.

With the current tendency for blastocyst transfer, there has been renewed interest in the development of methods for blastocyst biopsy, which has recently resulted in successful PGD cycles performed by blastocyst biopsy for translocations, yielding ongoing clinical pregnancies and births of healthy children [41]. Although blastocyst biopsy is not yet a method of choice in many centres, its potential is obvious, especially for additional testing required to confirm PB or blastomere diagnosis.

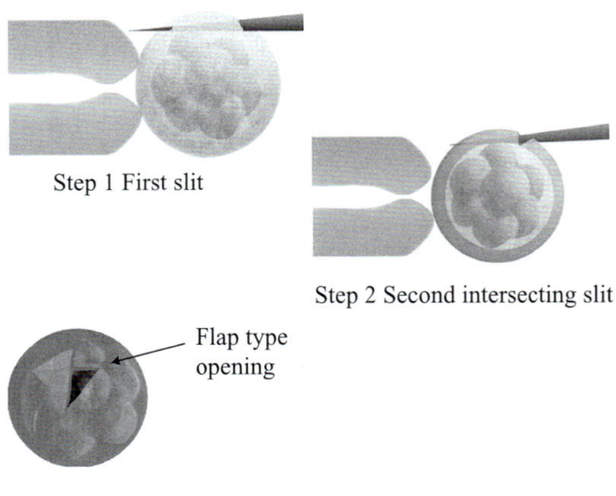

Figure 2.9. Three-dimensional partial zona dissection (3D-PZD) for embryo biopsy. Step 1: The embryo is held in position by gentle suction from the holding pipette. The embryo is rotated using the microneedle so that a blastomere with a well-visualized nucleus is present at 12 o'clock. The embryo is then rotated horizontally so that the blastomere of choice is facing the operator. The microneedle is passed tangentially under the zona pellucida through the perivitelline space moving from the 1–2 o'clock position to the 10–11 o'clock position. A first slit is created as described in Figure 2.3. Step 2: The embryo is rotated so that the slit opening is at the 1–2 o'clock position and a second intersecting slit is made by passing the microneedle through the first slit opening and out at the 10–11 o'clock position. The cut is accomplished as with all partial zona dissection previously described in Figure 2.3. This second intersecting slit creates a larger flap-type opening for the purpose of accessing PB2 as mentioned in Figure 2.5 or to utilize the required larger micropipette for embryo biopsy.

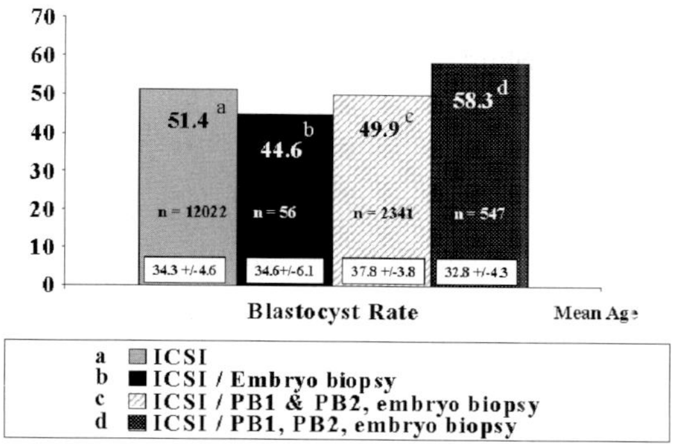

Figure 2.10. Effect of micromanipulations on embryo development. Bar graph demonstrating nondetrimental effect of biopsy procedures on the rate of blastocyst formation compaired to those following intracytoplasmic sperm injection (a control group). No significant difference was seen between embryos after one embryo biopsy procedure of a single blastomere and also embryos in which two biopsy procedures were performed in which the first and second polar bodies were removed simultaneously on day 1 (pronuclear stage) followed by the removal of a single blastomere on day 3, when compared with the control group. There was no decrease, but rather a significant increase in the rate of blastocyst formation when compared with the control group for the group of embryos in which three biopsy procedures were performed. The first polar body was removed prior to ICSI, the second polar body was removed at the time of fertilization assessment, and a single blastomere was removed on day 3. ($^{a,b}P = 0.316$ and $^{a,c}P = 0.209$, no significant difference. $^{a,d}P = 0.001$, significant difference.)

The follow-up studies of embryos after blastomere biopsy did not show any detrimental effects. No increase in congenital malformations has been reported among more than 1000 children born following PB or blastomere biopsy, although further systematic study will be needed to monitor the clinical outcomes of PGD using PB1 and PB2 sampling or embryo biopsy and collect further data on the safety of the procedures used in PGD for single-gene and chromosomal disorders. However, according to preliminary data presented in Figure 2.10, there seem to be no differences in developmental potential of embryos following one, two, or three biopsy procedures, compared with the rate of blastocyst formation following ICSI (control group). As can be seen from Figure 2.10, no significant differences were observed in blastocyst formation between the embryos with the embryo biopsy procedure of a single blastomere biopsy and the embryos with two biopsy procedures, including simultaneous PB1 and PB2 removal on day 1, followed by the removal of a single blastomere on day 3. There was no decrease, but rather an increase in the rate of blastocyst formation when compared with the control group, for the group of embryos in which three biopsy procedures were performed, including PB1 removal prior to ICSI, PB2 removal at the time of fertilization assessment, and a single blastomere removal on day 3. This is in agreement with pregnancy outcome data, which will be presented in Chapter 3, showing 38.8%, pregnancy rate in 562 PGD cycles for single-gene disorders, performed in majority of cases by sequential PB1 and PB2 analysis, followed by blastomere biopsy.

2.2 Single-Cell Genetic Analysis

Single-cell genetic analysis includes single-cell DNA and FISH analysis, which have become important tools for application of PGD in assisted reproduction and genetic services, providing a practical option for couples at genetic risk to avoid the birth of an affected offspring and have a healthy child of their own.

2.2.1 DNA Analysis

Because PGD for single-gene disorders is based on single-cell genetic analysis, its accuracy depends largely on the limitations of single-cell DNA analysis, which may potentially cause misdiagnosis. One of the key contributors to misdiagnosis is the phenomenon of preferential amplification, also known as allele-specific amplification failure (allele drop out, ADO), requiring the application of special protocols to ensure the highest ADO detection rate [29, 42, 43]. A few previously reported misdiagnoses, involving PGD for beta-thalassemia, myotonic dystrophy (DM), fragile-X syndrome (FMR1) and cystic fibrosis (CF), might have been due to this phenomenon, which has not initially been fully realized [21, 23, 28, 44].

It has been demonstrated that ADO rates in single cells might be different for different types of heterozygous cells [45]. The ADO rate may exceed 20% in blastomeres, compared with the ADO rate in single fibroblasts and PB1, which was shown to be under 10%. A high rate of ADO in blastomeres may lead to an obvious misdiagnosis, especially in compound heterozygous embryos. As mentioned, most misdiagnoses, especially those at the initial stage of application of PGD for single-gene disorders, were in the cases of blastomere biopsy from apparently compound heterozygous embryos.

The reliability of PGD depends on the controlling ADO. We have previously demonstrated that ADO rates in single-cell polymerase chain reaction (PCR) may vary with different lysis procedures, cell types, and loci analysed [29, 42, 46]. Therefore, reliable methods are needed to detect potential ADO, avoiding misdiagnosis in PGD for single-gene disorders. Our experience demonstrated feasibility of detection of ADO by a sequential analysis of oocytes, using PB1 and PB2, and by simultaneous amplification of mutant genes with linked polymorphic markers [29, 43, 46], which is described briefly below.

The biopsied single cells are placed directly into a lysis solution, consisting of 0.5 mcl 10 × PCR buffer, 0.5 mcl 1% Tween 20, 0.5 mcl 1% Triton × 100, 3.5 mcl H_2O and 0.05 mcl Protenase K (20 mg/ml in a 0.5 ml PCR tube). After spinning down at a low speed in a microfuge for a few seconds, the samples are covered with 1 drop of mineral oil and incubated at 45°C for 15 minutes in a thermal cycler. Proteinase K was then inactivated at 96°C for 20 minutes, which is also the beginning of the hot start of the first round PCR. Lower stringency and longer annealing time are used in the first-round PCR, with the introduction of the mixture of all outside primers for both mutant genes and polymorphic markers. Following the fist-round PCR, separate aliquots are amplified in the second-round PCR with specific inside primers for each site, using a higher stringency. Such a dual or multiple amplification reaction allows detection of most of the ADO cases. If there are pseudogenes, to eliminate false priming, the first-round primers are designed to anneal to the regions of nonidentity with pseudogenes [29]. In addition to STRs linked to the genes studied, STRs located on other chromosomes are also studied for testing of a possible contamination by extraneous DNA and identification of the origin of individual embryos in the established pregnancies. A list of STRs, their sequences and PCR reaction conditions for their analysis are presented in each respective section.

Fluorescent PCR (F-PCR) is used for a direct fragment size analysis of PCR product [47]. F-PCR is useful also for a direct sequencing of the PCR product in the detection of point mutations and for distinguishing preferential amplification from ADO. A considerable proportion of ADO is detected by sequential analysis of PB1 and PB2, which was demonstrated by data on sequential PCR analysis of 26 alleles in PB1 and PB2 obtained from 1047 oocytes, suggesting that this method may allow avoiding misdiagnosis due to ADO when no informative polymorphic markers are available. As mentioned, a high rate of ADO is observed especially in blastomeres, leading to an obvious misdiagnosis in compound heterozygous embryos, which was the case at the initial stage of application of PGD for single-gene disorders based on blastomere biopsy obtained from apparently compound heterozygous embryos [21, 28]. This may now be avoided by testing two or more linked polymorphic markers if available, making DNA testing in PB or single blastomeres a highly reliable procedure. Contrary to expectation, the application of F-PCR does not sufficiently improve detection of potential misdiagnoses in PGD of single gene disorders. Testing of 148 single fibroblasts by both conventional and F-PCR provided minor contributions to ADO detection rates. Based on this observation, it was postulated that simultaneous amplification of single cells for any causative gene, together with one linked polymorphic marker reduces the ADO rate by more than half, irrespective of the use of conventional or F-PCR. With the additional second marker in multiplex PCR, the ADO rate may further be reduced by half, being completely absent if three or more linked markers are simultaneously amplified.

We previously reported the results of 114 PGD cycles for couples at high risk for having children with single-gene disorders, resulting in preselection and transfer of a sufficient number of mutation-free oocytes in almost all cycles. Of 1047 oocytes with DNA results, 672 (64.1%) had heterozygous PB1, i.e., with both normal and mutant genes amplified, which is ideal for further testing, although their potential transfer depended entirely on the identification of mutant genes in the sequential analysis of PB2. Thus, a priority in preselection of embryos for transfer was given to the embryos resulting from the oocytes with heterozygous PB1 because in the absence of DNA contamination this indicates the absence of ADO of either normal or mutant alleles. Although most of the transferred embryos were preselected using this particular strategy, some preselected embryos still originated from homozygous normal oocytes, inferred from homozygous mutant status of PB1 and hemizygous normal status of PB2. These embryos were accepted for transfer only if ADO could have been excluded using linked polymorphic marker analysis. Otherwise, such embryos were excluded from transfer and exposed to follow-up confirmation analysis of the resulting embryos.

The follow-up analysis of the embryos excluded from transfer either because they were affected or there was insufficient information to preselect them for transfer, provided the data for evaluating the proportion of undetected ADO. Overall, 82 (7.8%) ADOs were observed, which included 75 detected and 7 undetected ones, suggesting that not 970 but actually 1052 oocytes were heterozygous. The genotype of 6 embryos appeared to be different from that predicted by mutation analysis, and 1 by STR, due to ADO in PB1, which were diagnosed as homozygous instead of their actual heterozygous status. The data indicated 98% accuracy, which is quite acceptable for clinical use of PGD for single-gene disorders.

As can be seen from these data, to avoid a misdiagnosis due to preferential amplification, a simultaneous detection of the mutant gene together with up to three or more highly polymorphic markers, closely linked to the gene tested, may be required [29, 43]. Each additional linked marker may reduce the misdiagnosis rate by half, so with one linked marker amplified together with mutation a misdiagnosis risk in blastomere analysis may be reduced from 20 to 10%; with two, from 10 to 5%; and with three, from 5% to practically zero. So a multiplex nested PCR analysis is performed, with the initial PCR reaction containing all the pairs of outside primers, so that following the first-round PCR, separate aliquots of the resulting PCR product may be amplified using the inside primers specific for each allele. Only when the polymorphic markers and the mutation agree, are embryos transferred. So multiplex amplification allows detecting ADO and prevents the transfer of misdiagnosed affected embryos.

Our data show that one of the efficient approaches for avoiding misdiagnosis may be a sequential genetic analysis of the PB1 and PB2 in PGD for maternally derived mutations. Detection of both mutant and normal alleles in the heterozygous PB1, together with the mutant allele in the corresponding PB2, leaves no doubt that the resulting maternal contribution to the embryo is normal, even without testing for the linked markers as a control. However, it will be ideal to test simultaneously at least for one linked marker to confirm the diagnosis. Alternatively, the mutation-free oocytes may be predicted also when corresponding PB1 is homozygous mutant, in which scenario the corresponding PB2 should be hemizygous normal, similar to the resulting maternal pronucleus. However, the genotype of the resulting maternal contribution may be quite opposite, i.e., mutant, if the corresponding PB1 is in fact heterozygous, but erroneously diagnosed as homozygous normal because of ADO of the normal allele. In the above scenario, the extrusion of the normal allele with PB2 would lead to the mutant allele left in the resulting oocyte. Therefore, the embryos resulting from the oocytes with homozygous mutant PB1 cannot be acceptable for transfer unless the heterozygous status of PB1 is excluded by the use of linked markers as described. The example of misdiagnosis due to ADO of the normal allele in PB1 has been described earlier in a PGD cycle performed for FMR1 [44]. To completely avoid misdiagnosis, a sequential PB1 and PB2 may be required to combine with multiplex PCR to exclude the possibility of an undetected ADO in heterozygous PB1. The accuracy of this approach may be demonstrated by the reports of PGD for thalassemia and familial dysautonomia (FD) (see below), resulting in the transfer of three unaffected embryos in each case, which were confirmed by the birth of the two sets of triplets free from thalassemia and FD [48, 49].

The other method with the proved potential for detecting and avoiding misdiagnosis due to preferential amplification is fluorescence PCR (F-PCR), which may allow detecting of some of the heterozygous PB1 or blastomeres misdiagnosed as homozygous in conventional PCR, therefore having the potential of reducing the ADO rates at least to some extent [43]. In addition, the method also allows a simultaneous gender determination, DNA fingerprinting, and detection of common aneuploidies. With further improvement and simplification, F-PCR combined with a multiplex system and sequential PB1 and PB2 analysis in cases of maternally derived mutations may allow excluding almost completely the risk for misdiagnosis due to preferential amplification.

The accuracy of PGD has been further improved with the application of fluorescent PCR with the expand long template (ELT) kit, which enabled reducing the undetected ADO rate from as high as 30–35% in both conventional and fluorescent PCR

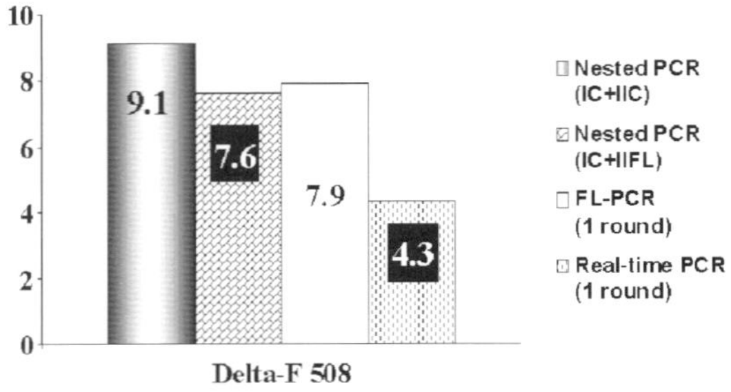

Figure 2.11. Reduction of undetected allele dropout (ADO) rates by real-time PCR. Undetected ADO rates are shown for nested conventional PCR (*left bar*), nested combined PCR (first-round conventional, second-round fluorescent (*middle left bar*), one-round fluorescent PCR (*middle right bar*) and one-round real-time PCR (*right bar*) with single human fibroblasts heterozygous for delta-F508 mutation in the CFTR gene. Comparable undetected ADO rates are seen in first three types of PCR, while application of one round of real-time PCR reduces ADO detection in single fibroblasts almost by half, demonstrating that real-time PCR is the most sensitive for detection of ADO, although it cannot completely prevent misdiagnosis.

to as low as 5% in testing for DM [50]. Another development in improving the accuracy of single-cell PCR analysis involves the application of real-time PCR, which was found to reduce the undetected ADO rate almost by half compared with conventional or fluorescent PCR (Figure 2.11). The application of these approaches together with simultaneous testing for the causative mutation along with at least one or two linked markers may allow reliably avoiding the risk for misdiagnosis.

Finally, because of the high rate of mosaicism at the cleavage stage, testing for the chromosome in which the gene in question is mapped is of obvious value to exclude the lack of mutant allele due to monosomy of this chromosome in the biopsied blastomere. Aneuploidy testing is technically feasible and may be done by adding primers for chromosome-specific microsatellite markers to the multiplex PCR protocols worked out for a specific genetic disorder [51]. The development of multiplex nested PCR systems will also allow performing PGD for different conditions simultaneously, as attempted for PGD for CFTR mutation together with FMR1 or gender determination [52, 53], and will be described in Chapter 3. The PCR-based strategy for aneuploidy testing is shown on Figure 2.12, listing polymorphic markers used for identification of copy number of chromosomes 13, 15, 16, 17, 18, 21, 22, and X.

Due to the need for development of a custom-made PGD design for each mutation and each couple, preparatory work has become an integral part of PGD for single-gene disorders to ensure avoiding potential misdiagnosis. For example, in some cases a particular set of outside primers has to be designed to eliminate false priming to the pseudogene, as described in PGD for long-chain 3-hydroxyacyl-Coa dehydrogenase deficiency [54]. Also, the preparatory work may frequently involve a single sperm typing needed for establishing paternal haplotypes, so that linked marker analysis could be performed in addition to mutation testing, especially in cases of paternally derived dominant conditions or PGD combined with preimplantation HLA matching (see below). The availability of the parental haplotypes allows not only confirming the absence of the mutant gene but also the presence of both maternal and paternal wild alleles in PGD by blastomere analysis, especially when only one informative marker is available (see Chapter 3).

2.2.2 FISH Analysis

PGD for the age-related aneuploidies is currently done by FISH analysis, using commercially available chromosome-specific probes (Vysis, Downers Groves, IL). It was first applied in 1991 for gender determination using DNA probes specific either for the X or Y chromosome [5]. Since testing for only one of the sex chromosomes could lead to misdiagnosis of gender due to a possible failure of hybridisation, a dual FISH was introduced, involving the simultaneous detection of X and Y, each in different colour [6]. Further, the dual FISH analysis was combined with a ploidy assessment to improve the accuracy by adding a centromeric probe specific for chromosome 18 [8, 55]. Testing was then extended to up to five autosomes, including chromosomes 13, 16, 21, and 22 (Figure 2.13) [29, 56, 57], although it is currently possible to analyze up to a dozen chromosomes using additional rounds of rehybridization.

As mentioned, the overall experience of preimplantation FISH analysis currently involves more than 5000 clinical cycles, resulting in an improved pregnancy rate in poor prognosis IVF patients [22, 23, 58–60]. Approximately half of these cycles were performed by FISH analysis of blastomeres and half by FISH analysis of PB1 and PB2, resulting in hundreds of unaffected pregnancies and healthy children born at the present time. The follow-up

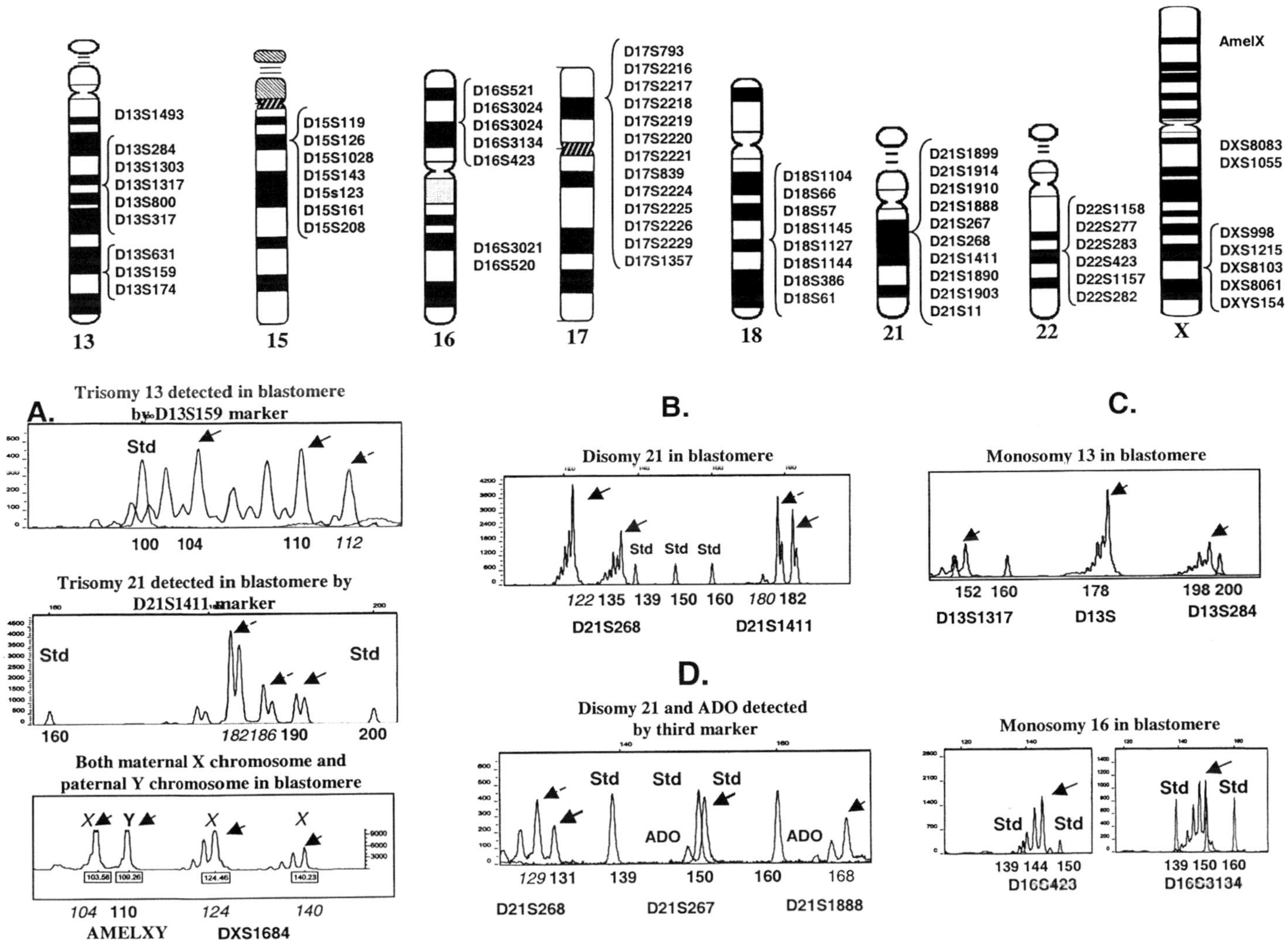

Figure 2.12. PCR-based aneuploidy testing. (Top) Positions of polymorphic markers applied for the PCR-based aneuploidy testing for selected chromsomes. To differentiate between allele dropout (ADO) and the absence of the whole chromosome (monosomy), at least three informative polymorphic markers for each chromosome are tested. (*Bottom*) Examples of single blastomere PCR results with trisomy 13, 21, and XXY (A); normal for chromosome 21 (B, D); monosomy 13 and 16(C); and ADO, distinguished from monosomy 21 by the use of three linked markers (D).

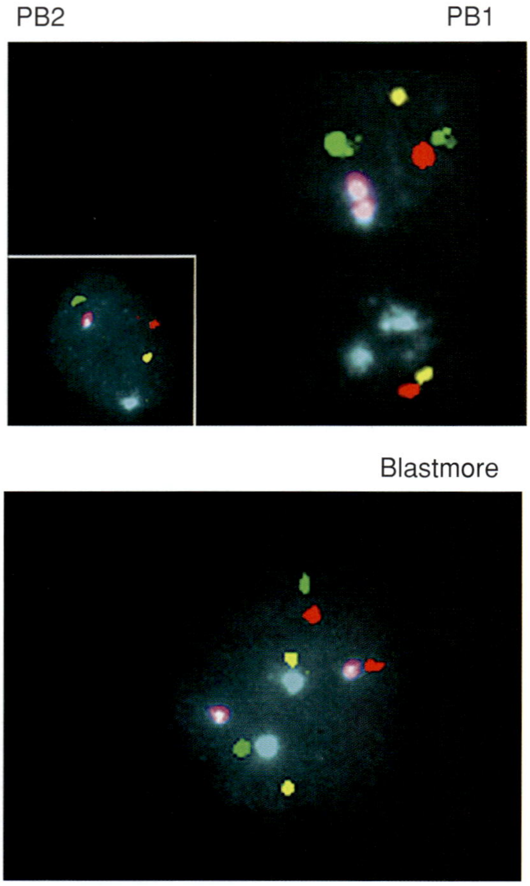

Figure 2.13. Normal pattern of FISH images of PB1 and PB2 and blastomeres after a 3-hr hybridization with MultiVysion PB panel probe for autosomes 13 (red), 16 (aqua), 18 (violet blue), 21 (green), and 22 (gold). PB1 and PB2 were removed simultaneously on day 1 at the pronuclear stage of development following fertilization assessment. (*Upper right*) PB1 chromosomes showing a normal number of single dot signals (two per chromosome, representing each chromatid). (*Upper left*) A normal number of signals (one per chromosome [*chromatid*]) are present in the PB2 interphase nucleus inferring a normal chromosome complement in the oocyte and resulting embryo. (*Bottom*) Normal FISH images of interphase nuclei isolated from blastomeres resulting from the oocytes shown on the left after embryo biopsy on day 3 of embryo development (two signals for each of the autosomes tested).

confirmation studies of the preselected abnormal embryos and the babies born following the procedure demonstrated an acceptable accuracy of the FISH analysis, which is described below.

The reliability of the FISH technique for aneuploidy detection in blastomeres has been extensively studied [9, 10, 61, 62]. By comparing the FISH results in the cleaving embryos to morphological abnormalities and maternal age, it was established that the observed chromosomal abnormalities were indeed not related to the limitations of the FISH technique, but were due to the embryo variables [10, 61]. However, a high rate of mosaicism was observed at the cleavage stage [9, 10, 61, 62], which was particularly high in slow embryos exhibiting an arrested development. Initially, an overall 12% mosaicism rate was suggested in cleaving embryos [12], but it has now been shown to occur in as much as half of preimplantation embryos, representing one of the major limitations of the FISH analysis for aneuploidies performed at this stage [17]. Clearly, mosaicism will affect the accuracy of the diagnosis, except for those cases when PGD detects the abnormal cell from a mosaic embryo, which will not be transferred. It was also shown that mosaicism may present diagnostic problems at the blastocyst stage [63], despite the initial prediction that the abnormal cells are deviated mainly to trophectoderm.

The first attempt to use FISH analysis for testing PB1 was undertaken in 1994. In this work, 130 unfertilized MII oocytes were tested simultaneously with their PB1, using X-chromosome– and chromosome-18–specific probes. It was demonstrated that PB1 FISH data allow an exact prediction of the chromosome set in the corresponding oocytes [64, 65]. Each chromosome in PB1 was represented by double dots (signals), corresponding to two chromatids in each univalent (see Figure 2.13). The data suggested that the number of signals (chromatids) in PB1 reliably predicts the corresponding number of signals (chromatids) in the MII oocytes, therefore providing an excellent tool for the genetic preselection of oocytes. It was also of interest that, in addition to a normal distribution of signals in PB1 and the corresponding MII oocytes, meiotic errors were also detected, confirming the accuracy of PB1 diagnosis for predicting the genotype of the corresponding oocyte. For example, in one PB1, four signals for chromosome 18 were detected, perfectly in accordance with the lack of the chromosome 18 signals in the corresponding MII oocyte (chromosome 18 nondisjunction). This suggested that the chromosomal complements of the oocyte could be inferred from the testing of PB1, which can be removed following its extrusion from the mature oocyte with no potential influence on the embryo viability. Another interesting phenomenon was the observation of chromatid malsegregation as a possible cause of chromosomal aneuploidy in the resulting mature oocytes. In four oocytes, instead of the expected two signals, three were found in the MII oocytes, which perfectly corresponded to a single signal in the corresponding PB1. Similar results were reported by another group, confirming diagnostic significance of the PB1 FISH analysis for predicting the genotype of the preimplantation embryo [66].

PB1 testing was one of the first approaches used for PGD of translocations [13], based on the fact that PB1 never forms an interphase nucleus and consists of metaphase chromosomes. It is known that PB1 chromosomes are recognizable when isolated 2–3 hr after in vitro culture, with degeneration beginning 6–7 hours after extrusion. Therefore, whole chromosome painting or the chromosome-segment–specific probes can be applied for testing of maternally derived chromosomal translocations in PB1. Although the method resulted in a significant reduction of spontaneous abortions in the patients carrying translocations, yielding unaffected pregnancies and births of healthy children, because of possible malsegregation and/or recombination between chromatids, a further follow-up analysis of PB2 may be required in order to accurately predict the meiotic outcome following the second meiotic division (see Chapter 5).

In contrast to PB1, PB2 is the by-product of the second meiotic division, extruded following fertilization of oocyte. The need for FISH analysis of PB2 for PGD of aneuploidies was first proposed in 1995, when it was demonstrated that PB1 testing alone does not allow predicting the resulting genotype of the oocyte [67]. It was shown that in contrast to the paired dots in PB1, each chromosome in PB2 was represented by a single dot (see normal pattern of FISH signals in Figure 2.13), so the lack of or addition of a signal for a particular chromosome provided evidence of a second meiotic division error. Although only 19 of 55 oocytes in this first study were tested by both PB1 and PB2, evidence of errors was observed not only in meiosis I, detected by PB1 analysis, but also in meiosis II, which may be detected by PB2 testing. These data suggested that some oocytes selected as normal, based on the PB1 FISH analysis, still could have been abnormal following nondisjunction in the second meiotic division. Therefore, FISH analysis for both PB1 and PB2 has become the basic requirement for PGD of aneuploidies, which allows detecting errors in both the first and second meiotic divisions. Currently, more than 15,000 oocytes have been analyzed by FISH analysis, showing the accuracy and reliability of PB1 and PB2 testing for predicting the karyotype of the embryo resulting from the corresponding oocyte. As will be described below, more than 50% of ooocytes from IVF patients of advanced reproductive age are abnormal, resulting from the errors in both the first and second meiotic divisions, in contrast to the previously believed concept that aneuploidies mainly originate from meiosis I.

As mentioned, PB2 testing is also an important component of PGD for maternally derived translocations. However, this is still done on interphase as, despite the progress in transforming PB2 into metaphase chromosomes via electrofusion of PB2 nucleus with foreign one-cell human embryo, the proportion of metaphase plates obtained was not sufficient to be useful in clinical practice. Much higher efficiency was observed in conversion of interphase nuclei of blastomeres, which is describred below.

Visualization of chromosomes of individual blastomere nuclei requires the application of nuclear transfer technique, which was initially attempted to conversion of interphase nuclei of PB2 into metaphase [68]. The original design was to fuse individual blastomeres with enucleated human oocytes. Although metaphases were obtained from two thirds of blastomeres treated by this method, its efficiency was not high enough to be applicable to PGD. This was due to inability of a replicating nucleus to form metaphase chromosomes after induction of premature chromosome condensation (PCC). However, because biopsied blastomeres may be at any stage of the cell cycle at the time of biopsy, there was the need to control the timing of mitosis of the blastomere nucleus, which can be achieved by introduction of blastomere into cytoplasm of a cell at a known cell cycle [69]. To achieve such reprogramming, the individual blastomeres were fused with intact or enucleated mouse zygotes at the pronuclear stage, known to be at the S-phase of the cell cycle.

So the method currently involves the use of frozen mouse zygotes, which can be purchased from Charles River Laboratories (Wilmington, MA), as recipient cytoplast to induce conversion of blastomere nucleus into metaphase. Also, there is no need for enucleation of mouse oocytes, as mouse and human chromosomes may be clearly distinguished. Blastomere biopsy is performed in the same way as described in Section 2.1.4, except for choosing only intact blastomeres with clearly visible nucleus. Also, several precautions have to be taken to ensure the integrity of the blastomere plasma membrane during the biopsy procedure. Although intact blastomeres may be inserted microsurgically into the perivitelline space, this appeared to be traumatic and was replaced by blastomere–zygote agglutination with phytohemagglutinin (Irvine Scientific, Santa Ana, CA). Before the procedure, the thawed mouse zygotes are freed of zonae pellucidae with acidic Tyrode's solution and pipetted through flame-polished Pasteur pipettes with an internal diameter of 80 μm. Then, using flame-polished Pasteur pipettes with an internal diameter of 100 μm, blastomere–zygote pairs are brought together and agglutinated in 300 μg/mL of phytogemagglutinin in protein-free human tubal fluid buffered with 20 mM of HEPES in a four-well plastic dish (Nunc).

Electofusion is induced in the same way as mentioned in the above section, except for substitution of 0.5% polyvinylpirrolidone in the electrofusion medium by 0.1% with molecular weight 360.000 (kd). Blastomere–zygote pairs are oriented between electrodes by hand, with the final orientation achieved with alternating current (500 kHz, 0.2 kV/cm for 2 seconds). Cell fusion is induced with a single direct current pulse (1 kV/cm for 500 μs), and the results are assessed in 20 minutes.

When human blastomeres are fused with intact mouse zygotes, the heterokaryons entering mitosis are identified under dissecting microscope. Because of the transparency of mouse cytoplasm, the disappearance of pronuclei and the formation of the joint metaphase plate are clearly visible. The heterokaryons with persisting pronucleus are exposed for 1 hr to 5 μM of okadaic acid (OA) in phosphate-buffered saline containing 3mg/nL of bovine serum albumin and 0.5 μg/nL of cytochalasin D. After a 10–15-min incubation in a hypotonic solution (0.1% sodium citrate and 0.6% bovine serum albumin) the resulting mitotic heterokaryons are fixed in a cold 3 : 1 solution of methanol and acetic acid in a four-well plastic dish. When the cytoplasm clears, heterokaryons are transferred onto slides and air dried. Chromosome plates are assessed by phase contrast and then used for standard chromosome analysis. For FISH analysis, the slides are pretreated with formaldehyde and pepsin (Abbott Inc., Downers Grove, IL).

Although the overall success rate of the procedure is as high as 83%, its efficiency can be further improved with experience [29, 18]. Similar results were obtained by using bovine ooplasts for fusion with human blastomeres [70]. Our data showed that some of the failures are simply due to the absence of the nucleus in biopsied blastomeres, or because the heterokaryons were fixed after they had already cleaved. It is also useful to perform blastomere biopsy not earlier than day 3 or 4, to avoid biopsy of 2- and 4-cell embryos, leading to the accelerated heterokaryon cleavage. However, the success rate did not depend at all on whether mouse zygotes were enucleated before fusion with blastomeres. This allows simplifying the procedure by using intact mouse zygotes.

So, the procedure is quite simple and includes the following components. Mouse zygotes are thawed, freed of zonae pellucidae and PB2 1–2 hr before electrofusion with human blastomeres. Four hours after fusion, heterokaryons are monitored for signs of the disappearance of pronuclei, and fixed at mitosis following hypotonic treatment. To avoid monitoring and a possible miss of mitosis, the heterokaryons may be cultured in the presence of microtubuli inhibitors, vinblastine or podophyllotoxin. All the embryos left in the culture by the ninth hour after fusion are fixed following 1-hr pretreatment with OA. The example of blastomere nucleus conversion applied for PGD of reciprocal translocation 46,XX, t(9;16)(q34.3;p13.1) is shown on Figure 2.14.

The method has been applied for PGD of paternally derived reciprocal translocations and for confirmation of PGD of chromosomal abnormalities performed by PB1 and PB2 FISH analysis. With the current success rate of blastomere nucleus conversion into metaphase it was applied to 183 clinical cycles for translocation carries, resulting in the transfer of balanced or normal embryos resulting in 45 unaffected pregnancies and birth of 39 healthy children (see Chapter 5).

It has recently been reported that the blastomere metaphase can be also obtained without application of specific conversion method [71]. To obtain analysable chromosomes, the embryos were monitored closely the second day after ICSI to identify the blastomere with nuclear breakdown, which was biopsied and fixed within 1 hr. This method was currently modified by 1-hr culture of the biopsied blastomere in medium with vinblastine, which resulted in harvest of metaphase chromosomes of good quality (Figure 2.15).

So, presented data show that the PB approach, based on PB1 and PB2 analysis, as well as genetic analysis of single blastomeres or blastocyst cells,

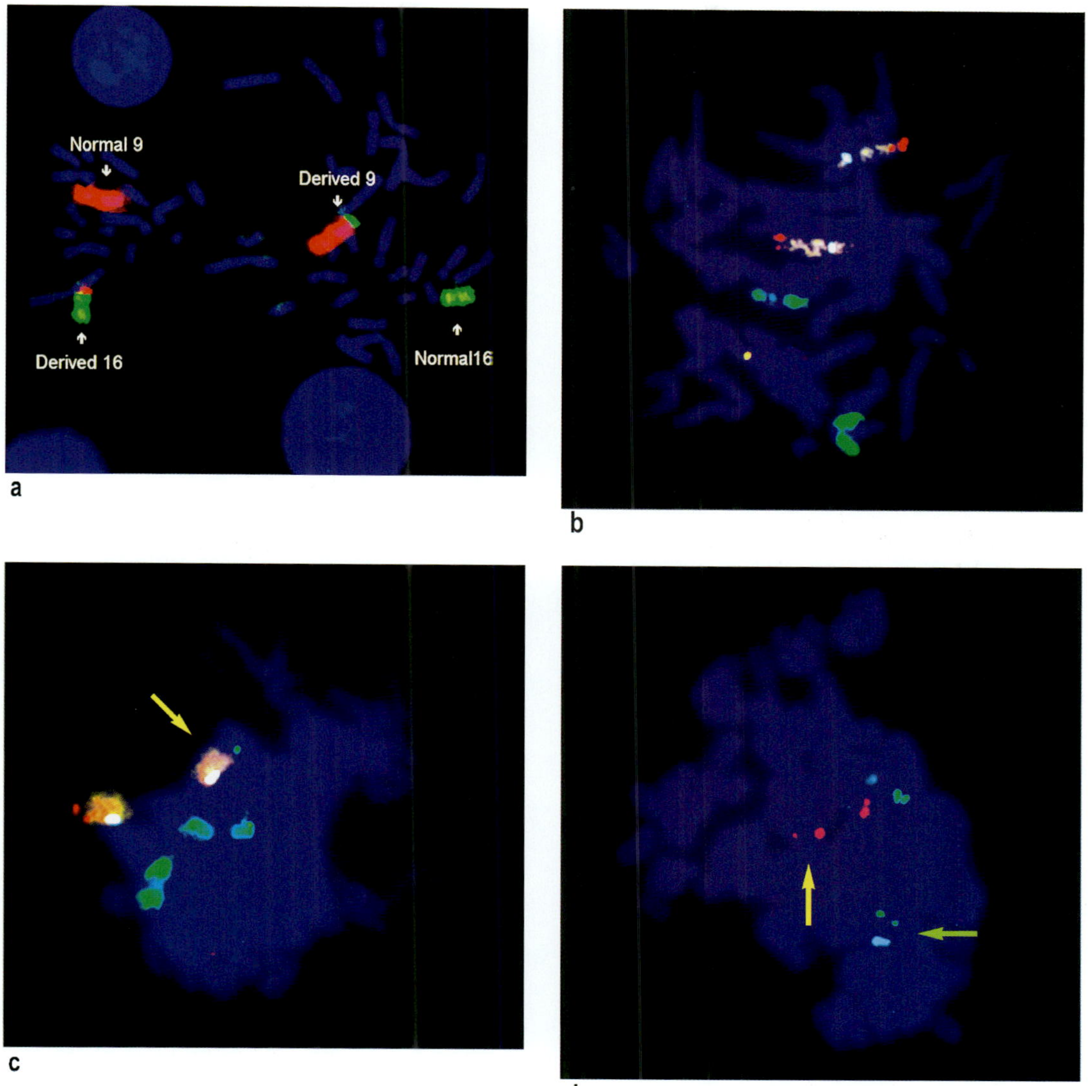

Figure 2.14. Blastomere nuclear conversion to metaphase for PGD of a maternally derived reciprocal translocation [46,XX,t(9;16)(q34.3;p13.1)] (*a*) FISH analysis of metaphase chromosomes of a peripheral blood lymphocyte from the carrier. Chromosome 9 is identified with WCP in orange (seen through a red single-bandpass filter) in conjunction with Tel 9q in red. Chromosome 16 is identified with WCP in green in conjunction with Tel 16p in green. Subtelomeric probes were added to this probe cocktail since the translocated portion of each chromosome is small. Derivatives are seen with both red and green fluorescence. (*b*) WCP in orange for chromosome 9 in conjunction with CEP 9 in aqua and Tel 9q in red along with WCP 16 (green) in conjunction with Tel 16p (green) on karyotypically normal, peripheral blood. (*c*) Analysis by embryo biopsy and blastomere nucleus conversion to metaphase chromosomes. Unbalanced chromosome complement [9,der (9),16,16] in which derivative (9) is present, evident by the presence of red and green fluorescence from Tel 16p (yellow arrow). This embryo was omitted from transfer. (*d*) Analysis by embryo biopsy and blastomere nucleus conversion to metaphase chromosomes. FISH analysis utilizing locus-specific and subtelomeric probes revealed a balanced chromosome complement by the presence of both derivatives. Derivative 9 is distinguished from the normal chromosome 9 by the presence of a CEP signal in aqua in conjunction with a Tel 16p signal in green (green arrow) and derivative 16 is distinguished from normal chromosome 16 by a Tel 9q signal (only) in red with the absence of a CEP 9 aqua signal (yellow arrow). This corresponding embryo was suitable for transfer barring any developmental problems.

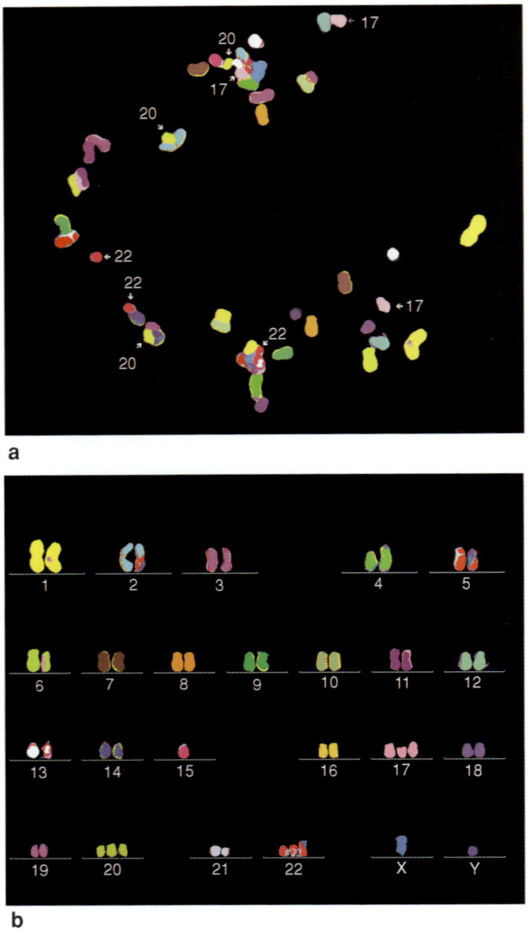

Figure 2.15. Karyotype of blastomere obtained without conversion technique. Embryo follow-up analysis after polar body analysis revealed second meiotic division error for chromosome 22 indicating a trisomy 22 embryo. (*a*) Testing was performed by embryo biopsy and "nonconversion" techniques used to obtain metaphase chromosomes from a single blastomere nucleus. Afterward, FISH was performed on the metaphase spread utilizing Metasystems' XCyte mFISH probe, which uses five fluorescent dyes to detect different painting probes at the same time, allowing for identification of all 24 different chromosomes with a single hybridization. Each chromosome is distinguished by a separate or combination of different fluorophores that are separated by appropriate filter sets. Based on the specific fluorochrome combination, the chromosomes are given a pseudo-color according to the 24XCyte labeling scheme. This allows for the analysis of complex numerical aberrations even when there is slight chromosome overlap present as seen in this metaphase spread. (*b*) Karyotype established by utilization of the Isis program revealing complex numerical aberrations consisting of trisomies 17, 20, and 22 and monosomy15.

may presently be applied for PGD of single-gene disorders, aneuploidies, and translocations. Based on this analysis, the embryos resulting from unaffected oocytes are transferred back to the patient within the implantation window, while those predicted to be affected are followed up by PCR or FISH analysis to confirm the PB or blastomere diagnosis. The results of genetic analysis using both sources of biopsy material is presented in Chapters 3 and 5, showing the accuracy and reliability of PGD for single-gene and chromosomal disorders.

References

1. Handyside AH, Kontogiani EH, Hardy K, Winston RML. Pregnancies from biopsied human preimplantation embryos sexed by Y-specific DNA amplification. Nature 1990;344:768–770.
2. Verlinsky Y, Ginsberg N, Lifchez A, Valle J, Moise J, Strom CM. Analysis of the first polar body: Preconception genetic diagnosis. Hum Reprod 1990;5:826–829.
3. International Working Group on Preimplantation Genetics. Current progress in preimplantation genetic diagnosis. J Assist Reprod Genet 1993;10:353–360.
4. International Working Group on Preimplantation Genetics. Preimplantation diagnosis of genetic and chromosomal disorders. J Assist Reprod Genet 1994;11:236–243.
5. Griffin DK, Handyside AH, Penketh RJA, Winston RML, Delhanty JDA. Fluorescent in-situ hybridization to interphase nuclei of human preimplantation embryos with X and Y chromosome specific probes. Hum Reprod 1991;6:101–105.
6. Griffin DK, Wilton LJ, Handyside AH, Winston RML, Delhanty JDA. Dual fluorescent in-situ hybridization for the simultaneous detection of X and Y chromosome specific probes for the sexing of human preimplantation embryonic nuclei. Hum Genet 1992;89:18–22.
7. Munné S, Lee A, Rozenwaks Z, Grifo J, Cohen J. Diagnosis of major chromosome aneuploidies in human preimplantation embryos. Hum Reprod 1993; 8: 2185–2191.
8. Munné S, Weier HUG, Stein J, Grifo J, Cohen J. A fast and efficient method for simultaneous X and Y in-situ hybridization of human blastomeres. J Assist Reprod Genet 1993;10:82–90.
9. Delhanty JDA, Griffin DK, Handyside AH. Detection of aneuploidy and chromosomal mosaicism in human embryos during preimplantation sex determination by fluorescent in situ hybridisation (FISH). Hum Mol Genet 1993;2:1183–1185.
10. Munné S, Weier HUG, Grifo J, Cohen J. Chromosome mosaicism in human embryos. Biol Reprod 1994;51:373–379.
11. Verlinsky Y. Preimplantation genetic diagnosis. J Assist Reprod Genet 1996;13:87–89.
12. International Working Group on Preimplantation Genetics. Current Status of Preimplantation Diagnosis. J Assist Reprod Genet 1997;14:72–75.
13. Munné S, Morrison L, Fung J, et al. Spontaneous abortions are reduced after preconception diagnosis of translocations. J Assist Reprod Genet 1998;15:290–296.

14. Verlinsky Y, Kuliev A. Preimplantation Genetics. J Assist Reprod Genet 1998;15:215–218.
15. Munné S, Sandalinas M, Escudero T. Outcome of preimplantation genetic diagnosis of translocations. Fertil Steril 2000;73:1209–1218.
16. International Working Group on Preimplantation Genetics. Preimplantation Diagnosis: An Alternative to Prenatal Diagnosis of Genetic and Chromosomal Disorders. Report of the 8th Annual Meeting of International Working Group on Preimlantation Genetics, in association with the 9th International Conference on Prenatal Diagnosis and Therapy, Los Angeles, June 7, 1998. J Assist Reprod Genet 1999;16:161–164.
17. Munné S. Preimplantation genetic diagnosis of numerical and structural chromosome abnormalities. Reproductive BioMedicine Online 2002;4:183–196.
18. Verlinsky Y, Cieslak J, Evsikov S, Galat V, Kuliev A. Nuclear transfer for full karyotyping and preimplantation diagnosis of translocations. Reprod BioMed Online 2002;5:302–307.
19. Kuliev A, Verlinsky Y. Thirteen Years' Experience of Preimplantation Diagnosis: Report of the Fifth International Symposium on Preimplantation Genetics. Reprod BioMed Online 2004;8:229–235.
20. International Working Group on Preimplantation Genetics. Preimplantation Diagnosis: An Integral Part of Assisted Reproduction. Report of the 9th Annual Meeting of International Working Group on Preimlantation Genetics, in association with the 11th IVF Congress, Sydney, May 10, 1999. J Assist Reprod Genet 2000;16:161–164.
21. International Working Group on Preimplantation Genetics. 10th Anniversary of Preimplantation Genetic Diagnosis. Report of the 10th Annual Meeting of International Working Group on Preimlantation Genetics, in conjunction with 3rd International Symposium on Preimplantation Genetics, Bologna, June 23, 2000. J Assist Reprod Genet 2001;18:66–72.
22. International Working Group on Preimplantation Genetics. Preimplantation Genetic Diagnosis: Experience of Three Thousand Clinical Cycles. Report of the 11th Annual Meeting of International Working Group on Preimlantation Genetics, in conjunction with 10th International Congress of Human Genetics, Vienna, May 15, 2001. Reprod BioMed Online 2001;3:49–53.
23. Verlinsky Y, Cohen J, Munné S, et al. Over a decade of preimplantation genetic diagnosis experience: A multicenter report. Fertil Steril 2004;82:292–294.
24. Verlinsky Y, Rechitsky S, Verlinsky O, et al. Preimplantation diagnosis for p53 tumor suppressor gene mutations. Reprod BioMed Online 2001;2:102–105.
25. Verlinsky Y, Rechitsky S, Schoolcraft W, Strom C, Kuliev A. Preimplantation diagnosis for Fanconi anemia combined with HLA matching. JAMA 2001;285:3130–3133.
26. Verlinsky Y, Rechitsky S, Sharapova T, Morris R, Tharanissi M, Kuliev A. Preimplantation HLA typing. JAMA 2004;291:2079–208.
27. Kuliev A, Verlinsky Y. Current features of preimplantation genetic disgnosis. Reprod BioMed Online 2002;5:296–301.
28. ESHRE Preimplantation Genetic Diagnosis (PGD) Consortium. Data Collecttion II (May 2002) Hum Reprod 2002;15,2673–2683.
29. Verlinsky Y, Kuliev A. Atlas of Preimplantation Genetic Diagnosis Parthenon, New York, 2000.
30. Verlinsky Y, Milayeva S, Evsikov S, et al. Preconception and preimplantation diagnosis for cystic fibrosis. Prenat Diagn 1992;12:103–110.
31. Verlinsky Y, Rechitsky S, Cieslak J, et al. Preimplantation diagnosis of single gene disorders by two-step oocyte analysis using first and second polar pody. Biochem Molec Med 1997;62:182–187.
32. Kaplan B, Wolf G, Kovalinskaya L, Verlinsky Y, et al. Viability of embryos following second polar body removal in a mouse model. J Assist Reprod Genet 1995;12:747–749.
33. KulievA, Rechitsky S, Verlinsky O, Strom S, Verlinsky Y. Preembryonic diagnosis for sickle cell disease. Molec Cell Endocrinol 2001;183:S19–S22.
34. Larsen WJ. Human Embryology. Churchill Livingstone, London 1994.
35. Willadsen S, Munné S, Schmmel T, Cohen J. Applications of nuclear sperm duplication. Fifth International Symposium on Preimplantation Genetics, June 5–7, Antalya, Turkey, 2003, p. 35. Istanbul Memorial Hospital, Istanbul Turkey 2003.
36. Munné S, Willadsen S, Schmmel T, Cohen J. Nuclear sperm duplication as a tool to study mosaicism. Fifth International Symposium on Preimplantation Genetics, June 5–7, Antalya, Turkey, 2003, pp. 55–56.
37. Lacham- Kaplan O, Daniels R, Trounson A. Fertilization of mouse oocytes using somatic cells as male germ cells. Reprod BioMed Online, 2001;3:205–211.
38. Tesarik J, Nagy ZP, Sousa M. et al. Fertilized oocytes reconstructed from patient's somatic cell nuclei and donor ooplast. Reprod BioMed Online 2001;2:160–164.
39. Galat V, Verlinsky Y. Haploidisation of somatic cell nuclei by cytoplasm of human oocytes. Reprod BioMed Online 2002;4(suppl 2):48–49.
40. Galat V, Ozen S, Rechitsky L, Kuliev A, Verlinsky Y. Cytogenetic analysis of human somatic cell paploidization. Reprod BioMed Online 2005;10:199–204.
41. McArthur S, Marshall J, Wright D, de Boer K. Successful pregnancies following blastocyst (day 5) biopsy and analysis for reciprocal and Robertsonial translocations. Fifth International Symposium on Preimplantation Genetics, June 5–7, Antalya, Turkey, 2003, p. 33.
42. Rechitsky S, Verlinsky O, Strom C, et al. Experience with single-cell PCR in preimplantation genetic diagnosis: How to avoid pitfalls. In: Hahn S, Holzgreve W. (eds.). Fetal Cells in Maternal Blood. New Developments for a New Millennium. 11th Fetal Cell Workshop, Karger, Basel, 2000, pp. 8–15.
43. Rechitsky S, Verlinsky O, Amet T, et al. Reliability of preimplantation diagnosis for single gene disorders. Molec Cell Endocrinol 2001;183:565–568.
44. Verlinsky Y, Rechitsky S, Verlinsky O, Strom C, Kuliev A. Polar body based preimplantation diagnosis for X-linked genetic disorders. Reprod BioMed Online 2002;4:38–42.
45. Rechitsky S, Strom C, Verlinsky O, et al. Allele drop out in polar bodies and blastemeres. J Assist Reprod Genet 1998;15:253–257.
46. Rechitsky S, Strom C, Verlinsky O, et al. Accuracy of preimplantation diagnosis of single-gene disorders by polar body analysis of oocytes. J Assist Reprod Genet 1999;16:192–198.
47. Findlay I, Quirke P. Fluorescent polymerase chain reaction: Part I. A new method allowing genetic diagnosis and DNA fingerprinting of single cells. Hum Reprod Update 1996;2:137–152.

48. Kuliev A, Rechitsky S, Verlinsky O, et al. Birth of healthy children following preimplantation diagnosis for thalassemias. J Assist Reprod Genet 1999;16:219–225.
49. Rechitsky S, Verlinsky O, Kuliev A, Ozen S, Masciangelo C, Lifchez A, Verlinsky Y. Preimplantation genetic diagnsis for familial dysautonomia. Reprod BioMed Online 2003;6:488–493.
50. Sermon K, Seneca S, De Rycke M, et al. PGD in the lab for triplet diseases-myotonic dystrophy, Huntington's disease and fragile-X syndrome. Molec Cell Endocrinol 2001;183:S77–S85.
51. Katz MG, Mansfield J, Gras L, Trounson A, Cram DS. Diagnosis of trisomy 21 in preimplantation embryos by single-cell DNA fingerprinting. Reprod BioMed Online 2002;4:43–50.
52. Harper J, Wells D, Piyamongkol W, et al. Preimplantation genetic diagnosis for single gene disorders: Experience with five single gene disorders. Prenat Diagn 2002;22:525–533.
53. Ray PF, Gigarel N, Bonnefont JP, Attie T, Hamamah S, Frydman N, Vekemans M, Frydman R, Munnich A. First specific preimplantation genetic diagnosis for ornithine transcarbamolase deficiency. Prenat Diagn 2000;20:1048–1054.
54. Verlinsky Y, Rechitsky S, Verlinsky O, Strom C, Kuliev A. Preimplantation diagnosis for long-chain 3-hydroxyacyl-CoA dehydrogenase deficiency. Reprod BioMed Online 2001;2:17–19.
55. Schrurs BM, Winston RML, Handyside AH. Preimplantation diagnosis of aneuploidy using fluorescent in-situ hybridization: Evaluation using a chromosome 18-specific probe. Hum Reprod 1993;8:296–301.
56. Munné S, Magli MC, Bahce M, Fung J, Legator M, Morison L. Preimplantation diagnosis of the aneuploidies most commonly found in spontaneous abortions and live births: XY, 13, 14, 15, 16, 18, 21, 22. Prenat Diagn 1998;18:459–1466.
57. Bahce M, Cohen J, Munné S. Preimplantation genetic diagnosis of aneuploidy: Were we looking at the wrong chromosomes? J Assist Reprod Genet 1999;16:154–159.
58. Gianaroli L, Magli MC, Ferraretti AP, Munné S. Preimplantation diagnosis for aneuploidies in patients undergoing in vitro fertilization with poor prognosis: Identification of the categories for which it should be proposed. Fertil Steril 1999;72:837–844.
59. Munné S, Magli C, Cohen J, et al. Positive outcome after preimplantation diagnosis of aneuploidy in human embryos. Hum Reprod 1999;14:2191–2199.
60. Munné S, Sandalinas M, Escudero T, et al. Improved implantation after preimplantation genetic diagnosis of aneuploidy. Reprod BioMed Online 2003;7:91–97.
61. Munné S, Alikani M, Tomkin G, Grifo J, Cohen J. Embryo morphology, developmental rates and maternal age are correlated with chromosome abnormalities. Fertil Steril 1995;64:382–391.
62. Harper JC, Coonen E, Handyside AH, et al. Mosaicism of autosomes and sex chromosomes in morphologically normal monospermic preimplantation human embryos. Prenat Diagn 1994;15:41–49.
63. Evsikov S, Cieslak J, Verlinsky Y. Survival of unbalanced translocations to blastocyst stage. Fertil Steril 2000;74:672–676.
64. Dyban A, Fredine M, Severova E, Cieslac J, Wolf G, Kuliev A, Verlinsky Y. Detection of aneuploidy in human oocytes and corresponding first polar bodies using FISH (about 97). 7th International Conference on Early Prenatal Diagnosis, May 22–27, Jerusalem, 1994.
65. Dyban A, Fredine M, Severova E. et al. Detection of aneuploidy in human oocytes and corresponding first polar bodies by FISH. J Assist Reprod Genet 1996;13: 72–77.
66. Munné S, Daily T, Sultan KM, Grifo J, Cohen J. The use of first polar bodies for preimplantation diagnosis of aneuploidy. Hum Reprod 1995;10:1014–1120.
67. Verlinsky Y, Cieslak J, Freidine M, Ivakhnenko V, Wolf G, Kovalinskaya L, White M, Lifchez A, Kaplan B, Moise J, Ginsberg N, Strom C, Kuliev A. Pregnancies following preconception diagnosis of commom aneuploidies by fluorescent in-situ hybridization. Hum Reprod 1995;10:1923–1927.
68. Verlinsky Y, Evsikov S, Karyotyping of human oocytes by chromosomal analysis of the second polar body. Mol Hum Reprod 1999;5:89–95.
69. Verlinsky Y, Evsikov S. A simplified and efficient method for obtaining metaphase chromosomes from individual human blastomeres. Fertil Steril 1999;72:1–6.
70. Willadsen S, Levron J, Munné S et al. Rapid visualization of metaphase chromosomes in single human blastomeres after fusion with in-vitro matured bovine eggs. Human Reprod 1999;14:470–474.
71. Tanaka A, Nagayoshi M, Awata S, Mawatari Y, Tanaka I, Kusunoki H. Preimplantation diagnosis of repeated miscarriage due to chromosomal translocations using metaphase chromosomes of a balstomere biopsied from 4–6 cell stage embryo. Fertil Steril 2004;81:30–34.

3

Preimplantation Genetic Diagnosis for Single Gene Disorders

As mentioned, preimplantation genetic diagnosis (PGD) was introduced 14 years ago with the purpose of performing genetic testing before pregnancy, in order to establish only unaffected pregnancies and avoid the need for pregnancy termination, which is the major limitation of traditional prenatal diagnosis [1, 2]. Despite the requirement for ovarian hyperstimulation and in vitro fertilization (IVF), needed to perform genetic testing of oocyte or embryo prior to transfer, PGD has been accepted in most parts of the world, with an overall number of approximately 7000 PGD cases performed by the present time, which have resulted in the birth of close to 2000 healthy children [3, 4]. At least 2000 of these PGD cycles were performed for single-gene disorders, and as will be shown below, is presently offered for some indications that have never been practiced in prenatal diagnosis, such as late-onset diseases with genetic predisposition and preimplantation HLA typing, making PGD not only an alternative but also a complement to prenatal diagnosis [4, 5]. The progress of PGD has been extensively reviewed, so the present book will mainly concentrate on those aspects of PGD that are useful for reproductive medicine and genetics practices, including available PGD approaches for different groups of genetic disorders, its accuracy, and major indications compared with prenatal diagnosis, and will present practical details useful for realization of PGD for each of the conditions described.

Indications for PGD were initially similar to those practiced in prenatal diagnosis, and applied for those at-risk couples that could not accept pregnancy termination, expected in 25–50% of cases following prenatal diagnosis, depending on the mode of inheritance. However, these indications have recently been extended beyond those for prenatal diagnosis, and currently include the conditions with a low penetrance, late-onset disorders with genetic predisposition, and HLA typing with or without testing for causative gene. The list of disorders for which PGD has been applied, according to our experience, now comprises close to 100 conditions (Table 3.1), with most frequent ones still being cystic fibrosis (CFTR), haemoglobin disorders, and some of the dynamic mutations. The choice between prenatal diagnosis and PGD mainly depends on the patient's attitude to termination of pregnancy, which is strongly influenced by social and religious factors. However, the majority of patients are still unaware of the availability of PGD, due to a relative novelty of the procedure. So there is obvious need for increasing awareness of PGD both for couples at risk and the medical profession, who require information about benefits, accuracy, safety, and expected risks of the procedure.

3.1 Autosomal Recessive Diseases

The most common indications for PGD were CFTR and haemoglobin disorders (HBB). As can be seen from the lists of CFTR and thalassemia mutations presented in Figures 3.1 and 3.2, 23 different mutations were tested for thalassemias and sickle cell disease, and 18 for CFTR. Overall, 113

Table 3.1. List of genetic disorders for which PGD was performed

Disease	mim Number	Inheritance	Gene Name/Symbol	Location	Protein Name
Achondroplasia (ACH)	100800	AD	FGFR3	4p16.3	Fibroblast growth factor receptor 3 [precursor]
Adenosine deaminase deficiency (ADA)	102700	AR	ADA	20q13.11	Adenosine deaminase
Adenomatous polyposis of the colon (APC)	175100	AD	APC	5q21-q22	Adenomatous polyposis coli protein
Adrenoleukodystrophy (ALD)	300100	XL	ABCD1	Xq28	Adrenoleukodystrophy protein
Alpha-1 antitrypsin deficiency (AAT)	107400	AR	SERPINA1	14q32.1	Alpha-1-antitrypsin [precursor]
Alport syndrome, X-linked (ATS)	301050	XL	AMMECR1	Xq22.3	AMME syndrome candidate gene 1 protein
Basal cell nevus syndrome (BCNS) (GORLIN)	109400	AD	PTCH	9q22.1-31	Patched protein homologue 1
Blood group—Kell-Cellano system	110900	AD	KEL	7q33	Kell blood group glycoprotein
Brain tumour, posterior fossa of infancy, familial	601607	AD	SMARCB1	22q11.2	SWI/SNF related, matrix associated, actin dependent regulator of chromatin subfamily B member
Ceroid lipofuscinosis, neuronal 2, late infantile (CLN2)	204500	AR	CLN2	11p15	Tripeptidyl-peptidase I [precursor]
Charcot-Marie-Tooth disease, axonal, type 2E	607684	AD	NEFL	8p21	Neurofilament triplet L protein
Charcot-Marie-Tooth disease, demyelinating, type 1A (CMT1A)	118220	AD	PMP22	17p12	Peripheral myelin protein 22
Charcot-Marie-Tooth disease, demyelinating, type 1B (CMT1B)	118200	AD	MPZ	1q23.3	Myelin P0 protein [precursor]
Charcot-Marie-Tooth disease, X-linked, 1 (CMTX1)	302800	XL	GJB1	Xq13.1	Gap junction beta-1 protein
Choroideremia (CHM)	303100	XL	CHM	Xq21.2	Rab proteins geranylgeranyltransferase component A 1
Citrullinemia, classic	215700	AR	ASS	9q34.1	Argininosuccinate synthase
Colon cancer, familial nonpolyposis, type 1	120435	AD	MSH2	2p22-21	DNA mismatch repair protein Msh2
Congenital adrenal hyperplasia (CAH)	201910	AR	CYP21A2	6p21.3	Cytochrome P450 XXIB
Craniofacial dysostosis, type I (CFD1)	123500	AD	FGFR2	10q26.13	Fibroblast growth factor receptor 2 [precursor]
Currarino syndrome	176450	AD	HLXB9	7q36	Homeobox protein HB9
Cystic fibrosis (CF)	219700	AR	CFTR	7q31.2	Cystic fibrosis transmembrane conductance regulator
Darier-White disease (DAR)	124200	AD	ATP2A2	12q23-q24.1	Sarcoplasmic/endoplasmic reticulum calcium ATPase 2
Diamond-Blackfan Anaemia (DBA)	205900	AD	RPS19	19q13.2	40S ribosomal protein S19
Dystrophia myotonica 1	160900	AD	DMPK	19q13.2-q13.3	Myotonin-protein kinase
Early-onset familial ALZHEIMER disease	104760	AD	APP	21q21.3	Amyloid beta A4 protein [precursor]
Ectodermal dysplasia 1, anhidrotic (ED1)	305100	XL	ED1	Xq13.1	Ectodysplasin A
Emery-Dreifuss muscular dystrophy autosomal recessive (EDMD3)	604929	AR	LMNA	1q21.2	Lamin A/C
Epidermolysis bullosa dystrophica, Pasini type	131750	AR	COL7A1	3p21.3	Collagen alpha 1(VII) chain [precursor]
Epiphyseal dysplasia, multiple, 1 (EDM1)	132400	AD	COMP	19p13.1	Cartilage oligomeric matrix protein [precursor]
Exostoses, multiple, Type 1	133700	AD	EXT1	8q24.11q24.13	Exostosin-1
Fabry disease	301500	XL	GLA	Xq22	Alpha-galactosidase A [precursor]

Table 3.1. *Continued*

Disease	mim Number	Inheritance	Gene Name/Symbol	Location	Protein Name
Fanconi anaemia, complementation group C (FANCC)	227645	AR	FANCC	9Q22.3	Fanconi anemia group C protein
Fancony anaemia, complementation group A (FANCA)	227650	AR	FANCA	16q24.3	Fanconi anemia group A protein
Fragile site mental retardation 1	309550	XL	FMR1	Xq27.3	Fragile X mental retardation 1 protein
Fragile site, folic acid type, rare, FRA(X)(q28), (FRAXE)	309548	XL	FMR2	Xq28	Fragile X mental retardation 2 protein
Galactosemia	230400	AR	GALT	9p13	Galactose-1-phosphate uridylyltransferase
Gaucher disease, type 1	230800	AR	GBA	1q21	Glucosylceramidase [precursor]
Glycogen storage disease type 4	232700	AR	PYGL	14q21-q22	Glycogen phosphorylase, liver form
Haemoglobin—alpha locus 1 (HBA1)	141800	AR	HBA1	16pter-p13.3	Hemoglobin alpha chain
Haemoglobin—beta locus (HBB)	141900	AR	HBB	11p15.5	Hemoglobin beta chain
Haemophilia A	306700	XL	F8	Xq28	Coagulation factor VIII [precursor]
Haemophilia B	306900	XL	F9	Xq27.1-q27.2	Coagulation factor IX [precursor]
HLA genotyping				6q21.3	
Huntington disease (HD)	143100	AD	HD	4p16.3	Huntingtin
Hurler syndrome	607014	AR	IDUA	4p16.3	Alpha-L-iduronidase [precursor]
Hydrocephalus, X-linked (L1CAM)	308840	XL	L1CAM	Xq28	Neural cell adhesion molecule L1 [precursor]
Hypophosphatasia, infantile	241500	AR	ALPL	1p36.1-34	Alkaline phosphatase, tissue-nonspecific isozyme [precursor]
Immunodeficiency with Hyper-IgM, type 1 (HIGM1)	308230	XL	TNFSF5	Xq26	Tumour necrosis factor ligand superfamily member 5
Incontinentia pigmenti (IP)	308300	XL	IKBKG	Xq28	NF-kappa-B essential modulator
Krabbe disease	245200	AR	GALC	14q31	Galactocerebrosidase [precursor]
Li-Fraumeni syndrome (LFS)	151623	AD	TP53	17p13.1	Cellular tumor antigen p53
Long-chain 3-hydroxyacyl-CoA dehydrogenase deficiency (HADHA)	600890	AR	HADHA	2p23	Trifunctional enzyme alpha subunit, mitochondrial [precursor]
Machado-Joseph disease (MJD)	109150	AD	ATX3	14q24.3-q31	Machado-Joseph disease protein 1
Marfan syndrome (MFS)	154700	AD	FBN1	15q21.1	Fibrillin 1 [precursor]
Metachromatic Leukodystropy	250100	AR	ARSA	22q13.31-qter	Arylsulfatase A [precursor]
Mucopolysaccharidosis type II (Hunter) Hunter-McAlpine craniosynostosis syndrome	309900	AD	IDS	Xq28	Iduronate 2-sulfatase [precursor]
Muscular Dystrophy, Becker type (BMD)	300376	XL	DMD	Xq21.2	Dystrophin
Muscular Dystrophy, Duchenne type (DMD)	310200	XL	DMD	Xq21.2	Dystrophin
Myotubular myopathy 1 (MTM1)	310400	XL	MTM1	Xq28	Myotubularin
Neurofibromatosis, type 1 (NF1)	162200	AD	NF1	17q11.2	Neurofibromin
Neurofibromatosis, type 2 (NF2)	101000	AD	NF2	22q12.2	Merlin
Neuropathy, hereditary sensory and autonomic, type 3 (HSAN3)	223900	AR	IKBKAP	9q31	I-kappa-B-kinase-complex–associated protein
Norrie disease (NDP)	310600	XL	NDP	Xp11.4-p11.3	Norrin
Oculocutaneous albinism, type 1 (OCA1)	203100	AR	TYR	11q14-q21	Tyrosinase [precursor]
Oculocutaneous albinism, type 2 (OCA2)	203200	AD	OCA2	15q11.2-q12	P protein

(*Cont.*)

Table 3.1. *Continued*

Disease	mim Number	Inheritance	Gene Name/Symbol	Location	Protein Name
Optic atrophy 1 (OPA1)	165500	AD	OPA1	3q28-q29	Dynamin-like 120 kDa protein, mitochondrial [precursor]
Ornithine transcarbamylase deficiency, hyperammonemia due to	311250	XL	OTC	Xp21.1	Ornithine carbamoyltransferase, mitochondrial [precursor]
Osteogenesis imperfecta congenita (OIC)	166200	AD	COL1A2	7q22.1	Collagen alpha 2(I) chain [precursor]
Osteogenesis imperfecta congenita (OIC)	166200	AD	COL1A1	17q21.31-q22	Collagen alpha 1(I) chain [precursor]
Osteopetrosis, autosomal recessive	259700	AR	TCIRG1	11q13.4-q13.5.	Vacuolar proton translocating ATPase 116 kDa subunit a isoform 3
Pelizaeus-Merzbacher-like disease (PMLD)	311601	XL	PLP1	Xq22	Myelin proteolipid protein
Phenylketonuria	261600	AR	PAH	12q22-q24.2	Phenylalanine-4-hydroxylase
Polycystic kidney disease 1 (PKD1)	601313	AD	PKD1	16P13.3	Polycystin 1 precursor
Polycystic kidney disease 2 (PKD2)	173910	AD	PKD2	4q22.1	Polycystin 2
Polycystic kidney disease, autosomal recessive (ARPKD)	263200	AR	PKHD1	6p12.3	Polycystic kidney and hepatic disease 1 [precursor]
Retinitis pigmentosa 2 (RP2)	312600	XL	RP2	Xp11.3	XRP2 protein
Retinoblastoma (RB1)	180200	AD	RB1	13q14.1-q14.2	Retinoblastoma-associated protein
Sickle cell anaemia	603903	AR	HBB	11p15.5	Haemoglobin beta chain
Sonic hedgehog (SHH)	600725	AD	SHH	7q36	Sonic hedgehog protein [precursor]
Spinal muscular atrophy, type 1 (SMa1)	253300	AD	SMN1	5q12.2-q13.3	Survival motor neuron protein
Spinocerebellar ataxia 2 (SCA2)	183090	AD	ATX2	12q24	SCA2 protein
Spinocerebellar ataxia 6 (SCA6)	183086	AD	CACNA1A	19p13	Voltage-dependent P/Q-type calcium channel alpha-1A subunit
Spinocerebellar ataxia 7 (SCA7)	164500	AD	SCA7	3p21.1-p12	Ataxin-7
Symphalangism, proximal (SYM1)	185800	AD	NOG	17q22	Noggin [precursor]
Tay-Sachs disease (TSD)	272800	AR	HEXA	15q23-q24	Beta-hexosaminidase alpha chain [precursor]
Torsion dystonia 1, autosomal dominant (DYT1)	128100	AD	DYT1	9q34	Torsin A [precursor]
Treacher Collins-Franceschetti syndrome (TCOF)	154500	AD	TCOF	5q32-q33.1	Treacle protein
Tuberous sclerosis type 1	191100	AD	TSC1	9q34	Hamartin
Tuberous sclerosis type 2	191100	AD	TSC2	16p13.3	Tuberin
von Hippel-Lindau syndrome (VHL)	193300	AD	VHL	3p26-p25	von Hippel-Lindau disease tumour suppressor
Wiskott-Aldrich syndrome (WAS)	301000	XL	WAS	Xp11.23-p11.22	Wiskott-Aldrich syndrome protein

PGD cycles were performed for haemoglobin disorders, representing currently the largest experience in the world, the majority of which were done in Cyprus, as routine procedure for those couples who cannot accept prenatal diagnosis and termination of pregnancy [6, 7]. Introduced for the first time in 1996, only 40 PGD cycles for haemoglobin disorders were performed before the year 2000, with almost 100 cycles performed since then. At the present time, the proportion of PGD cases for haemoglobin disorders in our overall experience of 567 PGD cycles for single-gene disorders is as high as 20%.

To improve accuracy of diagnosis of PGD for haemoglobin disorders, a set of polymorphic markers, listed in Figure 3.1, were used, which makes it realistic to select at least three closely linked informative markers in any case performed to analyse simultaneously with mutation testing. Of 113 clinical cycles performed, unaffected embryos for transfer were available in 104 (90.2%), resulting in 35 (33.6%) clinical pregnancies and

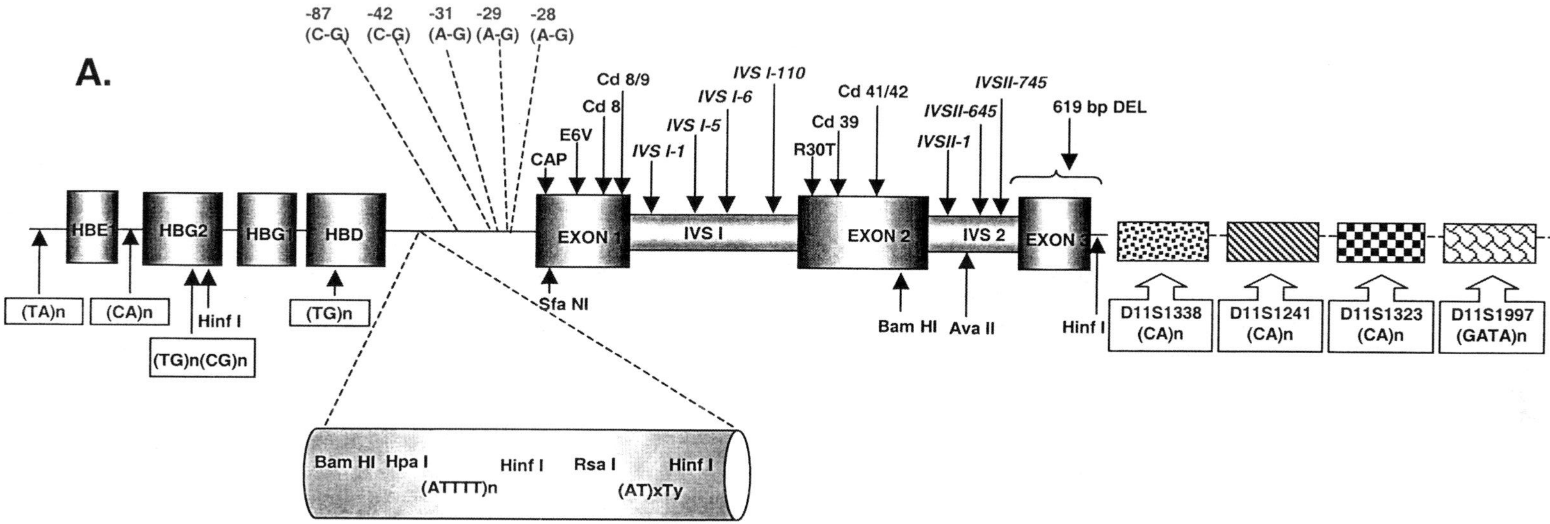

B. **Polymorphic markers**

List of mutations:

RNA Processing mutations:	Transcriptional mutations:	Nonfunctional mRNA:
IVS I – 1 (G – A)	- 87 (C-G	Codon 8 (-AA)
IVS I – 1 (G – T)	- 42 (C-G)	Codon 39 (C-T)
IVS I – 5 (G – C)	- 31 (A-G)	Codon 41/42 (-CTTT)
IVS I - 5 (G – T	- 29 (A-G)	
IVS I – 5 (G – A)	- 28 (A-G)	
IVS I – 6 (T – C)	E 6V (Sickle cell anemia)	
IVS I – 110 (G – A)	R30T	
IVS 2 – 1 (G – A)	**Deletion:**	**Cap site:**
IVS 2 – 654 (C – T)	619 bp	+1 (A – C)
IVS 2 – 745 (C – G)		+2 (T – C)

Figure 3.1. Mutations in beta-globin gene for which PGD was performed and polymorphic markers used in multiplex PCR analysis. Map of human beta-globin gene, showing sites and location of mutations (*A*), and linked polymorphic markers used for avoiding misdiagnosis (*B*). List of mutations is also presented in the lower panel.

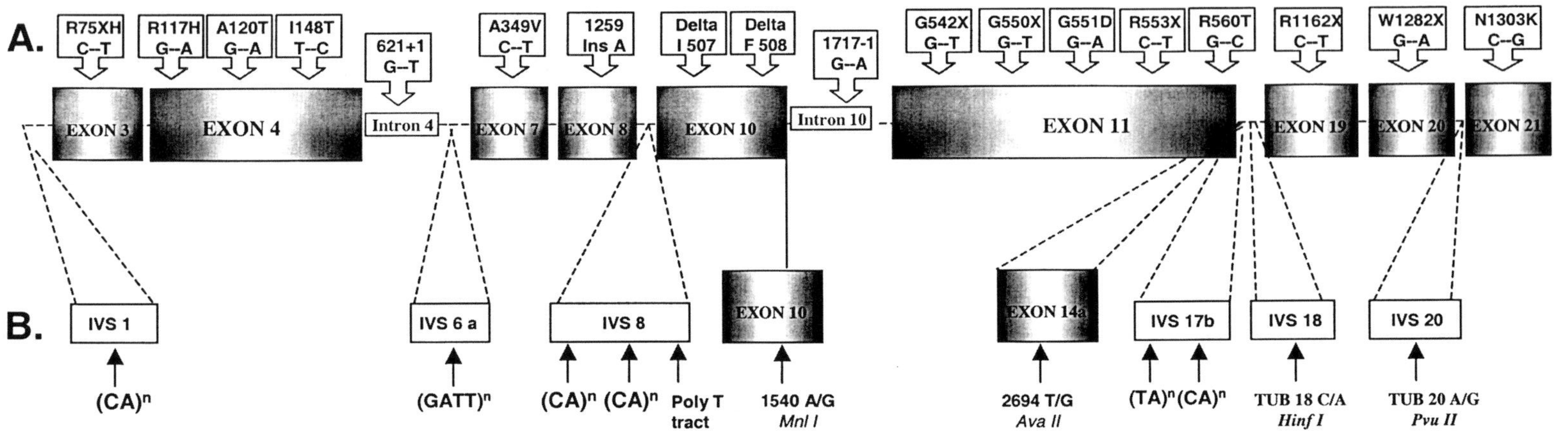

Figure 3.2. Mutations in CFTR gene for which PGD was performed and polymorphic markers used in multiplex PCR analysis. Map of CFTR gene, showing sites and location of mutations (*A*), and polymorphic markers used for avoiding misdiagnosis (*B*).

Table 3.2. Current experience in PGD for haemoglobin disorders

Mutation	Patient/ Cycle	ET/ Embryos	Pregnancy/ Birth
IVS I-110	32 / 55	51 / 130	20 / 16
IVS II - 1	1 / 2	1 / 3	1 / 2
IVS II - 745	4 / 7	6 / 7	2 / 2
IVS I-6	6 / 6	6 / 17	2 / 4
IVS I-1	5 /7	5 / 7	1/1
Codone 6	1 / 1	1 / 3	0
Codone 8	1 / 2	2 / 6	0
Codone 39	6 / 7	7 / 15	2 / 2
Deletion 619 bp	1 / 3	3 / 5	0
alpha-Thalassemia (45 kb deletion)	3 / 5	4 / 7	1 / 1
−29 (A-G)	1 / 1	1 / 1	0
Sickle cell anaemia	12 / 17	17 / 38	6 /2
TOTAL	*73 / 113*	*104 / 249*	*35 / 30*

birth of 30 healthy children. Because the majority of cases as mentioned were done for Cypriot patients, approximately half of the cases were performed for IVS I-110 mutation, which is the most common thalassemia mutation in the Mediterranean region (Table 3.2).

While mainly β-thalassemias were tested, it has recently become possible to test also for α-thalassemia mutations. The case of PGD for α-thalassemia 45K deletion is presented in Figure 3.3, showing the pedigree, PGD design, and the results of this world's first PGD for the disease resulting in birth of an unaffected child. As shown in the figure, both parents are carriers of this mutation, with the father also having haemoglobin H disease. The couple had one previous pregnancy resulting in spontaneous abortion caused by hydrops fetalis. To avoid misdiagnosis, the haplotype analysis with at least five polymorphic markers involved was performed. Of 15 oocytes and embryos tested, 8 were predicted affected, with the remaining 7 carrying one copy of the normal α-globin gene, of which 2 (embryos #7 and #15) were transferred, with the mutant embryos confirmed to be affected, showing the reliability of the approach.

PGD for CFTR has been the major indication from the very beginning of the application of PGD [8–11], and this is currently a routine procedure, which has been done in our experience in 106 cycles, involving testing for 18 different mutations in the CFTR gene (see Figure 3.2). Testing for CFTR mutation is usually performed simultaneously with at least three strongly linked polymorphic markers, which may realistically be selected from a set of 11 available markers listed in Figure 3.2. Of 106 PGD cycles performed by the present time, unaffected embryo transfer was possible in 91 (85.8%), resulting in 40 (44%) clinical pregnancies and birth of 41 unaffected children (four pregnancies are ongoing at the present time).

Because of high prevalence of CFTR mutations, there might be need to test simultaneously for two or three CFTR mutations in the same reaction, as presented in Figure 3.4. As can be seen from the pedigree, PGD for this couple seems to be the only choice, as the father is a double heterozygote for DF508 and R117H, and mother is a carrier of the DI507 mutation in CFTR gene. To avoid testing for three different mutations in blastomeres, taking into consideration approximately 20% risk for ADO for each of the three alleles, which may lead to a potential misdiagnosis, PB1 and PB2 testing was performed to limit the testing to the preselection of the mutation-free oocytes. As can be seen from Figure 3.4, testing for DI507 maternal mutation simultaneously with four closely linked markers allowed the identification of three mutation-free oocytes from the six oocytes available for testing. Two of these embryos resulting from oocytes #1 and 9 with acceptable development potential were transferred, yielding a singleton pregnancy and birth of a healthy boy, confirmed to be an unaffected carrier of the paternal mutation.

Although other autosomal recessive disorders are much rarer (see Table 3.1), overall, they have become the established indicators for PGD in genetic practices. The practical implications of PGD for these groups of disorders may be demonstrated by our experience of PGD for familial dysautonomia (FD) and spinal muscular atrophy (SMA), presented below.

Familial Dysautonomia (FD) is an autosomal recessive disorder, associated with the mutation affecting the donor splice site of intron 20 of the IKBKAP gene, assigned to chromosome 9q31 [12, 13]. It is the most common congenital sensory neuropathy present in 1/3600 live births in Ashkenazi Jews. FD is present at birth with characteristic features, including the absence of fungiform papillae on the tongue, absence of flare after injection of intradermal histamine, decreased or absent deep-tendon reflexes and absence of overflow of emotional tears. This is a devastating and debilitating disorder characterized by the poor development and progressive degeneration of the sensory and autonomous nervous system, gastrointestinal and respiratory dysfunctions, vomiting crisis,

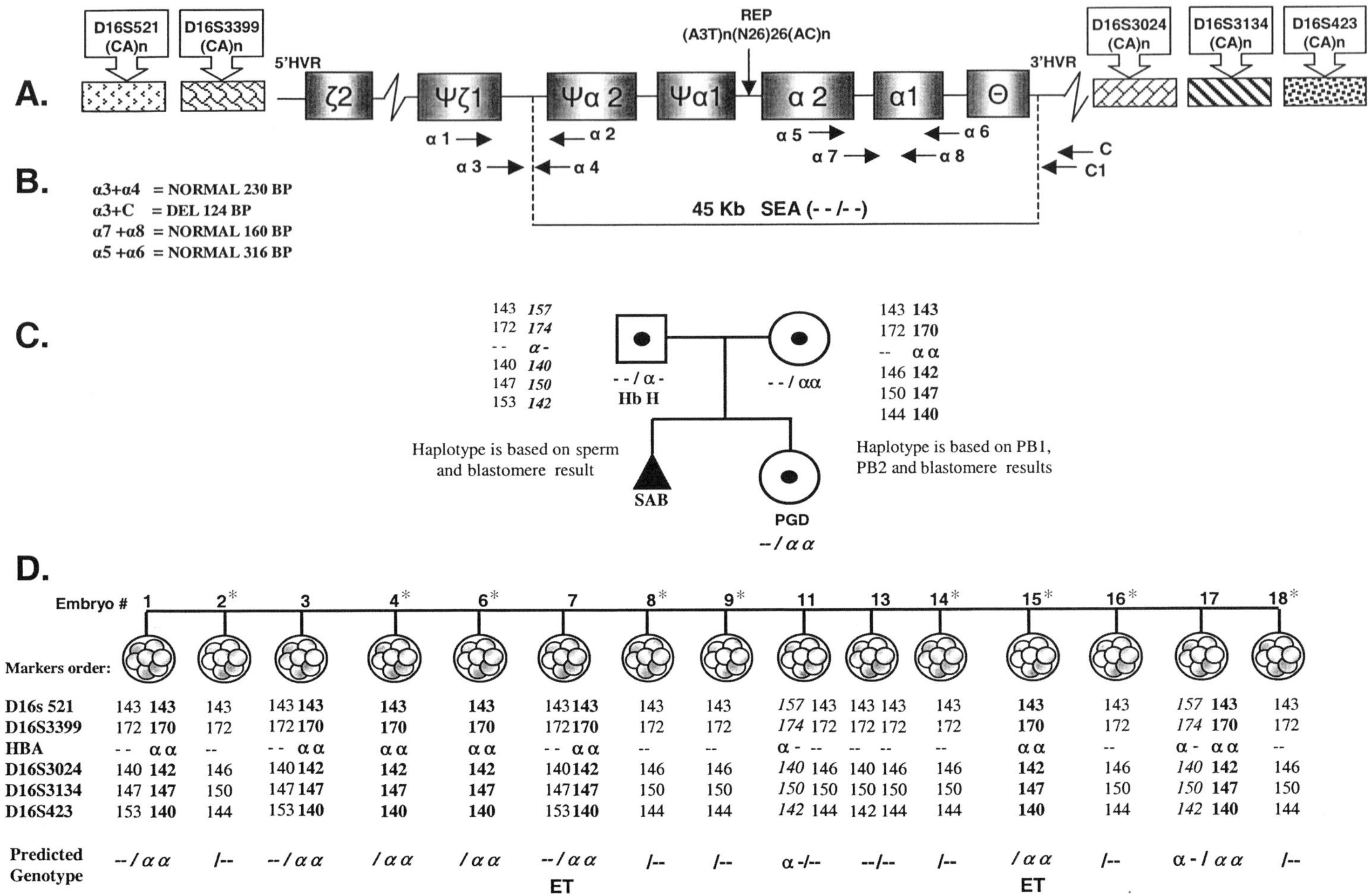

Figure 3.3. PGD for alpha-thalassemia. (*A*) Map of human alpha-globin gene, showing the position of 45 Kb SEA deletion and polymorphic markers used in multiplex PCR analysis; (*B*) size of fragments; (*C*) family pedigree showing both parents carrying SEA deletion, the father also having HbH disease; parental haplotypes are also shown, with paternal haplotypes obtained from sperm testing, and maternal haplotypes obtained from PB1 and PB2 analysis; (*D*) results of testing 15 oocytes and embryos showing the presence of 7 embryos with normal gene, of which two were transferred back to patient, resulting in birth of an unaffected child.

*Embryo was not biopsied; maternal haplotype is based on PB1 and PB2 analysis.

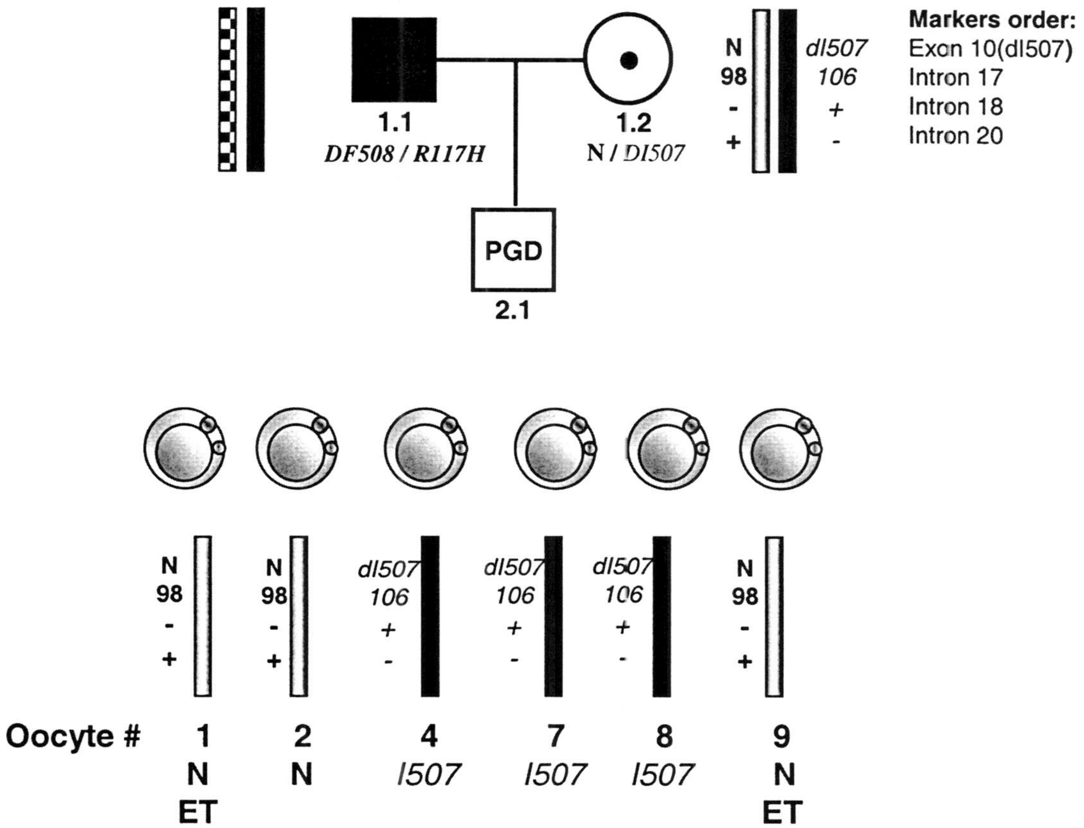

Figure 3.4. PGD for a couple with three different mutations in CFTR gene. (*Top*) Mother (1.2) is a carrier of the delta I507 mutation in CFTR gene. Father (1.1) is affected with cystic fibrosis and has delta F508 and R117H mutations in CFTR gene. (*Bottom*) PGD was performed by sequential PB1 and PB2 analysis. Mother is informative for three inside CFTR gene polymorphic markers (introns 17, 18, and 20). Multiplex heminested PCR of PB1 and PB2 revealed three normal (#1, 2, and 9) and three affected oocytes (#4, 7, and 8). Embryos #1 and 9 were transferred, and healthy boy was born.

excessive sweating, and postural hypotension. Despite a remarkable variability of the disease phenotype within and between families, expected to derive from different mutations producing inactivation of the gene, a single major mutation has been described [14]. It is also of interest that a single T→C change at the base pair 6 of the splice donor site, which is probably responsible for 99% of cases of FD, was shown to result in skipping of exon 20 (74 bp) from the IKBKAP mRNA only in the brain tissue, with varying level of the gene expression in other tissues. This may explain the severe progressive degeneration of the sensory and autonomous nervous system, leading to the continued neuronal depletion with age and early death. The product of the IKBKAP gene is a part of a multiprotein complex, hypothesized to play a role in general transcriptional regulation, so the complete inactivation of the gene might cause a lethal phenotype at any stage of embryonic development [14]. The remarkable variability of the disease phenotype may be explained by the presence of a partially functional gene product in some tissues, including brain, because even a small amount of the encoded protein expressed at critical developmental stages might permit sufficient neuronal survival. In addition, a very rare minor FD missence mutation was described (G→C change at base pair 17 in exon 19 of the gene), which is associated with a mild phenotype in patients with heterozygous status for the major mutation [14].

Despite the above progress in understanding the nature and pathogenesis of the disease, FD is still fatal, with no effective management available at the present time, making PGD a useful option for

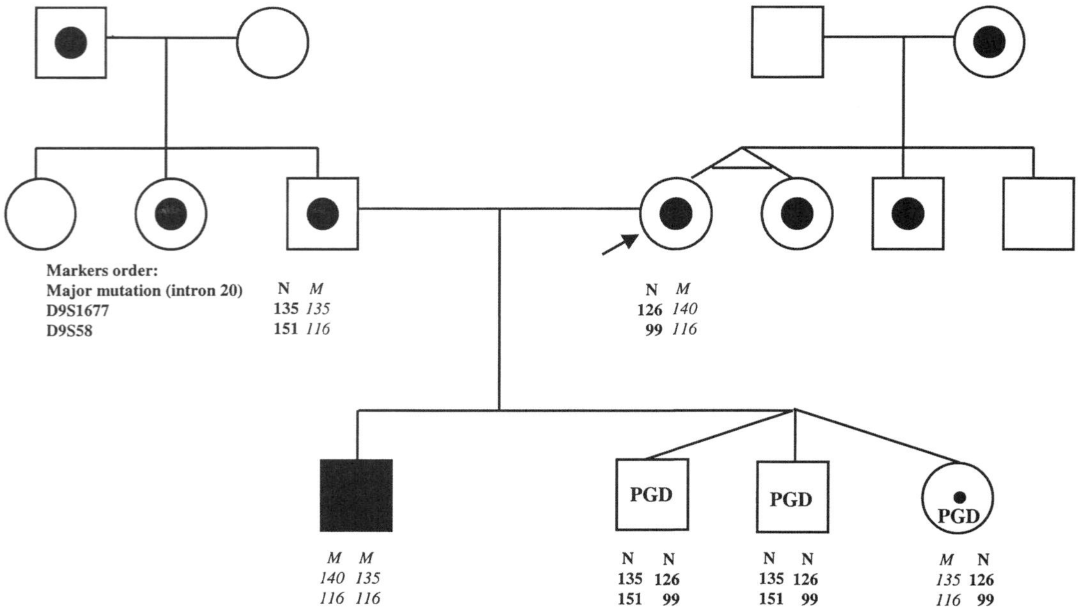

Figure 3.5. Family pedigree of couple undergoing PGD for FD. The father is a carrier of mutation of the donor splice site of intron 20 of the IKBKAP gene, which is linked to 135 bp repeat of D9S1677, and 116 bp repeat of D9S58, while the normal allele is linked to 135 pb and 151 bp repeat of the same polymorphic markers respectively. The mother is also carrier of the same mutation, linked to 140 bp and 116 bp repeat, the normal gene being linked to 126 and 99 bp repeats respectively. As seen from this panel, the paternal and maternal sisters and maternal brother are also carriers of the mutation, which was inherited from paternal father and maternal mother, respectively. Reproductive outcomes of this couple, including one previous affected child with FD and unaffected triplets born following PGD are shown at the bottom.

those at-risk couples who cannot accept prenatal diagnosis and termination of pregnancy as an option for avoiding FD in their offspring. One of such couples presented for PGD with one previous child diagnosed to be affected with FD. Both parents have an Ashkenazi Jews ancestor and, as seen from the pedigree (Figure 3.5), are carriers of the gene for FD, based on marker analysis, which has been available for all members of the extended family [15].

A PGD cycle was performed using a standard IVF protocol coupled with micromanipulation procedures for PB sampling, described previously in Chapter 2. PB1 and PB2 were removed following maturation and fertilization of the oocytes, and tested by the multiplex nested PCR analysis, involving the mutation testing simultaneously with different linked markers as described above. The mutation analysis involved the detection of T → C change in the donor splice site of intron 20, based on *Hha* I restriction digestion, which does not cut the normal allele, while creating two fragments of 60 and 94 bp in the mutant allele (Figure 3.6).

As described in Chapter 2, there may be three genetic possibilities for the PB1 genotype from a heterozygous mother. If no crossover occurs, PB1 will be homozygous (either normal or mutant), but in the event of a crossover, PB1 will be heterozygous. If crossover does not occur and the PB1 is homozygous for the mutant gene, the oocyte must contain two copies of the normal gene and any embryo resulting from this oocyte can be transferred, but this was not the case in any of the oocytes shown in Figure 3.6. If the PB1 is homozygous for the normal gene, the maternal contribution to the embryo must be the mutant gene, which was also not the case as seen from Figure 3.6. In both of these occasions, the extruded PB2 will have identical genotype to oocyte (ooposite to genotype of PB1). In the event of crossover, which has been observed in all cases shown in Figure 3.6, PB1 is heterozygous and the analysis of PB2 is required to predict which maternal allele has been extruded with PB2 and which left in the maternal pronucleus following fertilization. Accordingly, if the normal gene is extruded with PB2 (e.g., PB2 is hemizygous normal), the resulting maternal contribution to the embryos is the mutant gene, and, in reverse, if the mutant gene is extruded with PB2 (e.g., PB2 is

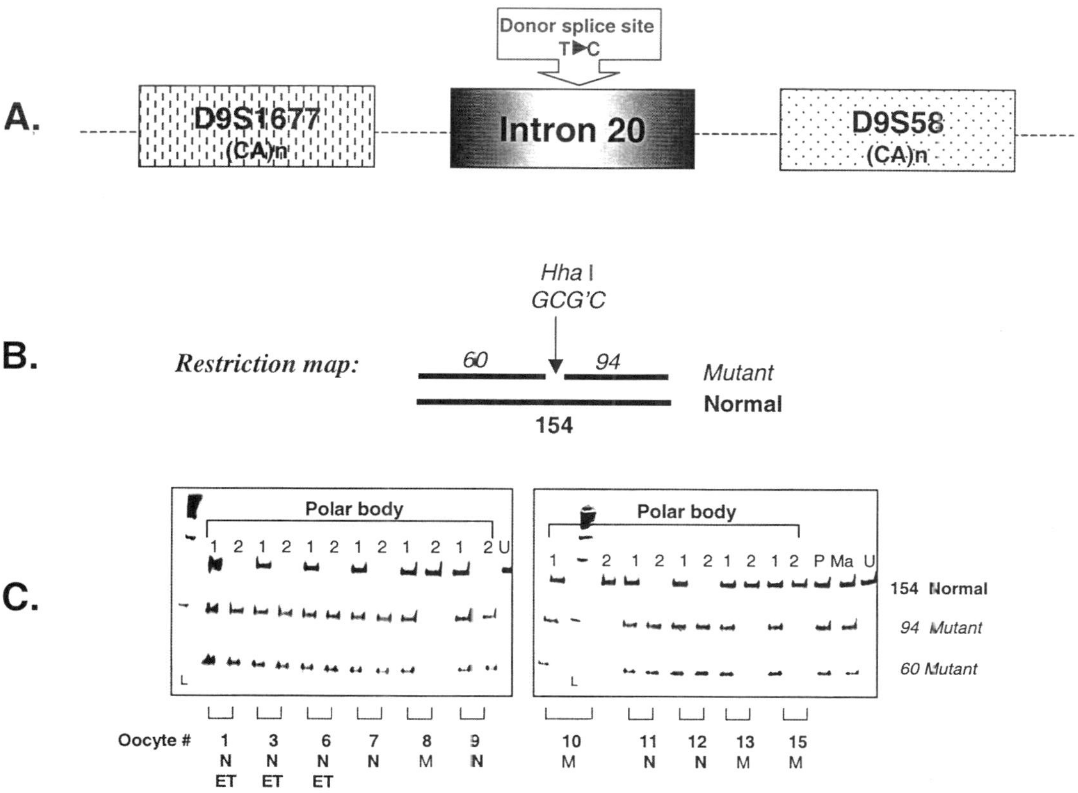

Figure 3.6. PGD for major mutation in IKBKAP gene causing FD. (*A.*) Position of major splice donor mutation T → C in IKBKAP gene and linked markers. (*B.*) Restriction map. Major mutation creates restriction site for *HhaI* enzyme. (*C.*) PB analysis of normal and mutant sequences of IKBKAP gene. Of 11 oocytes tested 7 were mutation-free based on heterozygous PB1 and affected PB2, of which oocytes #1, 3 and 6 were transferred resulting in unaffected triplets. (L, 100 bp, ladder; N, normal; M, mutant; ET, embryo transfer; U, undigested PCR product; Ma, maternal genotype; P, paternal genotype.)

hemizygous mutant), the resulting maternal contribution to the embryos is the normal gene. It is furthermore possible that even the oocytes predicted as mutant may further form unaffected heterozygous embryos following fertilization by mutation-free sperm. Therefore, with an insufficient number of mutation-free oocytes, preselected by PB1 and PB2 sequential analysis, further testing of the resulting embryos may allow the identification of heterozygous unaffected carrier embryos for transfer.

The preselection of mutation-free oocytes was performed based on simultaneous mutation detection and linked marker analysis, involving two strongly linked markers, D9S58 and D9S1677, which were shown not to be involved in recombination in the analysis of 435 FD chromosomes [15]. Therefore, prior to PGD, a single sperm testing was performed to identify the paternal haplotypes, which were as follows: the mutant allele was linked to 116 bp, and the normal to 151 bp repeat of D9S58 marker, while the D9S1677 marker was not informative. The maternal haplotypes were established based on PB analysis and were as follows: the mutant allele was linked to 116 bp repeat of D9S58, and 140 bp repeat of D9S1677 marker, while the normal allele to 140 bp repeat of D9S58 and 126 bp repeat of D9S1677 marker (see Figure 3.5). Primer sequences and reaction conditions are presented in Table 3.3.

A single PGD cycle was performed with 15 oocytes available for testing, of which 11 were with the information for both PB1 and PB2. Of these 11 oocytes, 4 were predicted to be mutant based on the heterozygous PB1 and hemizygous normal (mutation-free) PB2 (oocytes #8, 10, 13, and 15), while the remaining 7 oocytes were free of the mutant gene, as evidenced by the heterozygous PB1 and hemizygous mutant PB2. These results were in agreement with both markers, except for oocytes

Table 3.3. Primers and PCR conditions used in PGD for FD

Gene/Polymorphism	Upper Primer	Lower Primer	Annealing Tm (°C)
Intron 20 major mutation (T → C) *Hha I restriction* Heminested PCR	Outside 5′GTTGTTCATCATCGAGCCC3′	5′CTGATTGATGATATAGGTAATGAGG3′	62–55
	Inside 5′GTTGTTCATCATCGAGCCC3′	5′GCTTTTCATAATTTTAAGTTCTCG3′	55
D9S1677	Outside 5′TGGCTGTTTTGAGAAGT3′	5′ TGGGAGGATGAGTTGAG3′	62–55
	Inside 5′Hex ATGTAACCTGTCTCCACTG3′	5′CTGGGCAACATAGCAAG3′	57
D9S58	Outside 5′AAGCAATCCTCGCACCTCAG3′	5′CCAGGAGTTTGAGACCAGCC3′	62–55
	Inside 5′FamCCTGAGTAGCCGGGACTATA3′	5′TAGGCAACACATCAAGATCCT3′	57

#7 and 11, in which ADO of D9S1677 allele linked to the mutant gene was observed. Three embryos resulting from the above seven oocytes (embryos #1, 3, 6; see Figure 3.6), with the mutation-free status confirmed by both markers, reaching the blastocyst stage, were transferred back to patient, yielding a triplet pregnancy and birth of three unaffected children, including two homozygous normal and one heterozygous carrier. Two of the other embryos resulting from normal oocytes did not form blastocyst and the other two were further tested because of ADO of one of the markers (oocytes #7 and 11).

Three of four embryos deriving from the mutant oocytes were shown to be heterozygous carries of the mutant gene, and one homozygous mutant. One of the embryos deriving from normal oocytes was confirmed to be homozygous for the normal gene and the other was a heterozygous carrier. These embryos, as well as the other three embryos, which appeared to be heterozygous carriers, were frozen and are available for transfer in the future cycles, should the couple wish to have another unaffected child. Of course, the above normal noncarrier embryo could be given preference in transfer, but because a possible detrimental effect of removing blastomères from cleaving embryos cannot be completely excluded, we gave priority to the three nonbiopsied embryos resulting from the mutation-free oocytes. However, when more data are collected on the possible effect of different biopsy procedures on the outcome of pregnancy, and the accuracy of blastomere analysis is further improved, the possibility for parents to choose implanting normal or carrier embryos should be explored.

The results demonstrate a diagnostic accuracy of PGD for FD by sequential PB1 and PB2 analysis, as the follow-up analysis of the embryos resulting from either mutant or normal (mutation-free) oocytes was in agreement with sequential PB1 and PB2 analysis in all the embryos tested, which is in accordance with the extensive data on the sequential PB1 and PB2 analysis, described in Chapter 2. Although as mentioned, ideally at least three linked markers are considered necessary to completely exclude the risk for misdiagnosis due to ADO, the use of two linked markers or even one in the present study was reliable, because it was possible to follow both normal and mutant alleles through the first and second meiotic divisions in all but four oocytes, in which either no PB2 was available due to failure of fertilization or PB2 failed to amplify. In other words, the presence of both mutant and normal alleles in PB1 following meiosis I, and the detection of the mutant allele extruded with PB2 following meiosis II, leaves no doubt of mutation-free status of the maternal pronucleus, even if no linked markers are available for testing. However, the testing of a sufficient number of linked markers would be absolutely essential if the PB1 appears to be homozygous mutant, to exclude the possible ADO of the normal allele, because the failure of detecting ADO could lead to opposite interpretation of the results of PB2. With undetected ADO in PB1, the presence of normal allele in PB2 will erroneously suggest the normal (mutation-free) status of the resulting oocyte, which has apparently a mutant status.

The presented case is the first PGD for FD, demonstrating the clinical relevance of PGD in those couples who cannot accept prenatal diagnosis and termination of pregnancy [16]. Because of the high prevalence of FD in Ashkenazi Jews, with carrier frequency of 1 in 32, this approach may have practical implications, so the at-risk couples require information about the availability of PGD. The presented PGD design for FD may probably be applied without extensive preparatory work in different couples, due to the fact that a single major mutation is involved, although a sufficient number of informative linked markers should be selected, a variety of which are readily available.

As also shown by the above example of FD, it may be predicted that PGD may in the future be applied for gene expression abnormalities, which might be limited to particular tissue or a particular stage of embryonic development. This may also allow preselecting the embryos with the best potential to establish a viable pregnancy, based on the progress on the understanding of stage-specific gene expression.

The other example of autosomal recessive disease for which PGD is complex is spinal muscular atrophy (SMA) [17, 18], which is an autosomal recessive disorder with a newborn prevalence of 1/10,000 and a carrier frequency of 1 in 40–60 individuals. All types of SMA are caused by mutations in the survival motor neuron gene (SMN) locus mapped on chromosome 5q11.2-13 [12]. The SMN gene is present in two highly homologous copies, SMN1 and SMN2, of which homozygous loss of functional survival motor neuron 1 (SMN1) alleles results in SMA, while homozygous absence of SMN2 gives no clinical phenotype. Since 95% of SMA patients lack both copies of SMN1 in exon 7, PGD for SMA is based on the avoidance of SMN1 homozygous deletions, which is complex because of the sequence similarity between the SMN1 and SMN2 genes, requiring simultaneous linkage analysis involving STR markers. To test for paternal SMN1 and SMN2 deletion patterns on each chromosome by blastomere analysis, single sperm testing was required to establish the linkage between normal and deleted alleles with multicopy marker D15S1556 on promoter region of both gene copies.

Figure 3.7 presents PGD for SMA performed for a couple with both parents carrying the deletion in SMN1 gene. Paternal haplotype was predicted by multiplex heminested single sperm PCR analysis. Maternal haplotype was established by sequential PB1 and PB2 analysis. Of 16 embryos tested for deletion and four linked markers (primers are listed in Table 3.4), two embryos were mutant, 10 were predicted to be heterozygous for the deletion in SMN1 gene, and two with only one copy of the normal gene present. In addition, at least three STRs were amplified on each chromosome for aneuploidy testing of chromosomes 13, 16, 18, 21, and 22, which revealed five chromosomally abnormal embryos, including trisomy 13, monosomy 18, double trisomy 18 and 21, double monosomy 18 and 21, and triple monosomy 13, 16, and 22 (embryos #8, 10, 9, 14, and 13, respectively). Abnormalities predicted by PCR were confirmed by whole embryo fixation and FISH analysis. Embryos #2 and 11 were transferred, and healthy twins were born. All unaffected and chromosomally normal blastocysts were frozen for the future family use.

The data also show the importance of simultaneous testing for aneuploidy to avoid misdiagnosis and also to avoid the transfer of chromosomally abnormal embryos destined to be lost in preimplantation or postimplantation development.

3.2 Autosomal Dominant Disorders

Autosomal dominant conditions are important candidates for PGD, as couples have a 50% risk of producing an affected child. Autosomal dominant conditions for which we performed PGD have been presented in Table 3.1. Two of them, early-onset primary torsion dystonia and Charcot-Marie-Tooth disease, are described below.

3.2.1 Primary Torsion Dystonia

Primary torsion dystonia (PTD) is caused in majority of cases by a 3-bp deletion of DYT1 gene, located on chromosome 9q34 [12, 13, 19–21]. This is the most severe and common form of hereditary movement disorder, present in 1/15,000 live births. It presents between ages 4 and 44, characterised by sustained twisting contractures that begin in an arm or leg, spreading to other limbs

A.

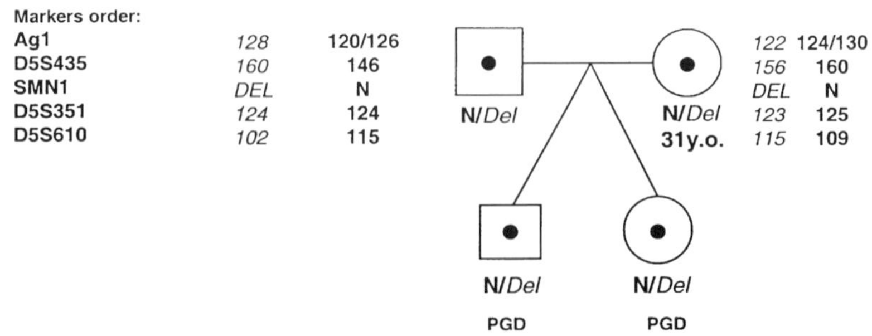

B.

Embryo #	1	2	3	5	6	7	8*	9	10*	11	12	13	14	15	16	18
Predicted SMN1 Genotype:	N/Del	N/Del	N/N	Del/Del	N/Del	N/Del	N/	N/Del	N/	N/Del	N/?	N/Del	Del/Del	N/Del	N/Del	N/Del
Predicted Karyotype:	13,13; 16,16; 18,18; 21,21; 22,22	13,13; 16,16; 18,18; 21,21; 22,22	13,13; 16,16; 18,18; 21,21; 22,22	13,13; 16,16; 18,18; 21,21; 22,22	13,13; 16,16; 18,18; 21,21; 22,22	13,13; 16,16; 18,18; 21,21; 22,22	*13,13,13;* 16,16; 18,18; 21,21; 22,22	13,13; 16,16; *18,18,18; 21,21,21;* 22,22	13,13; 16,16; *18, 0;* 21,21; 22,22	13,13; 16,16; 18,18; 21,21; 22,22	13,13; 16,16; 18,18; 21,21; 22,22	*0, 13; 0, 16;* 18,18; 21, 21; *0, 22*	13,13; 16,16; *0, 18; 0, 21;* 22,22	13,13; 16,16; 18,18; 21,21; 22,22	13,13; 16,16; 18,18; 21,21; 22,22	13,13; 16,16; 18,18; 21,21; 22,22
		ET								ET						

C.

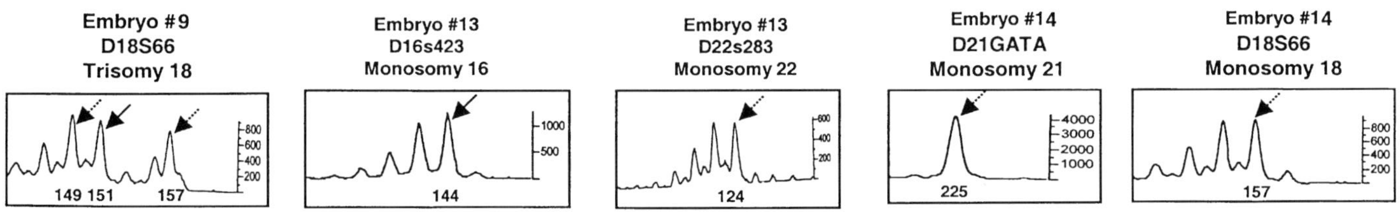

Figure 3.7. Combined PGD for SMA and aneuploidy. (*A*) Pedigree of the family undergoing PGD for SMA. Both parents are carriers of the deletion in SMN1 gene. Paternal haplotype was predicted by multiplex heminested single sperm PCR analysis. Maternal haplotype was established by sequential PB1 and PB2 analysis. As a result of IVF-PGD cycle, healthy twins were born. Positions of polymorphic markers linked to SMN1 gene and applied for improving accuracy of the mutation analysis are shown next to the paternal haplotypes. (*B*) Oocytes #3, 8, and 11 were predicted to be normal by polar bodies analysis. The embryos resulting from these oocytes were subjected to aneuploidy testing using five chromosome-specific probes. Trisomy for chromosome 13 was detected in embryo #8 and monosomy 18 in embryo #10. Blastomeres from the remaining 13 embryos were subjected to multiplex heminested PCR to perform simultaneous mutation, linked polymorphic marker, and aneuploidy analysis for chromosomes 13, 16, 18, 21 and 22. Of these, embryos #5 and 14 were affected, while embryos #1, 2, 6, 7, 9, 11, 13, 15, 16, and 18 were predicted to be heterozygous for the deletion in SMN1 gene. At least three STRs were amplified on each chromosome for aneupolidy testing, which revealed monosomy 13, 16, and 22 in embryo #13, monosomy 18 and 21 in embryo #14, and trisomy 18 and 21 in embryo #9. Chromosomal abnormalities predicted by PCR were confirmed by whole embryo fixation and FISH analysis. Embryos #2 and 18 were transferred, and healthy twins were born. All unaffected and chromosomally normal blastocysts were frozen for the future cycle. (*C*) Examples of chromosomal abnormalities (monosomy and trisomy) detected by PCR of different STRs. (N, normal allele; Del, deletion; ET, embryo transfer; FISH, fluorescent in situ hybridization.)

Table 3.4. Primers and PCR conditions used in PGD for SMA

Gene/Polymorphism	Upper Primer	Lower Primer	Annealing Tm (°C)
SMN (detection of Exon 7 deletion by *Hinf I* or *Dra* I digestion)	Outside 5′TGCAGCCTAATAATTGTTTTCTTTG3′	5′CCAACCAGTTAAGTATGAGAATTCTAG3′	62–45
	Inside for *Hinf I*: 5′CTTCCTTTTATTTTCCTTACAGGGATT3′	5′GTAGGGATGTAGATTAACCTTTTATCT3′	55
	Inside for *Dra I*[3]: 5′AAATGTCTTGTGAAACAAAATGC3′	Inside for *Dra I*[2]: 5′CCTTCCTTCTTTTTGATTTTGTTT3′	55
SMN (detection of Exon 8 deletion by *Dde* I digestion)	Outside 5′GTGGAATGGGTAACTCTTCTTGA3′	5′TAAACTACAACACCCTTCTCACAG3′	62–45
	Inside 5′TAATAACCAAATGCAATGTGAAAT3′	5′TAAACTACAACACCCTTCTCACAG3′	48
D5S1556 promotor area (Heminested)	Outside 5′TCCTTCCTGCCCCCATG3′	5′TCGGCATGTTGCTTAGGC3′	62–45
	Inside 5′FamAAATTCCTAGTAGGAGCTTACATTTAC3′	5′TCGGCATGTTGCTTAGGC3′	55
D5S435 (Heminested)	Outside 5′TTCCTCCATTGACAACTCCTT3′	5′CCTTAGATAGGGTTGATTTACACA3′	62–45
	Inside 5′HexGTCAAGAGCACAGTTTGGAGTG3′	5′CCTTAGATAGGGTTGATTTACACA3′	55
D5S351 (Nested)	Outside 5′ GAGTTTGAAGACCAGTCTATGG3′	5′GAGCATTGCCACTTTAGCC3′	62–45
	Inside 5′FamCACAGCGAGACCCCGTC3′	5′CCCGTGGAAGAAAGGCTAT3′	55
D5S610 (Heminested)	Outside 5′TCCAGTGAATTTTCATTTCAGATAC3′	5′CAAGTGACCCTCCCACCTT3′	62–45
	Inside 5′TCCAGTGAATTTTCATTTCAGATAC3′	5′HexCCAGCCTAAACTGAACTTTCAAAG3′	55

within about five years. Although the phenotypic expression of the disease is similar in all ethnic populations, the highest prevalence was reported among Ashkenazi Jews. Despite a low penetrance (30–40%), the disease phenotype varies greatly among families. In contrast to other neurodegenerative disorders, PTD does not show any distinct neuropathology. A 3-bp deletion in the coding sequence of the DYT1 gene is believed to result in loss of a pair of glutamic acid residues in a conserved region of an ATP-binding protein, torsinA, which has resemblance to the heat-shock proteins and may lead to imbalance of neuronal transmission in the basal ganglia implicated in dystonia. As low levels of dopaminergic metabolites in the cerebrospinal fluid of these patients show no response to dopa, it is probably caused by a defect in release rather than synthesis of dopamine.

The remarkable phenotypic variability of the disease may be explained by the interaction of the 3-bp deletion with modifying genetic, such as polymorphic variations in torsinA or mutations in the associated proteins, or with environmental factors, such as trauma, high body temperature or exposure to toxic agents. Although understanding of these relationships may allow elucidating neuronal mechanisms underlying loss of movement control, there is not yet effective treatment available. This makes PGD a useful option for those at-risk couples who cannot accept prenatal diagnosis and termination of pregnancy as an option for avoiding PTS in their offspring.

Two couples presented for PGD, both with the affected paternal partners carrying the DYT1 3-bp deletion. In one of the couples, the mutation was inherited from the paternal father (Figure 3.8), all of whose four sons were affected. In the other couple, the male partner inherited the mutation from his mother who did not have any other children (Figure 3.9).

PGD cycles were performed using blastomere biopsy, and tested by the multiplex nested PCR

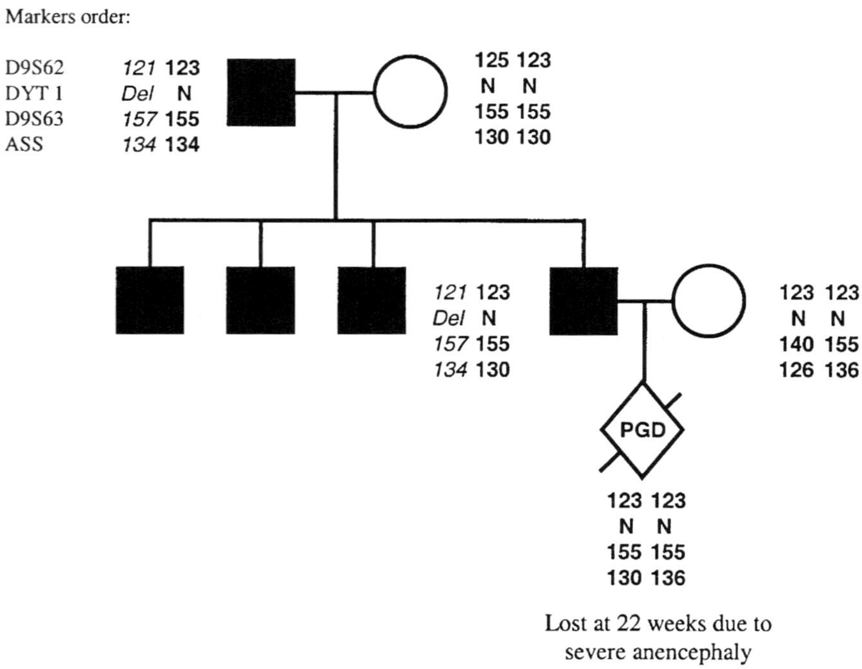

Figure 3.8. PGD for a couple at risk for torsion dystonia 1 resulting in miscarrige of unaffected child with anencephaly. (*Middle*) The father is a carrier of 3-bp deletion of DYT1 gene, which is linked to 121, 157 and 134 bp repeats of D9S62, D9S63, and ASS markers, respectively, while the normal allele is linked to 123, 155, and 130 pb repeats of the same polymorphic markers, respectively. The mother is normal, with one normal DYT1 allele linked to 123, 140, and 126 and the other to 123, 155, and 136 pb repeats of D9S62, D9S63, and ASS markers, respectively. As seen from this panel three paternal brothers are also affected obligate carriers of 3-bp deletion of DYT1 gene, inherited from their father *(top)*. (*Bottom*) Reproductive outcome of this couple, following PGD showing the 3-bp-deletion-free fetus, which was terminated due to anencephaly.

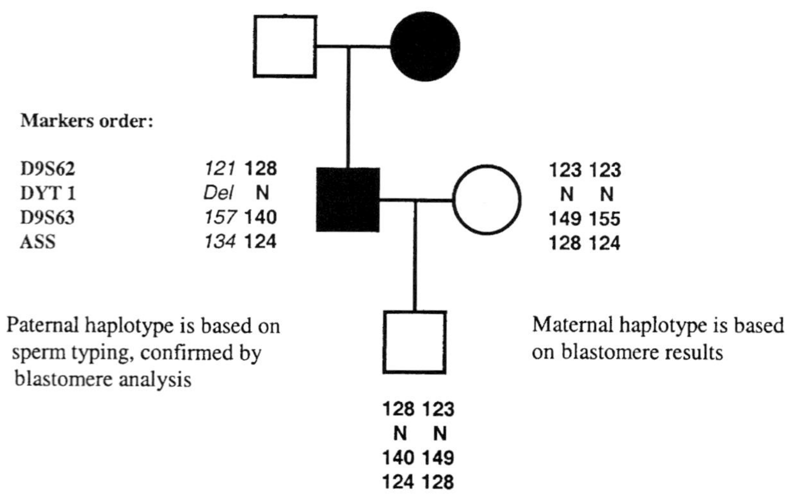

Figure 3.9. Pedigree of couple whose PGD for DYT1 resulted in birth of mutation-free baby. (*Middle*) The father is a carrier of 3-bp deletion of DYT1 gene, which is linked to 121, 157, and 134 bp repeats of D9S62, D9S63, and ASS markers, respectively, while the normal allele is linked to 128, 140, and 124 pb repeats of the same polymorphic markers, respectively. The mother is normal, with one normal DYT1 allele linked to 123, 150, and 128 and the other to 123, 155, and 124 pb repeats of D9S62, D9S63, and ASS markers, respectively. As seen from the Upper panel the mutation was inherited from the paternal mother (*top*) with no other family members available in the pedigree. (*Bottom*) Reproductive outcome following PGD, showing the 3-bp-deletion-free baby, which is in agreement with polymorphic markers, also suggesting the presence of both paternal and maternal normal genes. This embryo originates from the transfer of embryo #6, as shown in Figure 3.10.

analysis, involving the DYT1 mutation testing simultaneously with a set of linked polymorphic markers. The mutation analysis involved the detection of GAG deletion in the coding sequence of the DYT1 gene, based on either fragment size analysis using capillary elecrophoresis, or *BSeR* I restriction digestion, which creates three fragments of 161 bp, 24 bp, and 8 bp in the normal allele (Figure 3.10), in contrast to only two fragments of 185 and 8 bp in the mutant gene.

Three closely linked markers, D9S62, D9S63, and ASS (intron 14), which were shown not to be involved in recombination with DYT1 gene [19, 21–23], were used in the multiplex nested PCR system. To identify the paternal haplotypes, a single sperm testing was performed prior to PGD, which showed the linkage of the mutant allele in both couples to 121 bp, 157 bp, and 134 bp; the normal paternal allele in the first couple was linked to 128 bp, 140 bp, and 124 bp, and in the second to 123 bp, 155 bp, and 130 bp repeat of D9S62, D9S63, and ASS markers, respectively. The maternal haplotypes were, in the first couple 123/123 bp, 149/155 bp, and 128/124 bp, and in the second 123/123 bp, 140/155 bp, and 126/136 bp repeats of D9S62, D9S63, and ASS markers, respectively. Primer sequences and reaction conditions are presented in Table 3.5.

The embryos derived from the oocytes free of DYT1 3-bp deletion, in agreement with the information about the above polymorphic markers, were preselected for transfer back to patients, while those predicted to be mutant or with insufficient marker information were exposed to confirmatory analysis using genomic DNA from these embryos to evaluate the accuracy of single-cell based PGD.

Three PGD cycles were performed, including two for the first couple and one for the second couple, with a total of 19 embryos available for testing, of which 17 were with sufficient information on the mutation and marker analysis to predict the embryo genotype. Of these 17 embryos, 9 were predicted to contain a 3-bp deletion, while the remaining 8 were free of the mutant gene, as also confirmed by the polymorphic markers. Six of these embryos, which reached the blastocyst stage, were transferred back to the patients, two in each of the three cycles, yielding a singleton DYT1 mutation-free pregnancy in two of them.

In one couple, only the second cycle resulted in a clinical pregnancy in which eight embryos were available for testing (see Figure 3.10). Although the biopsied blastomere of one of these embryos (embryo #3) was free of the paternal mutant gene, no other paternally derived alleles were present, suggesting that this cell contained only maternal alleles, probably due to monosomy 9. In the other embryo (embryo #11), one of the biopsied blastomeres showed no amplification, so the second blastomere was removed, but did not show amplification of the DYT1 gene and ASS marker either, together with ADO of the D9S62 paternal allele. Despite this blastomere being informative for the D9S63 marker, which suggested the presence of both paternal and maternal normal alleles, the corresponding embryo was not transferred, to avoid the risk for misdiagnosis associated with the use of a single linked marker. Four of the remaining six embryos were predicted to be mutant (embyos #7, 10, 12, and 13), evidenced by the presence of the DYT1 3-bp deletion and all three markers linked to the mutant gene (see Figure 3.10). The remaining two embryos (embryos #6 and 8) were free of 3-bp deletion, with all three markers not only excluding a possible ADO of the mutation, but also confirming the presence of both maternal and paternal normal alleles. These two embryos were transferred back to the patient resulting in birth of a mutation-free boy. As can be seen from the inherited maternal normal chromosome, this baby originates from the transfer of embryo #6 (see Figures 3.9 and 3.10).

In the second patient, only four embryos were available for testing, two of which showed the presence of the DYT1 3-bp deletion confirmed by all three linked polymorphic markers. Of the remaining two embryos with no evidence for the presence of DYT1 3-bp deletion, the marker analysis also showed the presence of both paternal and maternal normal alleles, despite ADO of one of the paternal polymorphic markers in one of these embryos. The transfer of these embryos yielded a singleton pregnancy, which was terminated at 22 weeks of pregnancy due to a severe anencephaly (see Figure 3.8). The results of the mutation and marker analysis in the abortion material showed that the resulting fetus was free of mutation, with all three markers confirming the presence of both paternal and maternal normal alleles.

The presented results represent the first experience of PGD for TDY1, demonstrating the clinical usefulness of PGD in those couples who cannot accept prenatal diagnosis and termination of pregnancy. Because a single unique 3-bp deletion is involved in more than 70% cases of early-onset PTD,

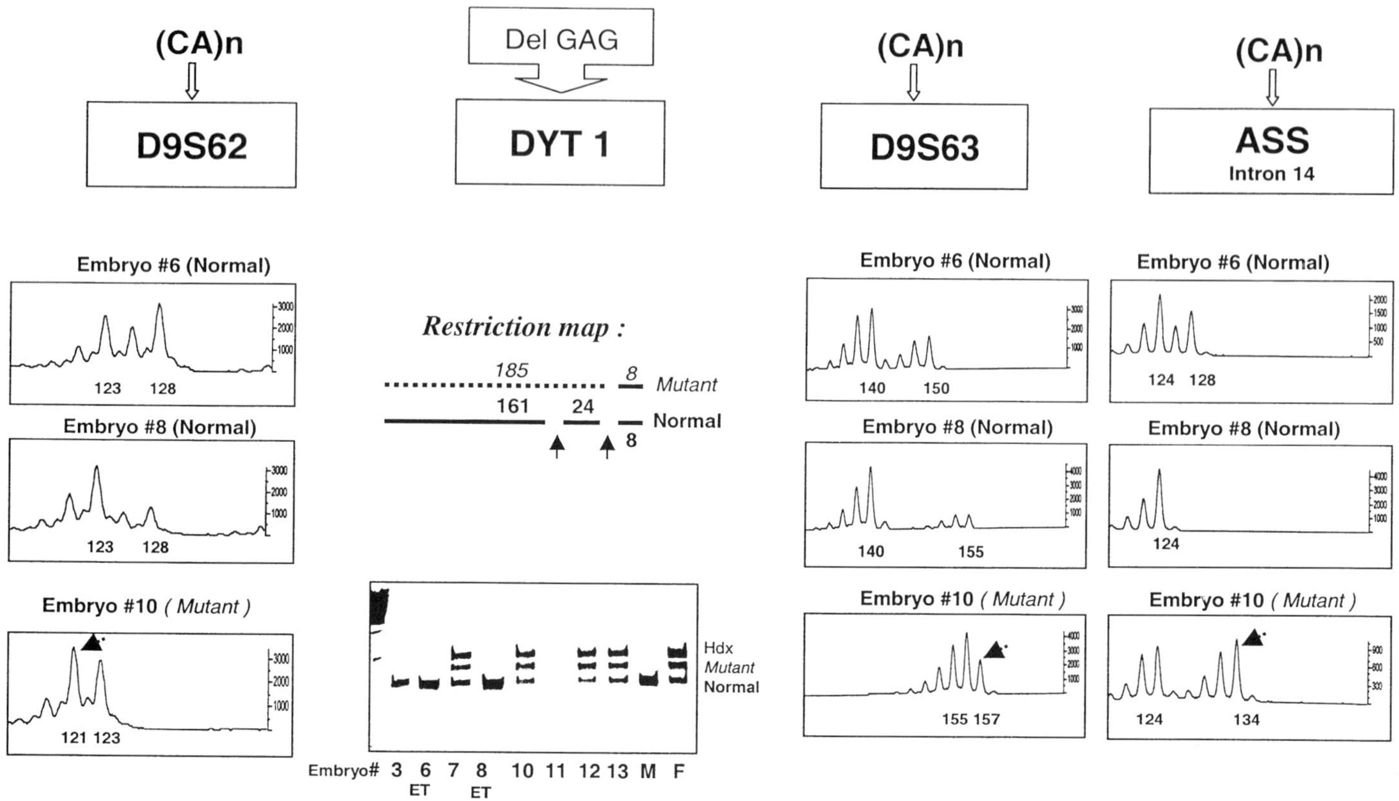

Figure 3.10. Preimplantation diagnosis for GAG deletion of DYT1 gene, resulting in an ongoing mutation-free pregnancy. Capillary electrophoregrams of fluorescently-labelled PCR products of linked markers D9S62 (*first on the left*) D9S63 (*third from the left*) and ASS (*first on the right*), scored by Genotyper™. The data of genotyping of only three embryos are shown as examples, including two transferred normal (embryos #6 and 8) and one affected (embryo #10). Paternally derived 128-, 140-, and 124-dinucleotides indicative of the DYT1 mutation are evident in blatomeres of embryos #6 and 8, together with the presence of maternal normal alleles. *Second panel from the left* shows the location (top) of the mutation in DYT1, restriction map for Dse RI digestion (second panel from the top), creating three fragments of 161, 24 and 8 bp in the normal allele, in contrast to only two fragments of 185 and 8 bp in the mutant gene. *Lower section of the same panel* shows the polyacrylamide gel electrophoregram of Dse RI-digested PCR products of eight blastomeres from one of the cycles of PGD, paternal DNA from sperm (F) and maternal (normal) DNA (M). Hdx—the extra fragment in the heterozygous mutant embryos as a result of heteroduplex formation. Of 8 embryos tested mutation is absent in embryos #6 and 8, in agreement with the linked marker analysis (see relevant panels on the left and right).

Table 3.5. Primers for detection of GAG deletion in DYT1 gene and linked polymorphic markers

Gene/Polymorphism	Upper Primer	Lower Primer	Annealing Tm (°C)
DYT 1	Outside		
(del GAG)	5′GCACAGCAGCTTAATTGACC3′	5′CGTAGTAATAATCTAACTTGGTGAA3′	62–55
(Heminested)	Inside		
	5′GCACAGCAGCTTAATTGACC3′	5′Hex TTTATCTGAGAAAACTCTCTCCTCT3′	56
D9S62	Outside		
(Heminested)	5′ACCTGTAATCCCAGTTGCT3′	5′TCACTTCTGACCCTCCTATCT3′	62–55
	Inside		
	5′Hex GGCAACAGGGCAAGACT3′	5′TCACTTCTGACCCTCCTATCT3′	56
D9S63	Outside		
(Heminested)	5′ATTCTGTGGGGGAATTATG3′	5′TTATAATGCCGGTCAACC3′	62–55
	Inside		
	5′Fam CACAAAAGAAAGTCACAATCC3′	5′TTATAATGCCGGTCAACC3′	56
ASS	Outside		
Intron 14 (CA)n	5′GGGAGCTATAAAAATGACAAT3′	5′TTAACAGGCTGTCTGGCA3′	62–55
(Heminested)	Inside		
	5′GGGAGCTATAAAAATGACAAT3′	5′Fam TAGGTCCGAAAACACAAAG3′	55

the presented PGD design for DYT1 mutation may probably be applied without extensive preparatory work in different couples, also taking into consideration the limited number of founder mutations [21]. As seen from the sperm haplotype analysis of our patients with GAG deletion of DYT1 gene, the same haplotypes appeared to surround the DYT1 gene. The availability of a sufficient number of highly variable and closely linked markers also allows testing for the mutation simultaneously with at least three markers to exclude misdiagnosis due to ADO, which may exceed 10% in blastomere analysis [24]. Because PTD is an autosomal dominant disorder, to ensure a reliable preselection of mutation-free embryos for transfer, PGD should include the detection of both paternal and maternal normal alleles in addition to the exclusion GAG deletion, which may be masked by ADO. This also allows identification of individual embryos that were implanted, as usually two or more embryos are transferred. As mentioned, the baby resulting from the transfer of two mutation-free embryos in one of the PGD cycles described has actually originated from implantation of embryo #6 [25].

Our data also showed that the above microsatellite markers used in this study were also useful in detection of the chromosomal number containing the DYT1 gene in single blastomeres, without which the accuracy of the predicted embryo genotype might not be sufficient. For example, without such information, one of the embryos could have wrongly been predicted to be normal, and which in fact contained no paternal linked markers either. This, therefore, may have suggested the presence of only maternal chromosome 9 due to mosaicism in this embryo, which in fact might have otherwise contained the paternal chromosome 9 with DYT1 mutation, so being affected.

Although prenatal diagnosis for PTD is also available, PGD may seem to be more attractive option. The fact that approximately 70% of the offspring will not develop the disease in the obligate carriers of the mutation makes the decision of what to do in case of a mutation carrier very difficult for the parents.

As for the neonatal outcome, the presented case with anencephaly detected in the second trimester of pregnancy is probably not related to the procedure. As will be described in Chapter 6, the analysis of the outcome of approximately 3000 PGD cases showed that the prevalence of congenital malformations (5%) was not different from the population prevalence. It is too early to analyze the prevalence of specific types of congenital malformations, but it has been demonstrated that the prevalence of anencephaly and other neural tube defects may be efficiently prevented by folic acid supplementation before pregnancy (see Chapter 1), so this should be recommended to all patients requesting PGD, as this is one of those rare occasions when the pregnancy is planned well ahead.

3.2.2 Charcot-Marie-Tooth Disease

Charcot-Marie-Tooth disease (CMT) represents a clinically and genetically heterogeneous group of hereditary peripheral neuropathies, affecting approximately 1 in 2500 in the United States. Although prenatal diagnosis is available, it may lead to termination of pregnancy, which is not acceptable for many couples. PGD has previously been applied in five couples with CMT1A, representing the most frequent autosomal dominant type of CMT caused by 1.5 Mb tandem duplication including the dosage-sensitive gene for peripheral myelin protein 22 (PMP22) on chromosome 17p11.2-12, presenting complexity of diagnosis requiring the application of multiple polymorphic markers [26]. We developed the PCR design, involving the application of 13 highly polymorphic microsatellite markers located within the duplicated area and closely linked to the PMP22 gene, some showing three alleles in patients with CMT1A duplication. Seven PGD cycles were performed for CMT type 1A and 1B. In two of three couples with paternally derived duplication, single sperms analysis was performed to determine normal and mutant haplotypes. In the remaining PGD cycles with maternal mutation, both PB1 and PB2 and blastomeres were analysed, using markers D17S1357, D17S2229, D17S2226, D17S2225, D17S839, D17S2224, D17S2221, D17S2220, D17S291, D17S2219, D17S2218, D17S2217, and D17S2216, which were amplified in the multiplex heminested PCR system, followed by fragment analysis on an ABI 3100 analyser. These cycles resulted in transfer of 11 embryos free of PMP 22 duplication, yielding one birth of an unaffected child and one ongoing pregnancy. Another PGD cycle was performed for maternally derived autosomal dominant CMT type 2E, caused by mutations in the light polypeptide neurofilament protein gene (NEFL). Testing of oocytes by PB1 and PB2 and also blastomeres for the presence of the mutation P8R, simultaneously with microsatellite marker D8S137, resulted in transfer of two unaffected embryos and birth of a child with the normal NEFL gene. Two PGD cycles were performed for the X-linked form of CMT, caused by mutations in the connexin-32 gene (Cx32), for which 28 oocytes were tested by sequential PB1 and PB2 for the presence of the V95M mutation in the Cx32 gene, simultaneously with STRs (DXS453, DXS8052, DXS8030, DXS559, and DXS441), which resulted in the transfer of four mutation-free embryos failing to yield a clinical pregnancy. The example of PGD for paternally derived CMT1A is presented in Figure 3.11, showing the outcome of testing of 10 embryos by blastomere analysis using 10 linked markers (primers and reaction conditions are listed in Table 3.6). Five unaffected embryos were identified, two of which were transferred, resulting in the birth of a healthy child. This is the largest experience of PGD for CMT, demonstrating the reliability and accuracy of the applied PGD designs.

3.3 X-Linked Disorders

More than half of PGD cycles for single-gene disorders were done by gender determination for X-linked conditions, the most straightforward indication from the very beginning of the introduction of PGD, either using PCR or FISH technique. This was not only because the sequence information was not always available, but also because it was technically straightforward to identify female embryos by DNA analysis or FISH technique, despite the obvious cost of discarding 50% healthy male embryos. So the couples at risk may chose either prenatal diagnosis at cost of 50% risk for pregnancy termination, or PGD by gender determination at cost of discarding 50% male embryos. However, because X-linked disorders are maternally derived, the most attractive option may be preselection of mutation-free oocytes through testing for specific causative mutations in PB1 and PB2. This allows avoiding any further testing of the resulting embryos, which may be transferred irrespective of gender or any contribution from the father.

PGD for the X-linked disorders was first attempted more than 14 years ago, and has been performed by gender determination [1]. Gender determination was initially done by single-cell PCR analysis, while now it is performed by the FISH technique in interphase cells. Although sufficiently accurate, as mentioned, this approach leads to a discard of 50% unaffected male embryos, which cannot be accepted by many patients. Thus, there has been a need for specific genetic tests allowing the identification of the unaffected healthy male embryos for transfer as an option to preselect the X-linked mutation-free embryos, irrespective of gender.

In an attempt to preselect the X-linked mutation-free embryos for transfer, a sequential

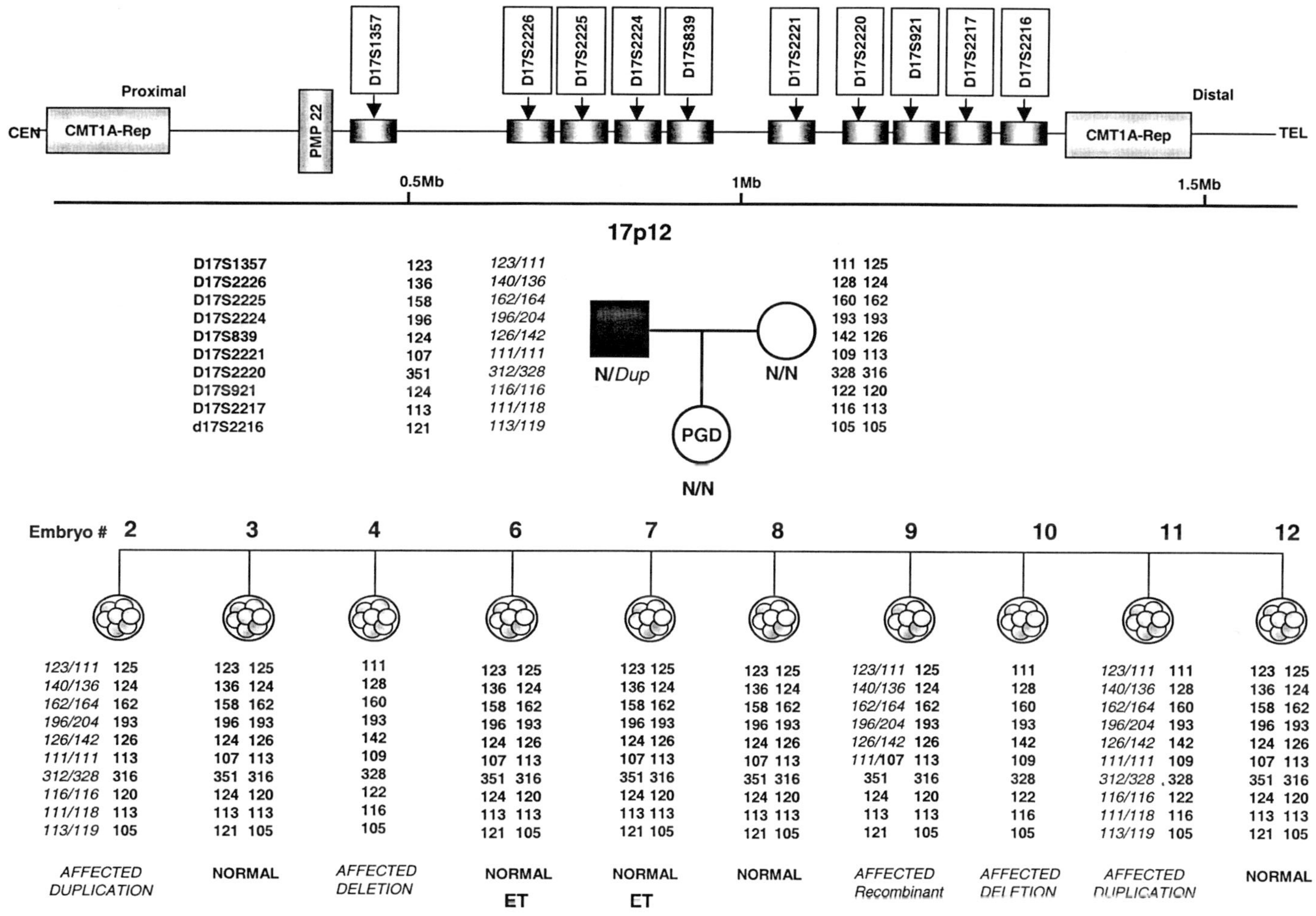

Figure 3.11. PGD for CMT1A resulting in birth of a healthy baby. (*Top*) Schematic representation of 1.5 Mb tandem duplication including dosage-sensitive gene for peripheral myelin protein 22 (PMP22) and linked markers, showing the complexity of the diagnosis. (*Middle*) family pedigree, showing the affected father with duplication and the results of haplotypes analysis established through sperm testing in the father and blastomere analysis in the mother. (*Bottom*) The results of multiplex PCR analysis for 10 markers, which identified five duplication free-unaffected embryos and five affected ones, including two with duplication, one recombinant containing partial duplication, and two with deletion. Two of the unaffected embryos were transferred, resulting in birth of an unaffected child.

Table 3.6. Primers and PCR conditions for PGD of CMT1 by linkage analysis

Gene/ Polymorphism	Upper Primer	Lower Primer	Annealing Tm (°C)
D17S2216 (Heminested)	Outside 5′TGGCAAAGAATAAGAAGGAAAGA3′	5′GTGTCTTGGACTCATTCGGAGACCCTT3′	62–45
	Inside 5′TGGCAAAGAATAAGAAGGAAAGA3′	5′Hex GACAGGCACGGGATTAGGA3′	53
D17S2217 (Heminested)	Outside 5′TCATATGCCCAGTGCCCACTC3′	5′GTGTCTTACACAGAGTTGATGAAAATCACTAA3′	62–45
	Inside 5′TCATATGCCCAGTGCCCACTC3′	5′FamCTTGTCCTGTTTCTCTCCTGGTTTC3′	55
D17S2218 (Heminested)	Outside 5′TTACAGAAGGACAAGGTAGCAAT3′	5′CTAAATGCTTGTGGATTAGTTGTA3′	62–45
	Inside 5′TTACAGAAGGACAAGGTAGCAAT3′	5′HexGTTGTATATTGGATTGGACCCTT3′	55
D17S2219 (Heminested)	Outside 5′GTGTCTTTGCTTTGCCAAGATTAC3′	5′CAACTGTAACACATAAAAGATGAGTTG3′	62–45
	Inside 5′GTGTCTTTGCTTTGCCAAGATTAC3′	5′FamGGGGCTTAGTCTCCATCTGGT3′	55
D17S2220 (Heminested)	Outside 5′ GTGTCTTTGGGCAACAGAGCA3′	5′CTAATAATTTTCCATTTGTAACCTTG3′	62–45
	Inside 5′GTGTCTTTGGGCAACAGAGCA3′	5′FamTTCTCTTGAAACCTCAGTCATCTTT3′	48
D17S2221 (Heminested)	Outside 5′GTGTCTTGAAAATTTCCCACCCAAAG3′	5′GCCTCTCCCTGAGTGTCTGGT3′	62–45
	Inside 5′GTGTCTTGAAAATTTCCCACCCAAAG3′	5′HexGCTGCTGTTCCTTCAAAATGTG3′	55
D17S839 (Heminested)	Outside 5′AACAACAGCGAAACTCTGTCTCA3′	5′CCAAACTTCCAAGTAAGAGAGGG3′	62–45
	Inside 5′AACAACAGCGAAACTCTGTCTCA3′	5′FamAGAGAGACCCTGGAAGATCAACTAC3′	55
D17S2224 (Heminested)	5′GTGTCTTGTTCATTCTATCGTCTCAAA3′	5′AGCAGAAATGAGAGAGGAGACATTA3′	62–45
	5′GTGTCTTGTTCATTCTATCGTCTCAAA3′	5′HexTACAAAGGATCATAAGGCTACCATA3′	55
D17S2225 (Heminested)	5′TGGTTAAATCTGTACTCATTATGGC3′	5′TGTCTTACAGTTGTAGAGGGCTTGTG3′	62–45
	5′Hex TTCATGTATCTGGGAGTATTCACG	5′TGTCTTACAGTTGTAGAGGGCTTGTG3′	55

PB1 and PB2 analysis was applied as an alternative to gender determination. One of such cases, involving PGD for ornithine transcarbamylase deficiency (OTC), was first reported more than five years ago, showing the feasibility of the approach [27].

The couples for whom PGD was applied now include those with previous male offspring affected by Duchenne muscular dystrophy (DMD), Charcot-Marie-Tooth disease X1 (CMTX1), fragile-X syndrome (FMR1), hemophilia A (F8) and B (F9), hyper IgM syndrome (HIGM1), Wiscott-Aldrich syndrome (WAS), X-linked hydrocephalus (L1CAM), Hunter syndrome (IDS), choroideremia (CHM), myotubular myotonic dystrophy (MTMD), Norrie disease (NDP), X-linked adrenoleukodysprophy (ALD), ornithine transcarbamylase deficiency (OTC), Pelizaeus-Merzbacher disease (PMLD), incontinencia pigmenti (IP), and Alport disease (ATS). Overall, 99 PGD cycles were performed for these couples as presented in Table 3.7, of which five were performed by blastomere biopsy only, with the remaining using PB1 and PB2 analysis, or both PB1 and PB2 followed by blastomere biopsy.

Following maturation of oocytes in a standard IVF protocol, PB1s were removed at least 3 hr after oocyte retrieval (38 hr after chorionic gonadothropin [hCG] administration), using micromanipulation procedures described previously in Chapter 2. Then oocytes were fertilized by intracytoplasmic sperm insertion (ICSI), followed by

Table 3.7. Results and outcomes of PGD for X-linked disorders

Disease	# Patients	# Cycles	Cycle by PB	Cycle by PB and BL	Cycle by BL	# ET	# Embryos	Pregnancy	Birth
DMD	8	14	1	12	1	13	36	5	6
CMT 1X	2	2	1	1	0	2	4	0	0
FRA X	29	43	16	24	3	37	86	16	15
Hemophilia A and B	4	5	0	5	0	5	11	2	2
HIGM syndrome	1	2	0	2	0	2	2	1	1
Wiscott-Aldrich	1	1	0	1	0	0	0	0	0
Hydrocephalus	3	8	2	6	0	8	18	0	0
Hunter syndrome	1	2	1	1	0	2	3	1	2
Choroideremia	1	1	1	0	0	1	2	1	1
MTM	1	2	2	0	0	1	3	0	0
Norrie disease	2	4	0	4	0	4	7	0	0
X-ALD	2	3	0	3	0	3	7	1	1
OTC	4	5	3	1	1	5	8	4	4
Pelizaeus-Merzbacher disease	2	2	0	2	0	2	5	1	2
Incontinencia pigmenti	3	4	0	4	0	4	8	1	1
Alport	1	1	1	0	0	1	2	0	0
TOTAL	*65*	*99*	*28*	*66*	*5*	*90*	*202*	*33*	*34*

PB2 removal, performed approximately 16–18 hr after ICSI. Although PB1 and PB2 were removed in sequence, they were amplified at the same time, using nested multiplex PCR, with the primer designs worked out for each of the conditions mentioned. Primers and PCR conditions for these conditions are presented in Table 3.8.

Based on the sequential PB1 and PB2 testing involving the mutation and/or linked marker analysis, results of which were already available on day 1 (approximately, 30–32 hr after oocyte retrieval), the embryos resulting from the oocytes predicted to be free from maternal mutation were transferred back to the patients on the day 3, while those predicted to be affected were further tested for confirmation of diagnosis, when available. Approximately, 7.7 oocytes per cycle on the average were obtained with PB1 results, which appeared to be heterozygous in 66% and homozygous normal or mutant in 34% of the oocytes. PB2 results were available in 79% of these oocytes (6.1 ooocytes per cycle on the average), so in the remaining oocytes, genotype prediction was not possible.

ADO, if undetected, remains the major problem in avoiding misdiagnosis, as observed in one of the cases of PGD for FMR1, presented below. ADOs were detected either by the use of the linked markers or by sequential PB1 and PB2 analysis. In the case of PGD for X-linked hydrocephalus, both PB1 and PB2 in one of the oocytes had identical homozygous mutant genotype, suggesting that PB1 was apparently heterozygous, with the normal allele not detected due to ADO. In eight PGD cycles performed for this condition, a total of 18 unaffected embryos were preselected for transfer (at least two embryos in each cycle). These embryos originated from the mutation-free oocytes preselected based on heterozygous status of PB1 and homozygous mutant PB2, which not only predicted the mutation-free status of the resulting embryos, but also reliably excluded any probability of ADO.

In two PGD cycles for MTMD in which oocytes were tested by both PB1 and PB2, three unaffected embryos were transferred, originating from the oocytes with heterozygous PB1 and homozygous mutant PB2, or with homozygous mutant PB1 and normal PB2, in which ADO in PB1 was excluded by two linked markers, but only chemical pregnancy was obtained. In five PGD cycles for F8 and F9, 11 embryos originating from the mutation-free oocytes predicted to be normal based on heterozygous PB1 and mutant PB2 were transferred in five cycles, yielding two clinical pregnancies and birth of two unaffected children.

The largest group of X-linked disorders for which PGD was offered was FMR1, which is tested indirectly using linkage analysis, as no test is available for direct analysis of the expanded allele [28].

Table 3.8. Primers to amplify mutation or linked markers for PGD of X-linked diseases

Gene/ Polymorphism	Upper Primer	Lower Primer	Anneal. Tm (°C)
FMR1 AC1	Outside 5′AGCTGCAAAGAGAAACAGACA3′	5′AATATCAGGCCAGGCACA3′	62–45
	Inside 5′GTTGATGCTGAACATCCTTATCG3′	5′Fam GGCTGAGGCATGATGAGAGTC3′	50
FMR1 AC2	Outside 5′GCCCTAATCAGATTTCCACA3′	5′GATGCGGTGGCTCAAG3′	62–45
	Inside 5′CAAAAAGAACCCAGATGTGA3′	5′Fam GGGAGGATAGCTCAAGCTC3′	50
DXs548	Outside 5′Fam CCTACATCAAAGTCCCAGCA3′	5′AAGCCTGCAACCAAACACTG3′	62–45
	Inside 5′Fam CCTACATCAAAGTCCCAGCA3′	5′GTACATTAGAGTCACCTGTGGTGC3′	50
DXs1193	Outside 5′GCCAAGGCATAGAAGACAAC3′	5′TGTCAGCACAAAAGGGCTTA3′	62–45
	Inside 5′ATTGTTCATGCAACTCTCCTC3′	5′Hex TCATCCAAGCTACTTATTTTAAG3′	50
Factor IX RFLP *MseI*	Outside 5′AGAGGGATAAATACATCAATGGC3′	5′AATATATTGTCTCCAGCCTGTAGC3′	62–45
	Inside 5′GATAGAGAAACTGGAAGTAGACCC3′	5′ATTAGGTCTTTCACAGAGTAGAATTT3′	60
Factor IX RFLP *MnlI*	Outside 5′TTGTAATACATGTTCCATTTGCC3′	5′GGGAATTGACCTGGTTTGG3′	62–45
	Inside 5′TTCTAGTGCCATTTCCATGTG3′	5′ATCTTCTCCACCAACAACCC3′	60
Factor IX VNTR	Outside 5′GGGACCACTGTCGTATAATGTG3′	5′GAAGAGACACTCCTGAACTCTGG3′	62–45
(50 bp repeat)	Inside 5′CCAAAATGTCATTGTGCAGC3′	5′TCTGAATCATATTTCTCCTTCCC3′	60
MTM1 R241C	Outside 5′TTGGATGTGGTGCTAATTAAG3′	5′CTCATCATCTTTATTTCGTTTC3′	62–45
(loses artificial *HhaI* site)	Inside 5′CGCAGTGAGATTGCAAGTG3′	5′CAAGAGGCTGACTGCAGC3′	55
MTM1 DXS1684	Outside 5′AGCACCCAGTAAGAGACTG3′	5′TGAATCAATCTATCCATCTCTC3′	62–45
	Inside 5′CAGGCCACTACCACTTATG3′	5′HexTACTGTTTTCCACTCTAATGC3′	55
X-Hydrocephalus Fr.168 (-C)	Outside 5′GAGTGTCAGCCCGTCTG3′	5′AGTAGAGGTTGCCGTTCTG3′	62–45
(loses *Hpa II* site)	Inside 5′GTGGCCAAAGGAGACAGTGAA3′	5′GAAGACACCCCCGCTAACA3′	60
PLP1 L86P	Outside 5′ATCTATGGAACTGCCTCTTTCTTCT3′	5′CGCTCCAAAGAATGAGCTTGAT3′	62–45
(*Msp I* cuts mutant)	Inside 5′ATCTATGGAACTGCCTCTTTCTTCT3′	5′GGCCCCCTGTTACCGTTG3′	60
5′PLP1 (CA)n	Outside 5′GGAAGGGACCTGAAGAAAAAGAT3′	5′GCTCTCATTTACCTGGCACACTA3′	62–45
(Heminested)	Inside 5′FamGGGGTGATTCTAGTAACCAGGC3′	5′GCTCTCATTTACCTGGCACACTA3′	55

Table 3.8. *Continued*

Gene/ Polymorphism	Upper Primer	Lower Primer	Anneal. Tm (°C)
DXS8020 (Heminested)	Outside 5′GATGGGTCGGTGATGAGAA3′	5′TCAGCATTACAATTCTGTATAGACT3′	62–45
	Inside 5′FamGGGAGGTAGAAAAGGGTTGGAAAGG3′	5′TCAGCATTACAATTCTGTATAGACT3′	60
DXS8096 (Heminested)	Outside 5′CCTAAGGTTTCCAGATTTAGCA3′	5′TGTGAGCCAGTTCTTGAAAAT3′	62–45
	Inside 5′FamACATCCAGAGAAACAGAACCAA3′	5′TGTGAGCCAGTTCTTGAAAAT3′	55
DXS8089 (Heminested)	Outside 5′ATGAATGGTGCTAGGATTTGG3′	5′ATTTTTACATTATGATGTGGTCAAA3′	62–45
	Inside 5′ATGAATGGTGCTAGGATTTGG3′	5′Fam TTCTTGTATGGCTTCTGGGTC3′	60
DXS1191 (Heminested)	Outside 5′TTGAAGGATGCACACTACAAA3′	5′GCCCCGTTTGATGCTTCTA3′	62–45
	Inside 5′TTGAAGGATGCACACTACAAA3′	5′FamCAGCAGTAAACTGTTTCCCTTT3′	60

We performed 43 PGD cycles for FMR1, which resulted in embryo transfer in 37 of them. As the mutant allele for this condition cannot be amplified, the preselection of the mutation-free oocytes was inferred from the presence of the linked markers for the normal allele. Only three of these cycles were performed by blastomere biopsy alone, with the remaining being done either by PB1 and PB2 only or PB1 and PB2 analysis followed by blastomere biopsy. Overall, 86 mutation-free embryos were transferred in 37 cycles resulting in 16 (43%) clinical pregnancies, an extremely high pregnancy rate observed in women carrying FRM1 mutation. Fifteen healthy children were born as a result of these PGD cycles, with only one misdiagnosis. The latter was due to the transfer of three embryos, two deriving from oocytes with heterozygous PB1 and mutant PB2, and one from the oocyte with presumably homozygous mutant PB1, which in fact turned out to be heterozygous because of an undetected ADO of both alleles linked to the normal gene (Figure 3.12). As only two of four available markers were informative for linkage analysis in this cycle, in the absence of the direct test for the expanded allele, the chance of ADO of the normal allele in this case was predicted to be in the range of 5%, based on our previous observations described in detail elsewhere.

The second largest group, for which 14 PGD cycles were performed, was DMD, resulting in the preselection and transfer of mutation-free embryos in all but one cycle. A total of 36 (2.8 per cycle) embryos resulting from mutation-free oocytes were transferred in 13 cycles, yielding five clinical pregnancies (38.5%) and four healthy mutation-free children born, with two pregnancies still ongoing.

Of five PGD cycles performed for four couples at risk for producing offspring with OTC, unaffected embryos for transfer were predicted in every cycle (1.6 embryos per cycle), resulting in four clinical pregnancies and birth of three healthy children free of mutation (one pregnancy is still ongoing following confirmation of PGD by prenatal diagnosis). The preselection of mutation-free oocytes was based on heterozygous PB1 and mutant PB2, except only a few with the homozygous mutant PB1 and normal PB2, but with the linked marker information available to confirm the diagnosis. Despite transferring only a single embryo or two from the mutation-free oocytes in each of these cycles, an extremely positive clinical outcome of PGD for OTC was observed.

Overall, 202 embryos resulting from the oocytes predicted to be free from mutation were transferred back to the patients in 90 cycles (approximately 2.2 embryos per cycle), yielding 33 clinical pregnancies, which have resulted in the birth of 33 unaffected children. In all embryos resulting from mutant oocytes available for follow-up analysis, the predicted results of sequential PB1 and PB2 analysis were confirmed. In those cycles performed by PB removal, both PB1 and PB2 results were obtained in 80% of oocytes,

A.

Oocyte	Cell Type	DXs1193	DXs548	Predicted Genotype
1	PB1	*154* / **156**	*245* / **247**	*AFFECTED*
	PB2	**156**	**247**	
3	PB1	154 / **156**	*245* / **247**	*AFFECTED*
	PB2	**156**	**247**	
5	PB1	154 / **156**	*245* / **247**	**NORMAL**
	PB2	*154*	*245*	
6	PB1	*245* / **247**	*245* / **247**	**NORMAL**
	PB2	*154*	*247*	
9	PB1	*154**	*245**	**NORMAL***
	PB2	**156**	**247**	
10	PB1	*245* / **247**	*245* / **247**	*AFFECTED*
	PB2	**156**	**247**	
Cord blood		*154*	*245*	*AFFECTED*

B.

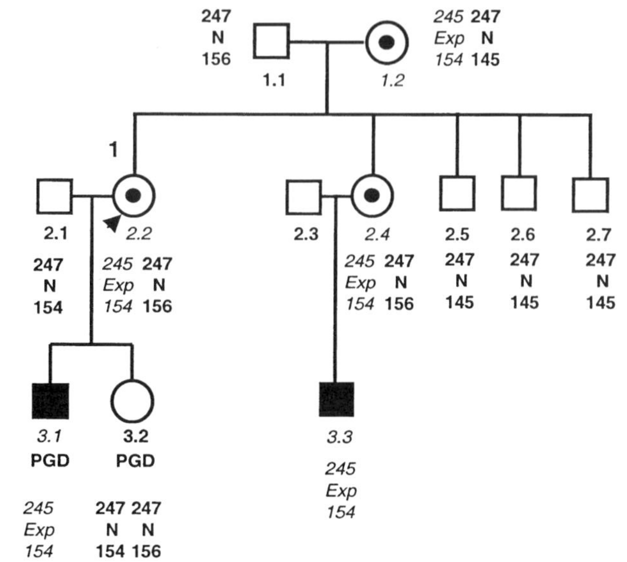

Figure 3.12. PGD for FRM1 with misdiagnosis due to undetected ADO in PB1. Family pedigree (*B*) and PGD results of haplotype analysis for CGG expansion in FMR1 gene using two linked markers (*A*). The mother (2.2) is carrier of an expanded allele linked to 154 and 245 markers. Her sister (2.4) also has expanded allele inherited from their mother (1.2). One of the sisters presented for PGD is shown by arrow. The other sister has an affected son (3.3), who carries only expanded allele (linked to 154 & 245 markers). On the left of the pedigree, an affected child carrying only expanded allele (linked to 154 & 245 markers) was born following PGD (3.1), due to the predicted 5% risk of misdiagnosis, resulting from undetected ADO of the normal allele in PB1 of the corresponding oocyte (shown by * in the table *B*), from which the transferred embryo derived. As can be seen from the table, misdiagnosis originates from undetected ADO of both markers linked to the expanded allele in PB1 of oocyte #9, which was erroneously considered homozygous mutant, while it was actually heterozygous; following the extrusion of PB2 with normal allele, the maternal contribution to the resulting oocytes should have been considered mutant.

in some cases making unrealistic the preselection of a sufficient number of embryos for transfer when only a few oocytes were available for analysis, so in more than half of such cases blastomere biopsy was required to avoid cancelling cycles. However, in some of such cases, due to a fertilization failure with no PB2 extruded, no embryos were available to perform blastomere biopsy either.

With progress in obtaining the sequence information for X-linked disorders, PGD for an increasing number of X-linked disorders may soon be performed by specific diagnosis to avoid discarding 50% healthy male embryos. On the other hand, as demonstrated above, testing for X-linked genetic disorders may be entirely limited to oocytes, because of the maternal origin of the mutations involved, making useless any further manipulation and testing of the resulting embryos, which may be transferred irrespective of gender or any contribution from the father. PGD using a specific diagnosis for X-linked disorders has been reported also based on blastomere biopsy [29].

Presented results further demonstrate the clinical usefulness of PGD for X-linked disorders by sequential PB1 and PB2 analysis, which resulted in the transfer of the embryos originating from the oocytes predicted to be free from X-linked mutation in all clinical cycles performed. As mentioned, these transfers yielded 33 clinical pregnancies, resulting in the birth of healthy unaffected children in all but one case, in which affected embryos were misdiagnosed to originate from the oocyte with the homozygous mutant PB1, predicted on the basis of only one additional polymorphic marker, which leaves at least a 5% chance for misdiagnosis. With two embryos having been reliably diagnosed for transfer, the couple may have had reason for accepting such misdiagnosis risk, to have three instead of two embryos for transfer, improving the chances to become pregnant. The results demonstrate once again that the embryos resulting from the oocytes with homozygous mutant PB1 should not be transferred unless sufficient number of linked markers is available to exclude the possible heterozygous status of PB1. This is especially relevant to testing for FMR1, which is entirely based on linkage analysis. To exclude completely the risk for misdiagnosis in the embryos resulting from homozygous affected PB1, testing for as many as four linked markers, overall, is required.

In conclusion, the data show that a sequential PB1 and PB2 analysis may be a practical option for PGD of X-linked disorders, providing an alternative to PGD by gender determination.

3.4 Homozygous or Double Heterozygous Recessive Conditions

PGD is of special value for dominant conditions and those couples with one homozygous or double heterozygous affected partner, when only a 50% chance of having an unaffected child may be expected. This was first applied for a couple with compound heterozygous male partner affected by phenylketonuria (PKU) and female partner carrying the third PKU mutation [30] One in every 10,000 infants in the United States is born with PKU, an inherited metabolic disorder that causes mental retardation if untreated. When infants are fed a strict diet from birth, they have normal development and a normal lifespan. However, a strictly controlled diet must be maintained, especially during pregnancy in female patients to avoid potential detrimental effects on the fetus. Because of considerable progress in screening newborns for PKU and dietary modification treatment, PKU has not often been included in prenatal diagnosis. With the introduction of PGD, which enables couples to have a healthy pregnancy, couples at risk may avoid giving birth to a child who needs life-long dietary treatment. PGD may be particularly useful for couples in which one partner is homozygous or compound heterozygous affected because there is a 50% chance of producing a child affected with PKU.

As mentioned, the first case of PGD for PKU was performed in a couple with compound heterozygous affected partner presented for PGD in connection with their first offspring with PKU, and their 50% risk of producing another affected child. The 31-year-old mother was a carrier of R408W mutation in exon 12 of phenylalanine hydroxylase (PAH) gene, while the affected father was compound heterozygous for R408W and Y414C mutations in the same exon. Following maturation of oocytes in a standard IVF protocol, PB1s were removed following maturation of oocytes. Then, following fertilization of oocytes by ICSI, PB2 were removed using the micromanipulation techniques described in Chapter 2. PB1 and PB2 were amplified by heminested multiplex PCR,

Table 3.9. Primers and reaction conditions used in PGD for PKU

Gene/Polymorphism	Upper Primer	Lower Primer	Annealing Tm (°C)
Y414C Nested PCR (*Pst I* cuts mutant allele)	Outside 5′ATGCCACTGAGAACTCTC3′	5′ AGTCTTCGATTACTGAGAAA 3′	62–45
	Inside 5′TACCTTTCTCCAAATGGTG3′	5′CCTTTAGGTGTGTAGGCTG3′	54
R408W Heminested PCR (*Sty* I cuts mutant allele)	Outside 5′ATGCCACTGAGAACTCTC3′	5′AGTCTTCGATTACTGAGAAA′3′	62–45
	Inside 5′TACCTTTCTCCAAATGGTG3′	5′AGTCTTCGATTACTGAGAAA3′	54
VNTR Nested PCR	Outside 5′GCTTGAAACTTGAAAGTTGC3′	5′GGAAACTTAAGAATCCCATC3′	62–45
	Inside 5′CTTGATTTAATCATTTTACAAT3′	5′CTCAGAGAAGCACATCTTTT3′	56
SNP (*Xmn* I) Heminested PCR	Outside 5′CTGTACTTGTAAGATGCAGC3′	5′GCAGTAACCACACTTCTGAA3′	62–45
	Inside 5′CTGTACTTGTAAGATGCAGC3′	5′ACTGTCCCAAGCAATCAAAG3′	58
Intron 3 STR $(TCTA)^n$ Heminested PCR	Outside 5′CAAATTGCCAGAACACATA3′	5′TCATAAGTGTTCCCAGACA3′	62–45
	Inside 5′TGTGGAAAGCAGAAAGAC3′	5′TCATAAGTGTTCCCAGACA3′	52

using a primer design and primer melting temperatures, presented in Table 3.9. Restriction digestion of PCR product for mutation was performed overnight with *Sty* 1 enzyme (Promega), followed by acrylamide gel electrophoresis. To detect possible ADO of mutant or normal allele, two linked markers were amplified simultaneously with PAH gene, one representing a STR in intron 3, and the other a restriction fragment length polymorphism (RFLP) *Xmn* I in intron 8. To detect potential contamination with extraneous DNA, an additional nonlinked VNTR was also amplified.

The embryos resulting from oocytes predicted to contain only normal maternal allele were cultured into the cleaving 6–8-cell embryos and transferred back to the patient at day 3 after aspiration. The resulting pregnancy was followed up by prenatal diagnosis to confirm the results of PGD. Accordingly, those embryos resulting from oocytes predicted to contain the mutant maternal gene were also analyzed further to investigate the accuracy of the results of the sequential PB1 and PB2 analysis.

Of 15 oocytes available for testing, the results for both PB1 and PB2 were obtained in 11 of them, allowing a reliable diagnosis. Six oocytes with both PB1 and PB2 data were predicted normal, based on the heterozygous PB1, containing both the normal and mutant maternal alleles, and the mutant PB2, suggesting that no mutant allele was left in the resulting oocytes (Figure 3.13). Another oocyte was predicted normal based on the presence of the mutant allele in PB1 and the normal allele in PB2. These results were in agreement with the linked marker analysis, showing that the homozygous status of PB1 was not due to ADO of the normal allele. Three oocytes were predicted affected, the diagnosis of two of which were based on the heterozygous status of PB1 and the normal PB2, and one on the homozygous normal status of PB1 and the mutant PB2. ADO of the mutant allele was observed in one oocyte, evidenced by the identical genotype of PB1 and PB2. This was confirmed by the presence of the linked polymorphic markers, suggesting that the resulting oocytes were mutant. The marker analysis also confirmed the mutant status of these three oocytes. ADO was observed in the STR analysis of two oocytes, evidenced by the identical patterns in PB1 and PB2. Three of four embryos resulting from these mutant oocytes were followed up by mutation and marker analysis, confirming the presence of the mutant allele in all of them. Four of seven embryos resulting from the mutation-free oocytes were selected for transfer back to the patient, yielding a clinical twin pregnancy, confirmed by the chorionic villus

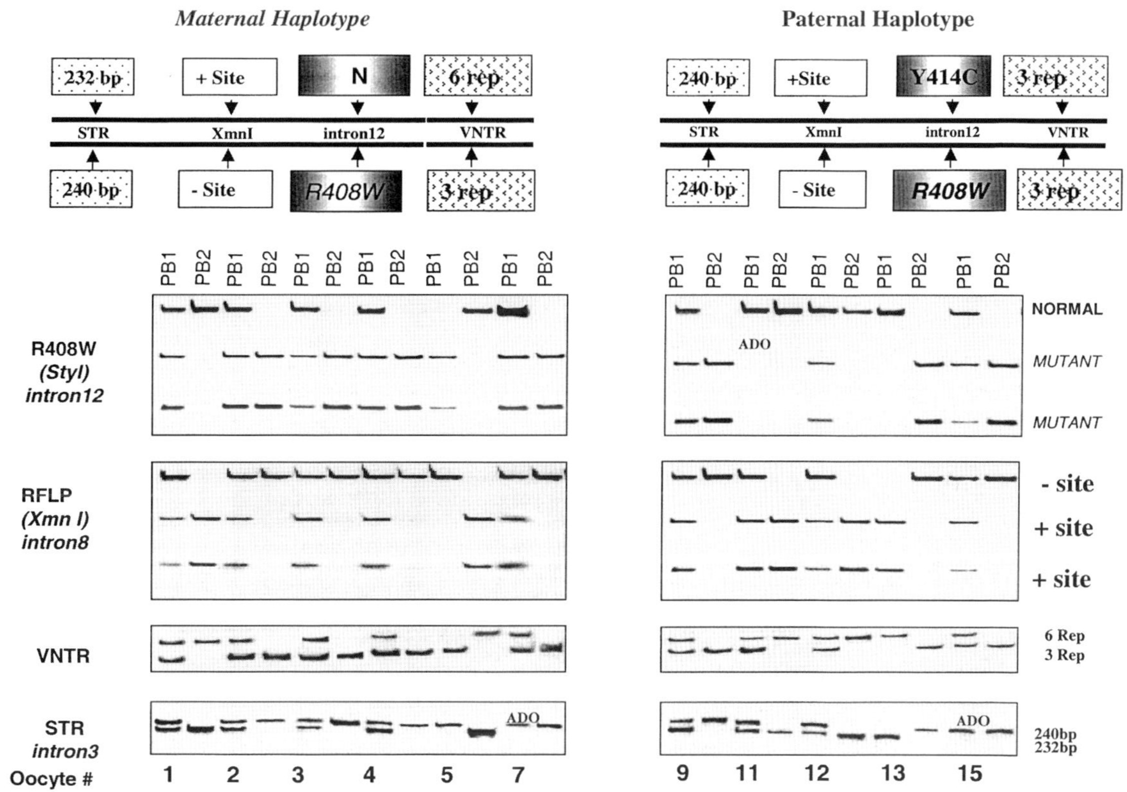

Figure 3.13. PGD for R408W mutation in phenylalanine hydroxylase gene in couple with male partner homozygous affected. (*Top*) Schematic representation of maternal (*left*) and paternal (*right*) haplotypes. Heterozygous mother has R408W mutation linked to short tandem repeats (STR) in intron 3, variable number of tandem repeats (VNTR) close to 3′ of gene (3 rep) and restriction fragment length polymorphism (RFLP) in intron 8 (— site). Homozygous affected father has two different mutations, one similar to maternal (R408W) mutation, and other, Y414C mutation, with its own linkage pattern. (*Bottom*) Genotyping oocytes by sequential analysis of PB1 and PB2 for R408W mutation and informative linked markers STR, VNTR, and RFLP. All series include PB1 followed by PB2 in lane to its immediate right, corresponding to 11 oocytes studied (oocytes are numbered at the bottom). As R408W mutation creates restriction site for *Sty* I enzyme, oocytes #2, 3, 4, 7, 9, and 15 were predicted normal based on heterozygous PB1 and homozygous PB2. Oocyte #5 was also predicted normal, but based on homozygous mutant PB1 and normal PB2, which was in agreement with marker analysis, excluding possibility for ADO in corresponding PB1. ADO of mutant allele is evident from identical genotype of both PB1 and PB2 in oocyte #11 (confirmed by all three markers), suggesting affected status of this oocyte. The other three affected oocytes were predicted based on heterozygous PB1 and normal PB2 (oocytes #1 and 12), and homozygous normal PB1 and mutant PB2 (oocyte #13). ADO was also detected in oocytes #7 and 15 (identical genotype of PB1 and PB2 for intron 3 STR). These are not in conflict with unaffected genotype of resulting embryos, which were transferred together with other two unaffected embryos (#2 and 4), resulting in twin pregnancy and birth of two healthy children, following confirmation of PGD by prenatal diagnosis.

sampling (CVS) to be unaffected. Healthy unaffected twins were born, presently 7 years of age, with normal patterns of mental development.

The presented case is the first experience of PGD for PKU in the experience of PGD for more than 100 different single-gene disorders, which were mainly performed for the at-risk couples with both unaffected partners who were heterozygous carriers of an autosomal recessive gene mutation, or one with a dominant mutation. Because, in the presented cases, it was a male partner who was compound heterozygous affected, the PGD strategy was based on the preselection of maternal mutation-free oocytes using a sequential PB1 and PB2 DNA analysis.

Based on the multiplex heminested PCR analysis, 6 of 11 oocytes with both PB1 and PB2 results were predicted to contain no mutant allele of PHA gene. Four embryos resulting from these zygotes were transferred, yielding an unaffected twin pregnancy and birth of healthy children, following the confirmation of PGD by CVS. The high accuracy of this strategy was obvious from the follow-up study of the embryos resulting from the mutant oocytes, confirming the results of the sequential PB1 and PB2 analysis.

With progress in treatment of genetic disorders, PGD will have an increasing impact on the decision of the affected and well-treated patients to reproduce. For example, the life expectancy has been significantly improved for such common conditions as CFTR and thalassemias, which presently may be treated radically by stem cell transplantation. In each of such cases, the strategy depends on whether the affected partner is male or female, because testing may be entirely based on oocytes if the male partner is affected, in contrast to embryo testing when the female partner is affected. We have performed PGD for couples with affected partners with both of these conditions, presented below.

In thalassemia cases, there may be male factor infertility involved, so the male partners need appropriate treatments before PGD, or testicular biopsy has to be applied. This was the case in one of our couples with male thalassemic partner, when a sufficient number of oocytes with homozygous mutant or heterozygous PB1 was detected, but no appropriated sperm can be found for fertilization, so matured oocytes were frozen for possible future use after treatment. In the other two cycles, because the female partner was affected, blastomere biopsy was utilized, resulting in preselection of a single unaffected embryo for transfer in each cycle, yielding no clinical pregnancy. This was due to a limited number of oocytes obtained from the thalassemic female partners. However, in the well-treated patients, the situation may be different as shown in the case with male affected partner with two different mutations, IVSI-110 and IVSII-745, and female partner with IVSI-110 presented in Figure 3.14. Of 25 embryos tested, 5 failed to amplify, 11 were mutant, and 9 heterozygous normal, providing a sufficient number of embryos for transfer.

We also performed PGD for two couples, one with female and one with affected male partner with CF. In a couple with maternal affected partner there were three different mutations in CFTR gene. The mother was affected with CF and had two different mutations, R117H and G542X (Fig. 3.15). The father was a carrier of 1717 mutation in CFTR gene. Paternal haplotypes were established using single sperm PCR, while maternal linkage was based on DNA amplification of PB1s. The couple had two previous pregnancies, the first one resulting in spontaneous abortion of twins, and the second pregnancy being terminated following prenatal diagnosis that identified an affected fetus with CF. So PGD was performed for three different mutations in CFTR gene, resulting in an unaffected pregnancy and birth of a healthy girl predicted and confirmed to be the carrier of maternal mutation G542X. It is of interest that multiplex PCR performed for three mutations in CFTR gene in this case was combined with age-related aneuploidy (see below) because of advanced reproductive age of the mother. Multiplex heminested PCR was performed on blastomeres from 14 embryos allowing simultaneous detection of the paternal and maternal CFTR haplotypes and nonsyntenic short tandem repeats (STRs) located on chromosomes 13, 16, 18, 21, 22, and XY. Six embryos were predicted to be carriers based on the presence of the maternal mutation and normal paternal CFTR gene. In addition to avoiding the transfer of the affected double heterozygous embryos, three aneuploid embryos were identified, involving aneuploidy for chromosomes 13, 18, 21, and X, and excluded from the transfer and freezing. Two unaffected embryos were transferred, resulting in pregnancy and birth of the healthy girl. DNA analysis of the newborn baby revealed the genetic profile identical to that of embryo #7, evidencing the usefulness and accuracy of the combined mutations analysis, linkage, and aneuploidy testing

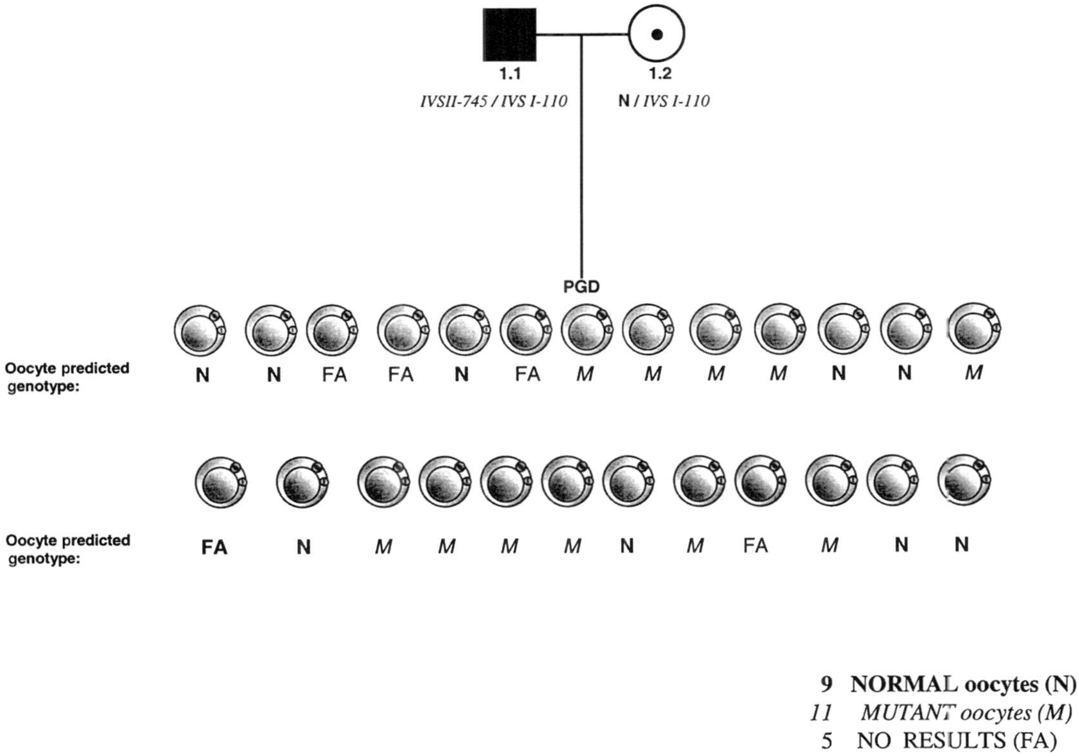

Figure 3.14. PGD for a couple with double heterozygous affected male partner with thalassemia. (*Top*) Family pedigree with father double heterozygous IVSI-110 and IVSII-745, and mother carrier of IVSI-110. (*Bottom*) Genotyping of 20 of 25 available oocytes for maternal IVSI-110 mutation by sequential PB1 and PB2, resulting in prediction of nine mutation-free oocytes. N-normal; M-mutant; FA-failed amplification.

in PGD for single-gene disorders in patients with advanced reproductive age (Figure 3.15).

PGD for another couple with three different mutations in CFTR gene was performed in which the paternal partner was affected, carrying both delta F508 and R117H mutations in CFTR gene, which was described previously (see Figure 3.4).

3.5 Conditions with No Available Direct Mutation Testing

Although the availability of the sequence information is one of the major conditions in undertaking PGD for Mendelian diseases, it is also possible to perform PGD when no exact mutation is known. One of the examples demonstrating feasibility of the approach may be the most recent report of PGD for autosomal dominant polycystic kidney disease (ADPKD), caused by either PKD1 or PKD2 genes, for which no direct testing is yet available in most of the couples at risk, making linkage analysis the method of choice.

ADPKD is a common genetic disorder present in 1/1000 individuals worldwide, which causes progressive cyst formation and may eventually lead to renal failure by late middle age, requiring renal transplantation or dialysis [12]. The overall health implications of ADPKD are obvious from the fact that approximately 10% of all patients at need for renal transplantation or dialysis have this disease. ADPKD is caused by either PKD1 or PKD2 genes [31–33]. Because of an extremely variable expression and the age of onset, much higher in PKD2, only half of the patients carrying these genes may present with severe clinical manifestation and renal failure, making prenatal diagnosis and pregnancy termination highly controversial. PGD may, therefore, be more attractive for the at-risk couples, because an ADPKD-free pregnancy may be established from the onset,

A.

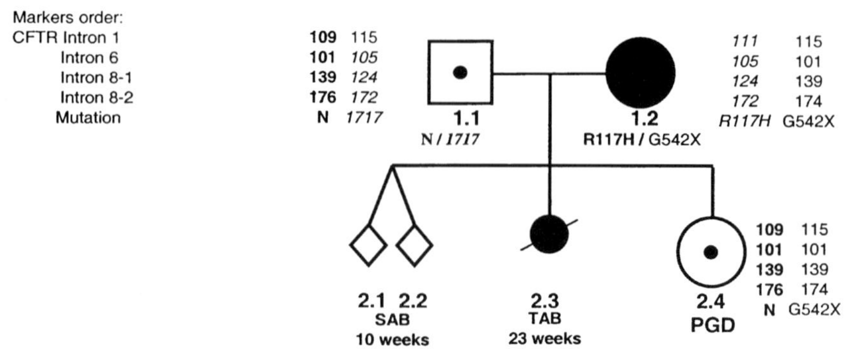

B.

Embryo #	2	3	4	5	6	7	8	9	10	11	13	14	15	17
	115 115 105 101 124 139 172 174 1717 G542X	109 115 101 101 139 139 176 174 N G542X	115 111 105 105 124 124 172 172 1717 R117H	115 111 105 105 124 124 172 172 1717 R117H	109 115 101 101 139 139 176 174 N G542X	109 115 101 101 139 139 176 174 N G542X	115 115 105 101 124 139 172 174 1717 G542X	115 115 105 101 124 139 172 174 1717 G542X	115 115 105 101 124 139 172 174 1717 G542X	109 111 101 105 139 124 176 172 N R117H	115 111 105 105 124 124 172 172 1717 R117H	115 115 105 101 124 139 172 174 1717 G542X	109 111 101 105 139 124 176 172 N R117H	109 111 101 105 139 124 176 172 N R117H
Predicted genotype:	1717/ G542X	N/ G542X	1717/ R117H	1717/ R117H	N/ G542X	N/ G542X	1717/ G542X	1717/ G542X	1717/ R117H	N/ R117H	1717/ R117H	1717/ G542X	N/ R117H	N/ R117H
Predicted karyotype:	13,13; 16,16; 18,18; 21,21; 22,22 XX	13,13; 16,16; 18,18; 0,21; 22,22 XY	13,13; 16,16; 18,18,18; 21,21; 22,22 XXY	13,13,13; 16,16; 18,18; 21,21; 22,22 XY	13,13; 16,16; 18,18; 21,21; 22,22 XY	13,13; 16,16; 18,18; 21,21; 22,22 XX	13,13; 16,16; 18,18; 21,21; 22,22 XY	13,13; 16,16; 18,18; 21,21; 22,22 XX	13,13; 16,16; 18,18; 21,21; 22,22 XY	13,13; 16,16; 18,18; 21,21; 22,22 XX	13,13; 16,16; 18,18; 21,21; 22,22 XY	13,13; 16,16; 18,18; 21,21; 22,22 XX	13,13; 16,16; 18,18; 21,21; 22,22 XX	13,13; 16,16; 18,18; 21,21; 22,22 XX
					ET	ET								

Figure 3.15. PGD for a couple with female affected double heterozygous partner with two different CFTR mutations combined with aneuploidy testing. (*A*) Family pedigree of a couple with three different mutations in CFTR gene. Mother is affected with cystic fibrosis and had two different mutations (R117H and G542X). Father is a carrier of 1717 mutation in CFTR gene. Paternal haplotype was established by multiplex heminested single sperm PCR. Maternal linkage was based on DNA amplification of polar bodies. Marker order of the mutations and polymorphic markers in CFTR gene are shown on the upper left. First pregnancy with twins resulted in spontaneous abortion; second pregnancy with affected fetus was terminated; PGD was performed for three different mutations in CFTR gene, resulting in an unaffected pregnancy and birth of a healthy girl predicted and confirmed to be the carrier of maternal mutation G542X. (*B*) Outcome of PGD cycle for three mutations in CFTR gene combined with age-related aneuploidy testing. Multiplex heminested PCR performed on blastomeres from 14 embryos allowing simultaneous detection of the paternal and maternal CFTR haplotypes and nonsyntenic short tandem repeats located on chromosomes 13, 16, 18, 21, 22, and XY. Six embryos (#3, 6, 7, 11, 15, and 17) were predicted to be carriers based on the presence of the maternal mutation and normal paternal CFTR gene. Monosomy 21 was found in the blastomere from embryo #3, which was excluded from the transfer and freezing. Trisomy 18 and XXY were observed in the blastomere from embryo #4, and trisomy 13 was detected in the blastomere from embryo #5. Embryos #6 and 7 were transferred, resulting in pregnancy and birth of the healthy girl. DNA analysis of newborn baby revealed the genetic profile identical to that of embryo #7.

without risk for pregnancy termination after prenatal diagnosis.

The majority (85%) of ADPKD is caused by PKD1, assigned to chromosome 16p13.3, the rest being attributed to PKD2, located on chromosome 4q21-q23 [12]. Because both PKD1 and PKD2 are characterised by enlargement of the kidneys due to the formation of bilateral or multiple unilateral fluid-filled cysts, presymptomatic diagnosis is available by abdominal ultrasound examination of young adults at risk, which also allows improving hypertension management in these patients, appearing long before the actual manifestation of renal disease. Although mutation rate is believed to be high, especially in PKD1, approximately half of the cases are still ancestry-related, allowing the diagnosis by linkage analysis. PKD1 gene contains 46 exons encoding the membrane protein polycystin 1 involved in cell-to-cell interaction, while PKD2 has at least 15 exons and encodes polycystin 2, which is a channel protein. Both of these proteins interact to produce new calcium-permeable nonselective cation currents, contributing to fluid flow sensation by the primary cilium in renal epithelium. Therefore, a loss or dysfunction of any of these proteins leads to polycystic kidney disease (PKD) due to the inability of cells to sense mechanical cues that normally regulate renal tubular morphology and function.

Because direct mutation testing is not presently available for the majority of patients either for PKD1 or PKD2 and their clinical manifestations are almost identical, except for the severity and earlier onset of PKD1, the linkage analysis is usually applied for both of these genes to trace the disease gene from the affected person through the family to the patient. DNA sequencing may in the future be applied for PKD2, but this may not be expected to be useful for PKD1 gene, because a large part of this gene is a duplication of multiple pseudogenes, which produce mRNA but are not translated.

PGD was performed for a woman with a family history of ADPKD, who inherited the disease from her father with severe ADPKD and also had a brother with clinical symptoms of ADPKD (Figures 3.16 and 3.17). Initial linkage analysis could not exclude either PKD1 or PKD2 as a cause of ADPKD in this family, so a set of linked markers have been designed to trace both PKD1 and PKD2 in the same reaction.

PGD cycle was performed using PB sampling and blastomere biopsy, described previously. PB1 and PB2 were removed in sequence and tested by multiplex nested PCR analysis, involving the above markers simultaneously in a multiplex heminested system. The preselection of mutation-free oocytes was performed based on the linked marker analysis, involving at least three closely linked markers to PKD1, Kg8, D16S664, and SM7, and at least four closely linked markers to PKD2, D4S2922, D4S2458, D4S423, and D4S1557 (primers and reaction conditions are presented in Tables 3.10 and 3.11). The maternal haplotypes were 90 bp, 110 bp, and 133 bp repeats for Kg8, D16S664, and SM7 markers, respectively, to trace PKD1 (see Figure 3.16), and 123 bp, 91 bp, 133 bp, and 100 bp repeats for D4S2922, D4S2458, D4S23, and D4S1557 markers, respectively, to trace PKD2 (see Figure 3.17).

The embryos derived from the oocytes free of PKD1 and PKD2, based on the information about polymorphic markers, were preselected for transfer back to the patient, while those predicted to be mutant or with inconclusive marker information were further tested by blastomere analysis removed from the resulting embryos either to confirm the diagnosis or identify additional mutation-free embryos for transfer.

Of 14 oocytes available for testing, only 11 were with the information for both PB1 and PB2. Two of these oocytes were predicted to contain PKD1 (oocytes #1 and 4), three to contain PKD2 (oocytes #2, 8, and 9), three neither PKD1 nor PKD2 (oocytes #5, 7, and 10), and three were inconclusive for one or both genes, due to failed amplification of PB2 (oocytes #3 and 6) or ADO in PB1 (oocyte #11) (Table 3.12). This last oocyte, however, was shown to be normal following blastomere analysis, although there was still a probability of recombination between Kg8 marker and PKD1 gene. The other two embryos (resulting from oocytes #3 and 6) with inconclusive results were left with insufficient information because of the amplification failure of embryo #6, or shared parental markers for PKD2 in embryo #3. Blastomere biopsy from the embryos resulting from the oocytes predicted to be mutant confirmed them to be affected, including one (embryo #9) in which, according to linkage analysis, neither PKD1 nor PKD2 could be excluded.

Three embryos resulting from oocytes #5, 7, and 10 were transferred, yielding a twin pregnancy and birth of two children confirmed to be free of both PKD1 and PKD2. As seen from the pattern of markers used to exclude PKD1, these mutation-free children are dizygotic twins,

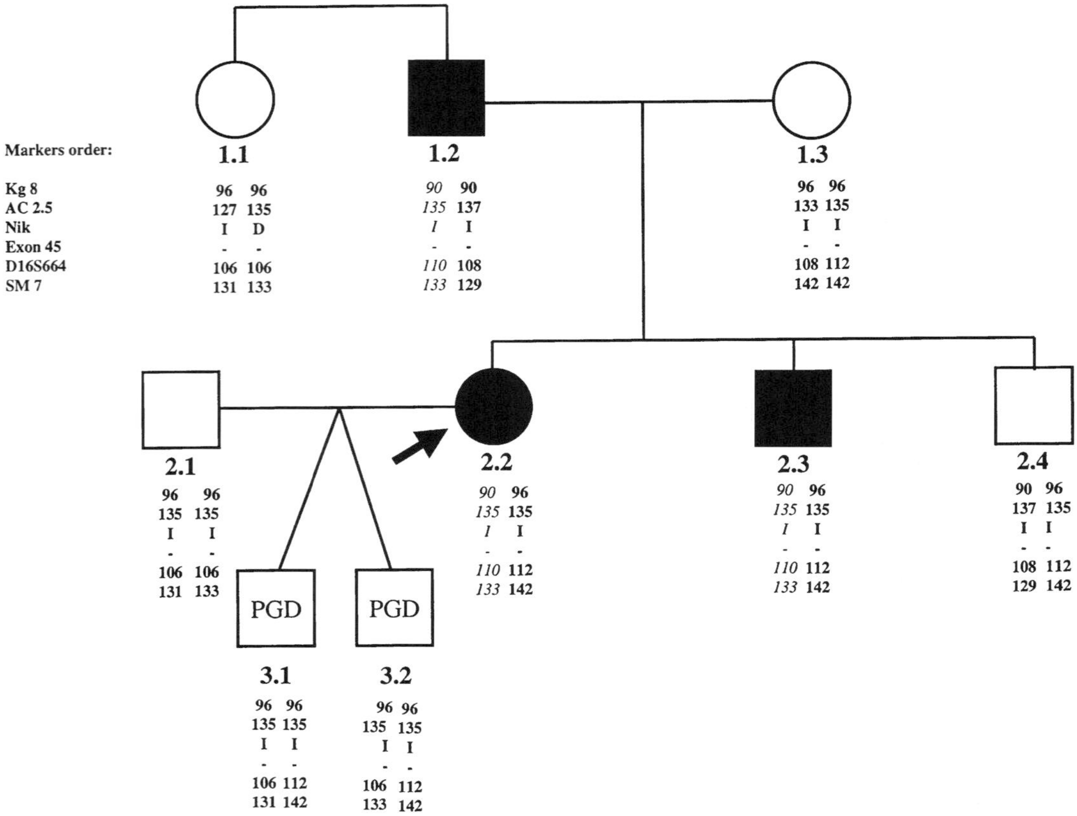

Figure 3.16. Preimplantation linked marker analysis for PKD1 resulting in birth of unaffected twins. ADPKD mutation originated from the affected maternal father (1.2), from whom the patient (2.2) inherited the mutant chromosome 16 (italic); Mutation-free chromosomes of maternal mother (1.3) are shown in **bold** face. The affected patient (2.2) had one affected brother (2.3), who also inherited the mutant chromosome from his father, and the same normal chromosome 16 from mother. Her unaffected husband's (2.1) normal chromosomes 16 differ by one of the markers (131 bp vs. 133 bp), which makes possible to identify twins, resulting from PGD (3.1 and 3.2).

as they inherited different normal chromosomes 16 from the father. The follow-up analysis of the embryos resulting from the mutant oocytes also confirmed the accuracy of sequential PB1 and PB2 by linked marker analysis, which is in accordance with the data presented in Chapter 2. In addition, in cases of inconclusive results by PB analysis, the blastomere biopsy of the resulting embryo may still allow preselecting additional embryos for transfer, as in embryo #11, mentioned above. However, as for avoiding a potential recombination between PKD1 and Kg8 alleles, closer linked markers were required; this embryo was frozen for possible future use by the couple.

The presented case is the first known PGD for ADPKD [34]. The clinical relevance of PGD for ADPKD is evident from the fact that prenatal diagnosis and termination of pregnancy may not be acceptable for a considerable proportion of ADPKD cases, which might be benign with no clinical manifestation during the whole lifetime.

The practical implication of PGD for ADPKD is also obvious from the high worldwide prevalence of this disease. So the information about the availability of PGD for ADPKD will be useful for practicing physicians, to provide at-risk couples with appropriate advice regarding the available options for avoiding the risk of producing a child with ADPKD. Because of a straightforward clinical manifestation of the disease and the availability of a simple ultrasound screening, with no need for the mutation identification, the PGD strategy described may be applied without extensive preparatory work usually required for PGD for many other conditions.

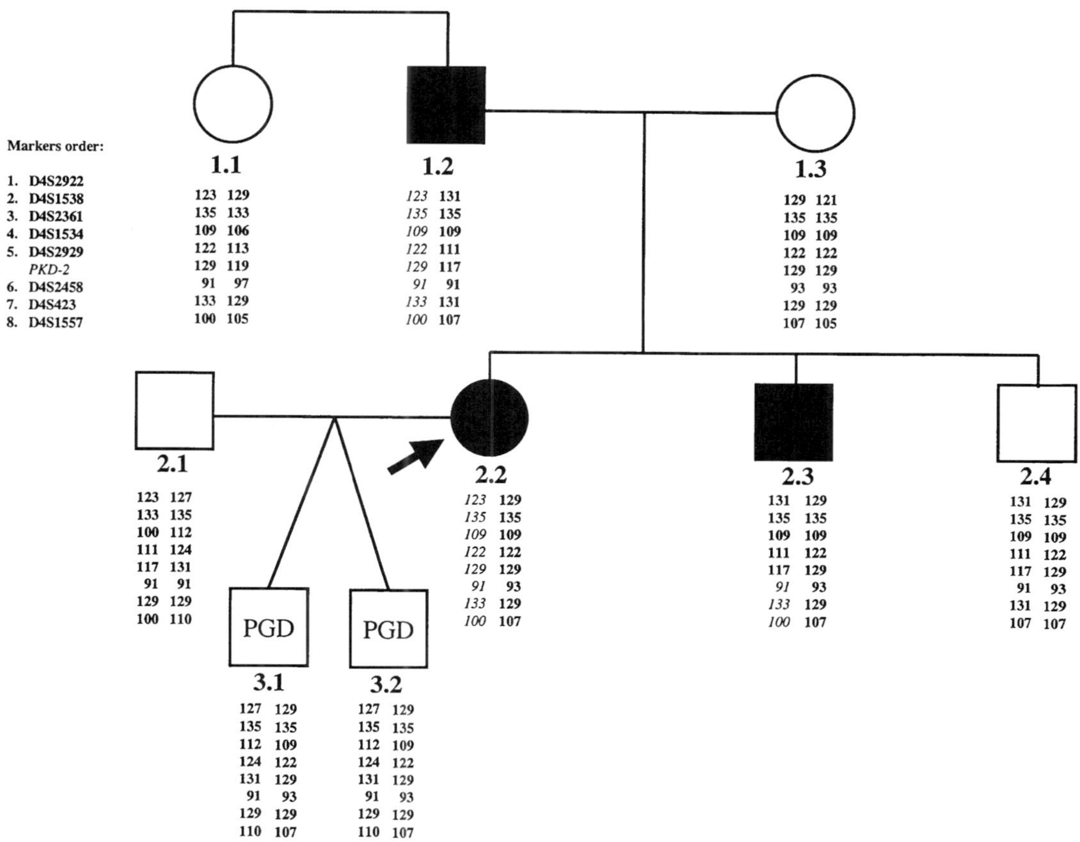

Figure 3.17. Preimplantation linked marker analysis for PKD2 (same pedigree as in Figure 3.16). ADPKD mutation (mutant chromosome 4 is shown in italic face) originated from the affected maternal father (1.2), from whom the patient (2.2) inherited the mutant chromosome 4 (italic face); Mutation-free chromosomes of maternal mother (1.3) are shown in **bold**. The affected patient (2.2) had one affected brother (2.3), who also inherited the mutant chromosome from his father, but following recombination between D4S2929 and D4S2458, which resulted in a recombinant chromosome 4 (**bold** and italic), and the same normal chromosome 4 from mother. Her unaffected husband's (2.1) normal chromosomes 4 cannot be distinguished due to sharing the same markers, so the fact that the resulting PGD twins are dizygotic cannot be demonstrated (3.1 and 3.2).

The presented case also demonstrates the feasibility of PGD in cases without exact information on the causative gene involved, and, therefore, may have practical implications for PGD of other conditions, in which the mutation might not be known but tracing of the mutant chromosome is possible using highly variable linked markers.

3.6 Late-Onset Disorders with Genetic Predisposition

The diseases with genetic predisposition have not traditionally been considered an indication for prenatal diagnosis, as this would lead to pregnancy termination, which may hardly be justified on the basis of genetic predisposition alone. On the other hand, the possibility of choosing the embryos free of genetic predisposition for transfer would obviate the need for considering pregnancy termination, as only potentially normal pregnancies are established. PGD for such conditions appeared to be acceptable on ethical grounds because only a limited number of the embryos available from hyperstimulation are selected for transfer.

The first PGD for inherited predisposition has been performed for couples carrying p53 tumour-suppressor gene mutations [35], known to

Table 3.10. Primers for the detection of linked markers to PKD 1 (16p13.3)

Gene/Polymorphism	Upper Primer	Lower Primer	Annealing Tm (°C)
D16S3252	Outside		
Kg 8	5′CGGCCATGCCCACAGAAG3′	5′CCTGGGGGCTGGCTCC3′	62–45
(Nested)	Inside		
	5′CACAGCCAGCTCCGAGG3′	5′Fam CAGGGTGGAGGAAGGTGAC3′	58
D16S291	Outside		
AC2.5	5′TGCAGCCTCAGTTGTGTTTC3′	5′TGCTGGGATTACAGGCATG3′	62–45
(Heminested)	Inside		
	5′TGCAGCCTCAGTTGTGTTTC3′	5′Fam AAGGCTGGCAGAGGAGGTGA3′	58
Nik 2.9	Outside		
(Nested)	5′GGCCCCAGGTCTCTTTCC3′	5′TCCGTGAGTTCCACTTGTC3′	62–45
	Inside		
	5′GTGGACGGGCATACATCAG3′	5′CCAGGCCGATGATGTGCAGC3′	58
EXON 45	Outside		
Ava II and *Bsp1286 I*	5′CATCCTGGTAGGTGACTG 3′	5′GAACAACTCCACCATCTCG3′	62–45
(Nested)	Inside		
	5′TACGCCCTCACTGGTGTC3′	5′ACAGCTCTCCACGCAAGG3′	58
SM7	Outside		
(Heminested)	5′CTCCGTCTCAAACAAACAAC3′	5′TTGTGGCCCAAATATATCA3′	62–45
	Inside		
	5′CTCCGTCTCAAACAAACAAC3′	5′Hex TAGTCCTGGTCCCTTCCA3′	58
D16S664	Outside		
CW3	5′TCATCGTTAGTGGGAGTCTG3′	5′TGCCCGGTCATAAATTG3′	62–45
(Heminested)	Inside		
	5′FamTCAATGAGATTTCGGGTAA	5′TGCCCGGTCATAAATTG3′	58

Table 3.11. Primers for the detection of linked markers to PKD 2 (4q21–q23)

Gene/ Polymorphism	Upper Primer	Lower Primer	Annealing Tm (°C)
D4S2922	Outside		
(Heminested)	5′AATAATAGAAAACAATCTCCTCA3′	5′TACACATCTTCTTTACTCAAACTAC3′	62–45
	Inside		
	5′Hex AGTTCTTCCTCTCCTGCTTAT3′	5′TACACATCTTCTTTACTCAAACTAC3′	55
D4S1538	Outside		
(Heminested)	5′AGCCTGGGTGACAGAGACTC3′	5′AACAATAAGTACATAAAAGAATGAAAC3′	62–45
	Inside		
	5′AGCCTGGGTGACAGAGACTC3′	5′Hex ATAGAAGCCTCAGTGCCTCTT3′	55
D4S2361	Outside		
(Heminested)	5′TCCTCCTCCCCACTGAAGT3′	5′GCACATGTACCCTAGAACTTAAAG3′	62–45
	Inside		
	5′Fam AAAGCCACGTGACTTTCATTA3′	5′GCACATGTACCCTAGAACTTAAAG3′	55
D4S1534	Outside		
(Heminested)	5′GTATTCAGTTTCAGCCCCATC3′	5′CCAAGGTAGAGGAGGGAAGG3′	62–45
	Inside		
	5′HexGTGAATATGCAGGTTACTCCCA3′	5′CCAAGGTAGAGGAGGGAAGG3′	55
D4S2929	Outside		
(Heminested)	5′TGTATTCAGTTTCAGCCCCATC3′	5′AGACCAGCCCAAGGTAGAGG3′	62–45
	Inside		
	5′Fam GAATATGCAGGTTACTCCCAGGA3′	5′AGACCAGCCCAAGGTAGAGG3′	55
D4S2458	Outside		
(Heminested)	5′CGAATGTTTCTATGTCATTGGAC3′	5′TGTCTTTCATTACAGAGATTCCTTC3′	62–45
	Inside		
	5′CGAATGTTTCTATGTCATTGGAC3′	5′Hex CCCCTGTAGCTGCCCTGTA3′	55
D4S423	Outside		
(Heminested)	5′TATTTGTTTCTCTTGAGTAGTTCCT3′	5′TGATCTCTTCTTTGGAAATATAAAT3′	62–45
	Inside		
	5′TATTTGTTTCTCTTGAGTAGTTCCT3′	5′Hex GTACATTGCAAAGTCCTCCATC3′	55
D4S1557	5′TTAATTTCAGACTTTTCTATTTTCA3′	5′GGAGATATTGGAAAGGCTGT3′	62–45
(Heminested)	5′FamTATATTCTCGATATGAGATTGTGG3′	5′GGAGATATTGGAAAGGCTGT3′	55

Table 3.12. Results of polar body and blastomere testing for PKD1 and PKD2 by linked marker analysis

		PKD 1			PKD 2					
Oocyte #	**Cell type**	**Kg8**	**D16S664**	**SM7**	**D4S2922**	**D4S2458**	**D4S423**	**D4S1557**	**Predicted genotype**	**ET**
1	PB1	96	112	142	123 / 129	ADO / 93	133 /129	100 / 107	Oocyte: PKD1 AFFECTED	NO
	PB2	90	110	133	123	91	133	100	PKD2 NORMAL	
2	PB1	90 / 96	110 / 112	133 / ADO	123 / 129	91 / 93	129	100	Oocyte: PKD1 NORMAL	NO
	PB2	90	110	133	129	93	129	107	PKD2 AFFECTED	
3	PB1	90 / 96	110 / 112	133 / ADO	129	91 / 93	133 / 129	100 / 107	Oocyte: PKD1 NORMAL	NO
	PB2	FA	110	133	123	91	133	100	PKD2 INCONCLUSIVE	
	Blast.	96 / 96	106 / 112	133 / 142	127 /123	91 / 93	129 / 129	110 / 107	Embryo: PKD1 NORMAL	
									PKD2 INCONCLUSIVE	
4	PB1	ADO / 96	110 / 112	ADO / 142	123 / 129	91 / 93	133 / 129	100 / 107	Oocyte: PKD1 AFFECTED	NO
	PB2	96	112	142	123	91	133	100	PKD2 NORMAL	
	Blast.	96 / 90	ADO / 110	133 / 133	123 / 129	91 / 93	129 / 129	100 / 107	Embryo: PKD1 AFFECTED	
									PKD2 NORMAL	
5	PB1	90 / 96	110 / 112	133 / ADO	123	91	133	100	Oocyte: PKD1 NORMAL	YES
									PKD2 NORMAL	
	PB2	90	110	133	129	93	129	107		
6	PB1	96	110 / 112	133 / 142	123	91 / 93	133 / 129	100 / 107	Oocyte: PKD1 INCONCLUSIVE	NO
	PB2	FA	110	133	129	91	133	100	PKD2 INCONCLUSIVE	
	Blast.	FA	FA	FA	FA	FA	FA	FA	Embryo: INCONCLUSIVE	
7	PB1	90 / 96	110 / 112	133 / 142	123 / 129	91 / 93	133 / 129	100 / 107	Oocyte: PKD1 NORMAL	YES
	PB2	90	110	133	123	91	133	100	PKD2 NORMAL	
8	PB1	96	110 / 112	142	123 / 129	91 / 93	133 / 129	100 / 107	Oocyte: PKD1 INCONCLUSIVE	NO
	PB2	FA	110	FA	129	93	129	107	PKD2 AFFECTED	
	Blast.	96 / ?	106 / 112	131 / 142	127 /123	91 / 91	129 / 133	110 / 100	Embryo: PKD1 INCONCLUSIVE	
									PKD2 AFFECTED	
9	PB1	96	ADO / 112	142	123 / 129	91 / 93	133 / 129	100 / 107	Oocyte : PKD1 INCONCLUSIVE	NO
	PB2	FA	112	FA	129	93	129	107	PKD2 AFFECTED	
	Blast.	96 / 90	106 / 110	133 / 133	123 / 123	91 / 91	ADO / 133	100 / 100	Embryo: PKD1 AFFECTED	
									PKD2 AFFECTED	
10	PB1	90 / ADO	110 / 112	133 / 142	123 / 129	91 / 93	133 / 129	100 / 107	Oocyte: PKD1 NORMAL	YES
	PB2	90	110	133	123	91	133	100	PKD2 NORMAL	
11	PB1	96 / ?	110 / 112	ADO / 142	123	91	133	100	Oocyte: PKD1 INCONCLUSIVE	YES
	PB2	90	110	133	129	93	129	107	PKD2 NORMAL	
	Blast.	96 / ?	106 / 112	133 / 142	123 / 129	91 / 93	129 / 129	100 / 107	Embryo: PKD1 NORMAL	
									PKD2 NORMAL	
Affected mother		90 / 96	110 / 112	133 / 142	123 / 129	91 / 93	133 / 129	100 / 107		
Affected maternal brother		90 / 96	110 / 112	133 / 142	131 / 129	91 / 93	133 / 129	100 / 107		
Unaffected father		96 / 96	106 / 106	131 / 133	123 / 127	91 / 91	129 /129	100 / 110		

PB1, first polar body; PB2, second polar body; Blast, blastomere; FA, failed amplification; ADO, allele drop out; ET, embryo transfer.
?, Inconclusive results.

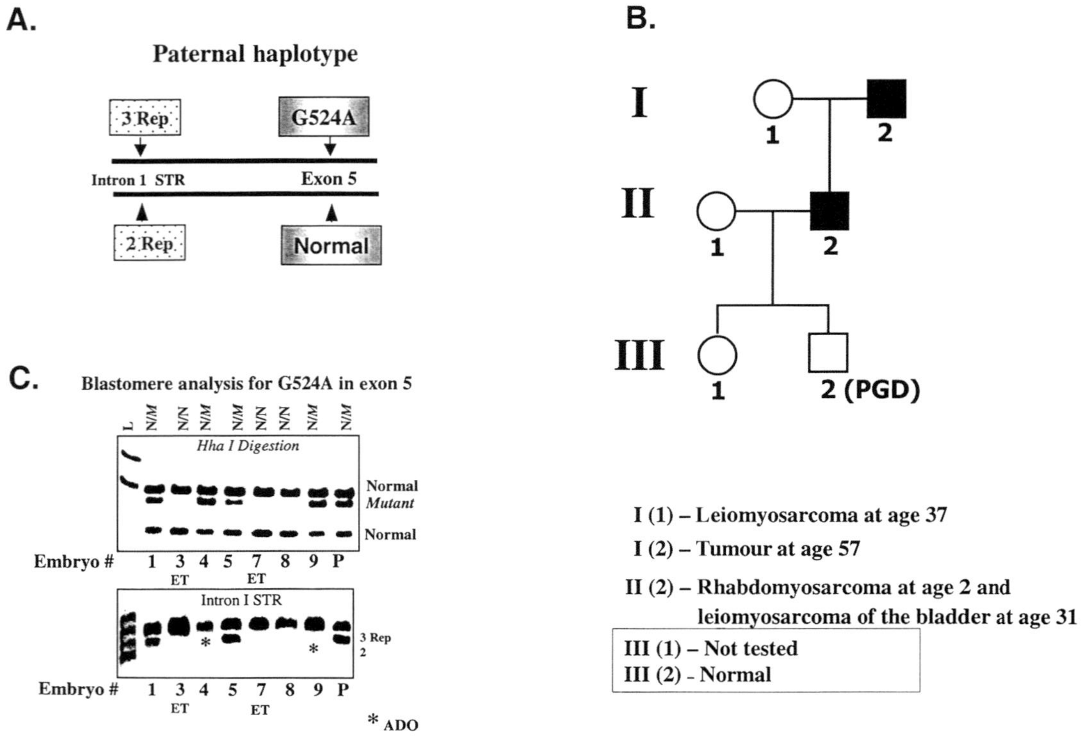

Figure 3.18. PGD for paternally derived mutation in p53 tumour suppressor gene mutation. (*A*) Paternal haplotypes are based on single sperm analysis, showing linkage of affected allele to 3 repeats and normal allele to 2 repeats in intron 1 STR. (*B*) Family pedigree, also showing a medical history on the bottom. (*C*) Duplex blastomere analysis for G524A mutation in exon 5 and liked STR in intron 1, showing that embryos #1, 4, 5, and 9 are affected, while embryos #3, 7, and 8 are mutation-free and suitable for transfer. ADO of three repeats in intron 1 STR was detected in blastomere from embryos #4 and 9.

determine a strong predisposition to the majority of cancers, which is described below.

3.6.1 p53 Tumour-Suppressor Gene Mutations

Two couples presented for PGD, one with the maternally and one with the paternally derived p53 tumour-suppressor mutations. The paternal one was a missence mutation due to a transversion of a G to A in exon 5 of the p53 tumour-suppressor gene, resulting in a change from arginine to histidine at the 175 amino acid residue of the protein [12, 36]. The carrier was a 38-year-old proband with Li-Fraumeni syndrome (LFS), diagnosed with rhabdomyosarcoma of the right shoulder at age 2 followed by right upper extremety amputation. At age 31, he was also diagnosed with a high-grade leimyosarcoma of the bladder and underwent a radical cystoprostatectomy. His mother was diagnosed with leimyosarcoma at age 37 (Figure 3.18).

In the other couple, the 39-year-old mother with LFS was a carrier of 902insC mutation of the p53 tumour-suppressor gene, representing an insertion of C in exon 8. She was diagnosed with breast cancer at the age of 30, followed by bilateral mastectomy. She also had thyroid cystocarcinoma, which was also removed. Her mother died from stomach cancer at age 51. One of her sisters, diagnosed with breast cancer at age 48 followed by mastectomy, also died at age 51. Two of her four brothers were diagnosed with bone or brain tumours at their teens (Figure 3.19).

Prior to PGD cycles, linkage analysis had been performed for each couple in order to establish the maternal and paternal haplotypes, needed for performing a linked marker analysis in addition to the mutation testing. Thus, in both cases, a multiplex nested PCR was performed by the testing of the mutations simultaneously with the linked

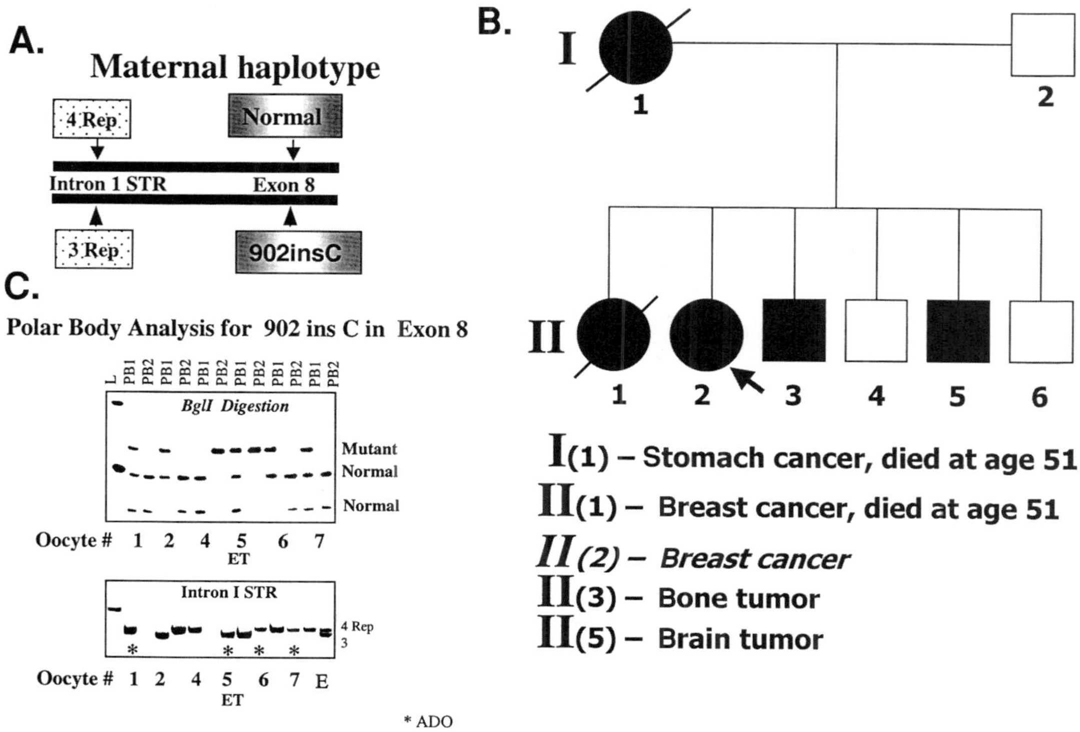

Figure 3.19. PGD for maternally derived mutation in p53 Tumour-suppressor gene mutation. (*A*) Maternal haplotypes are based on PB1 and PB2 analysis, showing linkage of affected allele to 3 repeats and normal allele to four repeats in intron 1 STR. (*B*) Family pedigree, also showing a medical history on the bottom. (*C*) Duplex PB1 and PB2 analysis for 902insC mutation in exon 8 and linked STR in intron 1, showing that all but one oocyte (oocyte #5) are affected, based on heterozygous PB1 and homozygous normal PB2. Thus, only embryos deriving from oocyte #5 were predicted normal and suitable for transfer, as shown by heterozygous PB1 and affected PB2. ADO of three repeats in intron 1 STR linked to 902insC mutation was detected in four of seven PB1 studied, all of which were heterozygous in mutation analysis.

polymorphic markers, representing the short tandem repeats (STR) in the intron 1. To establish the paternal haplotypes, a single sperm analysis was performed, showing a linkage of the affected allele to the three repeats and the normal one to the two repeats. The maternal haplotypes were established by PB1 and PB2 analysis, demonstrating the linkage of the affected allele to the three repeats, and the normal one to the four repeats. The outside and inside primer sequences, the restriction sites, and the primer melting temperature for DNA analysis of both mutations are shown in Table 3.13. The PCR products were identified by restriction digestion, using *Hha* I for G524A and *Bgl* I for 902insC mutation.

Two PGD cycles were performed for each couple as described above. Testing for the maternal 902insC mutation was done by DNA analysis of PB1 and PB2, removed sequentially following maturation and fertilization of oocytes. The paternal G524A mutation was tested by DNA analysis of single blastomeres, removed from the eight-cell embryos. Based on both mutation and STR analysis, unaffected embryos were preselected for transfer back to the patients, while those predicted mutant were exposed to confirmatory analysis using the genomic DNA from these embryos to evaluate the accuracy of the single-cell based PGD.

A total of 18 day-3 embryos were tested in the couple with the paternally derived G524A mutation. Eleven embryos were heterozygous, with both alleles of the exon 5 present, while seven contained only the normal allele in agreement with the STR analysis in the intron 1. Figure 3.18 demonstrates the results of the testing of eight embryos in one of the cycles performed for this couple, showing that embryos #1, 4, 5, and #9 were affected, and embryos #3, 7, and 8 were unaffected. These results were in agreement with the

Table 3.13. Primers and reaction conditions for PGD of p53 tumour-suppressor–gene mutations

Gene/Polymorphism	Upper Primer	Lower Primer	Annealing Tm (°C)
G524A	Outside		
-Hha I	5′GTGCAGCTGTGGGTTGA3′	5′GGAATCAGAGGCCTGGG3′	62–45
	Inside		
	5′CCATGGCCATCTACAAGCA3′	5′GGGACCCTGGGCAACC3′	56
902insC	Outside		
-Bgl I	5′AAGAGAATCTCCGCAAGAA3′	5′GAGGCAAGGAAAGGTGATAA3′	62–45
	Inside		
	5′GCCTGTCCTGGGAGAGAC3′	5′GCTTACCTCGCTTAGTGCG3′	60
(TAAA)n	Outside		
(Heminested)	5′CATTTGAATCCGGGAGGAG3′	5′ACAAAACATCCCCTACCAAACA3′	62–45
	Inside		
	5′GCCTGGGCAATAAGAGCTG3′	5′ACAAAACATCCCCTACCAAACA3′	60

marker analysis in all but two embryos (embryos #4 and 9), in which the three-repeat STR marker linked to the mutant gene was missing, probably due to ADO, because these embryos were clearly heterozygous according to the mutation analysis.

A total of 10 oocytes were tested in two PGD cycles from the couple with the maternally derived 902insC mutation in the exon 8. Mutation analysis was performed simultaneously with the linked STR in the intron 1, using a sequential PB1 and PB2 analysis. Six oocytes with the available PCR results in both PB1 and PB2 were predicted affected, based on the heterozygous PB1 and normal PB2, and four unaffected, based on the heterozygous PB1 and the mutant PB2. As in the previous case, ADO of the three-repeat marker linked to 902insC mutation was observed in four of seven heterozygous PB1 (see Figure 3.19).

The unaffected embryos were preselected and transferred back to the patients in each of the four PGD cycles performed for these two couples. Neither transfer yielded clinical pregnancies in the two cycles performed for the maternal mutation 902insC, while one of the transfers resulted in a singleton pregnancy and birth of a mutation-free child in a couple with the paternally derived G525A mutation, following the confirmation of PGD by prenatal diagnosis. The follow-up PCR analysis of the embryos predicted affected were possible for the embryos resulting from the six affected oocytes for the maternal mutation 902insC, and showed a concordance with the results of the PB analysis.

In one of our couples, the maternal carrier of the p53 tumour-suppressor mutation was age 39, thus her elevated age-related risk for aneuploidy may have contributed to the failure of establishing pregnancy in the two subsequent clinical cycles. Although testing for chromosomal abnormalities could be done for this couple, only one or two embryos, respectively, were available for the transfer in each of the two cycles, preventing further embryo preselection for the age-related aneuploidies. On the contrary, the other patient, with the established pregnancy, was neither a carrier of the p53 tumour-suppressor mutation nor had the risk for the age-related aneuploidy.

As seen from the presented cases, indications for PGD are being extended steadily compared with the practice of prenatal diagnosis. This is due to the potential of PGD for the preselection of mutation-free embryos and the establishment of an unaffected pregnancy, instead of the testing and termination of the pregnancies already in process. In fact, many at-risk couples have had such unfortunate experiences of repeated prenatal diagnoses and termination of affected pregnancies that they regard PGD as their only hope for having children of their own, despite having to undergo IVF. This has made PGD also attractive for couples at risk for the late-onset disorders with the genetic predisposition, although such conditions have never been an indication even for prenatal diagnosis.

The presented cases demonstrate practical feasibility of PGD for the late-onset disorders, providing a principally new option for a large group of couples who wish to avoid the risk of having children with a strong inherited predisposition to common disorders. The application of PGD for avoiding the establishment of pregnancy with neurofibromatosis is presented below.

3.6.2 Neurofibromatosis

Neurofibromatosis (NF) is a common autosomal dominant neurological disorder, with at least two distinct major forms: NF type I (NF1), which is more common (1/4000) and characterized by fibromatous skin tumours with cafe-au-lait spots, known also as Von Recklinghausen disease, and NF type II (NF2), which is less common (1/100,000) and characterized by bilateral acoustic neuromas, meningeomas, schwannomas, and neurofibromas [12].

The NF1 gene is located on chromosome 17q11.2, while the NF2 gene is mapped on chromosome 22q12.2. Alterations in the sequence of these genes affect the tumour-suppressor function of their products (neurofibromin and merlin, respectively), leading to a strong predisposition to malignancies. Different mutations in these genes have been described, resulting in a variety of clinical manifestations. Approximately half of these mutations are sporadic [37, 38], with the rest representing germ-line mutations, which may be detected before the establishment of pregnancy to ensure unaffected pregnancy and birth of a healthy child without an inherited predisposition to malignancy.

Although PGD for inherited cancer predisposition is still a controversial issue [39, 40], the possibility of establishing only mutation-free pregnancies makes PGD an attractive option for the late-onset disorders with genetic predisposition because of the possibility for preselection of the genetic-predisposition-free embryos for transfer, avoiding the risk for pregnancy termination, as only potentially normal pregnancies could be established.

Three couples presented for PGD, two at risk for producing a child with NF1 and one with NF2 mutation. In one of the couples at risk for producing a child with maternally derived NF1, presented in Figure 3.20, the 23-year-old mother with cafe-au-lait spots, axillary freckling, and Lisch iris nodules had a nonsense mutation (Trp → Ter) resulting from TGG → TGA substitution in exon 29 of the NF1 gene coding sequence. This was a de novo mutation, as no clinical symptoms of NF1 were known in the extended pedigree of the patient.

In the couple at risk for producing an offspring with NF2, the 32-year-old father had NF2 mutation, represented by the DNA sequence alteration at codon 141 of the merlin gene, due to a T-C base substitution in the nucleotide position 422, leading to the change from leucine (L) to proline (P) (Figure 3.21).

For testing of the maternal NF1 mutation, PB1 and PB2 were removed sequentially following maturation and fertilization of oocytes, while single blastomeres were removed from the eight-cell embryos for testing the paternal NF2 mutation.

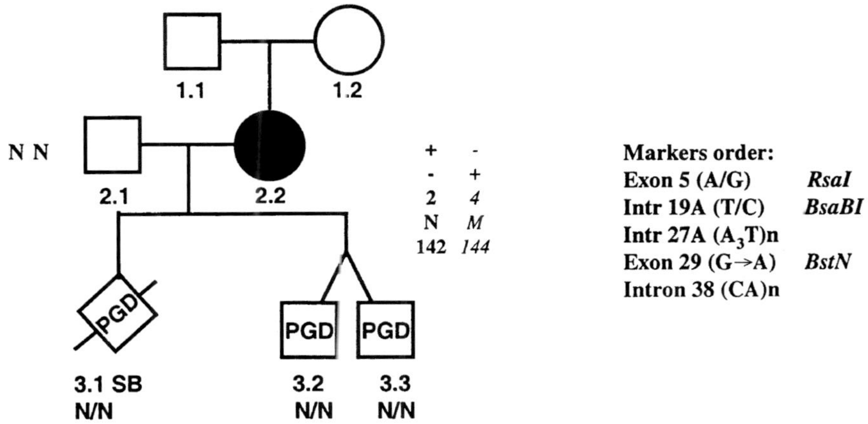

Figure 3.20. PGD for NF1: family pedigree. Haplotype analysis in PGD couple with maternal de novo nonsense NF1 mutation Trp → Ter (TGG → TGA) in Exon 29 of NF1 gene (based on sequential polar body analysis) *Rsa*– in exon 5, *Bsa*+ in intron 19A and AAATup in intron 27A markers (2.2) (*middle*). (*Bottom*) Sibling (SB) (3.1) representing a stillborn baby born after the first PGD cycle, with confirmed mutation-free status in agreement with linked presented markers, while 3.2 and 3.3 represent unaffected twins resulting from next PGD cycle.

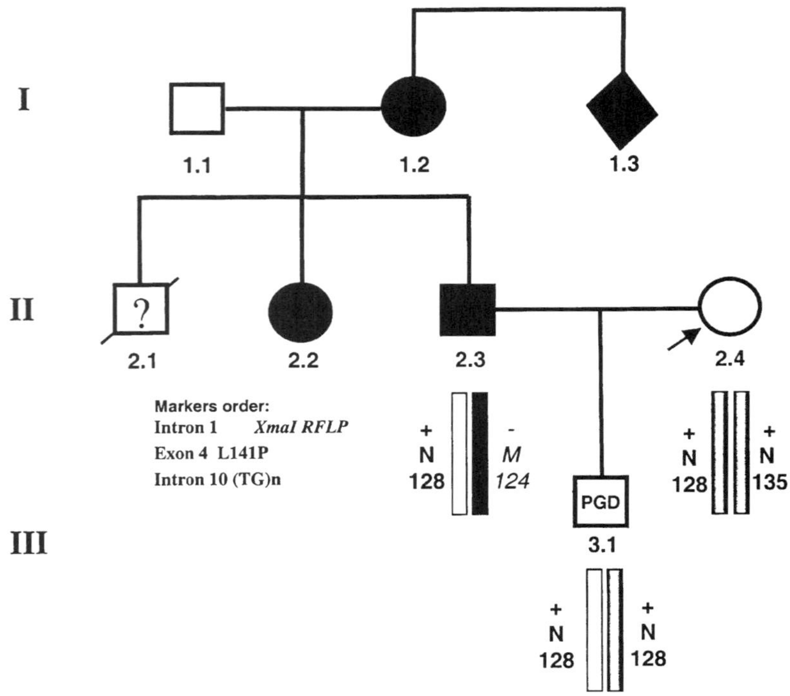

Figure 3.21. PGD for NF2: family pedigree. Haplotype analysis in couple with paternal L141P mutation in exon 4 of NF2 gene. Paternal haplotype (2.3) was established by multiplex heminested PCR analysis of single sperm. Maternal haplotype was based on embryo genotypes. PGD cycle resulted in delivery of the normal boy (3.1).

Linkage analysis had been performed for each couple, in order to establish the maternal and paternal haplotypes and to amplify these markers simultaneously with the mutation testing. In the NF1 case, three markers were informative, including *BsaB* 1 restriction site in intron 19 A [41], short tandem repeat (STR) (AAAT) in intron 27A [42] and *Rca* I restriction site in exon 5 [43]. In the NF2 case, two markers were informative, one representing single nucleotide polymorphism (SNP) in intron 1, following *Xma* I restriction digestion, and the dinucleotide repeat (TG) in intron 10 [44, 45]. Thus, in both cases a multiplex nested PCR was performed, amplifying the mutations simultaneously with the linked polymorphic markers [46, 47]. The primers designed for amplification of NF1 and NF2 genes and their linked markers mentioned are listed in Table 3.14.

Testing for NF1 was performed, using *BstN* I restriction digestion, which produces two fragments of 20 bp and 52 bp in the second-round PCR product of the normal allele, in contrast to a single undigested fragment of 72 bp in the mutant gene. The maternal haplotypes were established by PB1 and PB2 analysis, demonstrating the linkage of the affected allele to *Rsa*−, *Bsa*+, and AAATup markers (Figure 3.22)

Testing for NF2 mutation was done by blastomere analysis, using MspI restriction digestion, which produces two fragments of 36 bp and 86 bp in the second-round PCR product of the mutant allele, in contrast to a single 122 bp fragment in the normal gene. The single sperm haplotype analysis demonstrated the linkage of the affected allele to the *Xma* I undigested fragment, and to the 124 bp dinucleotide repeat in intron 10, the latter being tested using fluorescent PCR (Figure 3.23).

A total of 57 oocytes were tested in five PGD cycles from the patient carrying NF1 mutation. Based on simultaneous mutation and marker analysis in PB1 and PB2, 20 oocytes with available PCR results in both PB1 and PB2 were predicted affected. Of 26 oocytes containing no mutant gene, eight (2 in each cycle) with sufficient marker information available resulted in embryos of acceptable quality for embryo transfer, and three

Table 3.14. Primers and reaction conditions for PGD of NF1 and NF2

Gene/Polymorphism	Upper Primer	Lower Primer	Annealing Tm (°C)
NF1	Outside		
exon 29 (TGG-TGA)	5′GTTTAATTCTTCTCCACTTCACC3′	5′CAACACTGCATACCTTCCA3′	62–45
Bst NI cutsnormal	Inside		
	5′TTCATATCCGGACCCCC3′	5′CTTTTGGCCGAATCTTGG3′	53
NF1	Outside		
(AAAT)n	5′TGGTGGCACATACCTGTAC3′	5′TTACAGATTAAGGCAATTCTGA3″	62–45
	Inside		
	5′TGCATTCTAGCCTGAGTGA3′	5′AAACAAGCAAGAATAGAAAAAG3′	48
NF1	Outside		
SNP	5′ATTAGTGGGTTTTACTGTG3′	5′CTGAGGCTTTATGTATCTTA3′	62–45
Bsa BI	Inside		
	5′TGTGTATTTAAACTTTTGAG3′	5′TTCCCAATAACTGTAGAC3′	53
NF1	Outside		
SNP	5CATGTGGTTCTTTATTTATAGG3′	5′TTGACACCAGTTGACAATAG3′	62–45
Rsa I	Inside		
	5′GGTAGAAAATTATCCAGATGA3′	5′AACTTGGAAAACGATGATAG3′	55
NF2	Outside		
Msp I	5′GGCAGCCCTCATTAGAAC3′	5′AGAATACAGAAAACCCAAAG3′	62–45
cuts mutant	Inside		
	5′AAGATCTACTGCCCTCCTG3′	5′TGATCCCATGACCCAAATTA3′	54
NF2	Outside		
SNP	5′AAGAATATTCGCCGTGTGTC3′	5′GACTTTCCCGCTCCGTC3′	62–45
Xma I	Inside		
	5′CAACGAAGGACCCAATTCC3′	5′AAGCAGGCCTAGGCTCG3′	54
NF2	Outside		
Intron 10	5′GGAGAAAATTGGAGAAGAACT3′	5′CCACTCTGGTCATACAACG3′	62–45
(TG)n	Inside		
	5′TTCACTGTTTTATTGCTTGTC3′	5′FamGACTGCTGCTTTTTCTAAATC3′	

reached blastocyst stage and were frozen. Due to the close location of the gene to centromere, which might explain the presence of only two oocytes with heterozygous PB1, the information on the linked markers was of particular importance to avoid misdiagnosis due to potential ADO of one of the alleles in PB1.

Two of these transfers resulted in clinical pregnancies. In one of these cycles, of 17 tested oocytes, 9 contained no mutant allele, from which only 2 with sufficient linked marker information to exclude the risk for misdiagnosis resulted in embryos of acceptable quality for transfer, yielding a singleton clinical pregnancy. Prenatal diagnosis by chorionic villus sampling (CVS) confirmed the mutation-free status of the fetus, but the pregnancy ended in a stillbirth at 24 weeks. Similar results were obtained in the next cycle, resulting in preselection and transfer of two mutation-free oocytes, yielding a twin pregnancy and birth of healthy twins (see Figures 3.20 and 3.22). As seen from polyacrylamide gel electrophoresis of the restriction-digested PCR products of PB1 and PB2 from six oocytes from this cycle, mutation-free status of oocytes # 2, 4, and 5 is evidenced by a 72 bp undigested fragment in PB1 and a 52 bp digested fragment in PB2, in agreement with all three markers. The most recent cycle performed for this couple, resulting in the establishment of embryonic stem cell line from the affected NF1 embryos, will be describe in Chapter 7.

Of 18 day-3 embryos tested in one cycle from the couple with the paternally derived NF2 mutation, 8 were predicted to be mutation-free, including 4 with no or only one linked marker information available to exclude the risk for misdiagnosis. Three of four normal embryos with both linked markers in agreement with mutation analysis (see Figure 3.23) were transferred back to patient, resulting in a clinical pregnancy and birth of

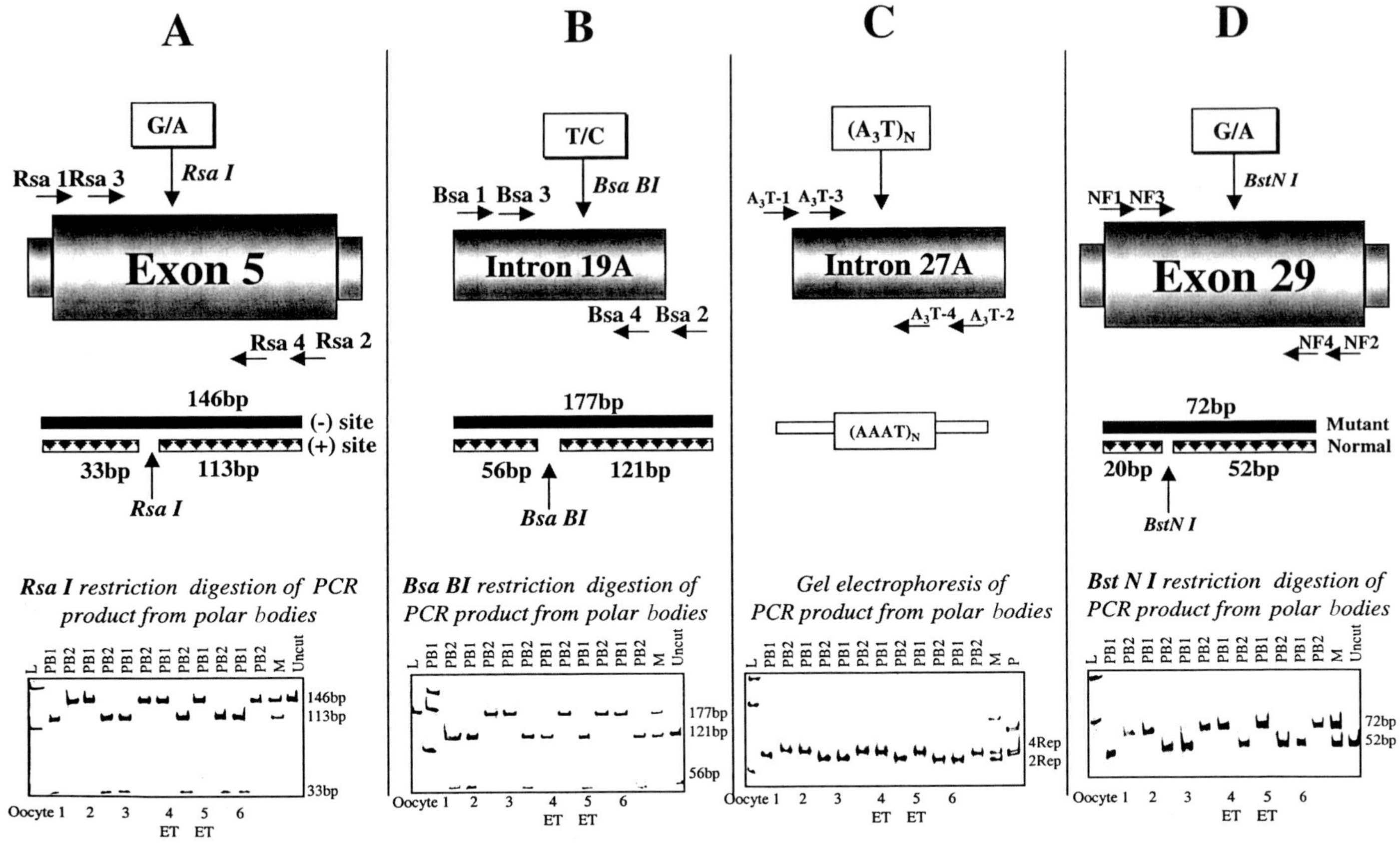

Figure 3.22. PGD for NF1 nonsense mutation Trp ⟶ Ter (TGG ⟶ TGA) in exon 29: mutation and polymorphic marker analysis. (*Top*) Schematic diagram of the mutation (*D*), and linked markers (*A, B, C*) on chromosome 17q11.2. (*Middle*) Restriction maps for *BstN* I (*D*), *Rsa* I (*A*) and *BsaB* I (*B*) restriction fragment length polymorphism; and short tandem repeat in intron 27A (*C*). (*Bottom*) Polyacrylamide gel electrophoresis of the restriction-digested PCR products of PB1 and PB2 from six oocytes in one of the cycles of PGD for NF1, showing mutation-free oocytes #2, 4, and 5, evidenced by 72 bp undigested fragment in PB1 and a 52 bp digested fragment in PB2 (*D*), in agreement with all three marker analyses (*A, B, C*) (for linkages to mutant and normal alleles).

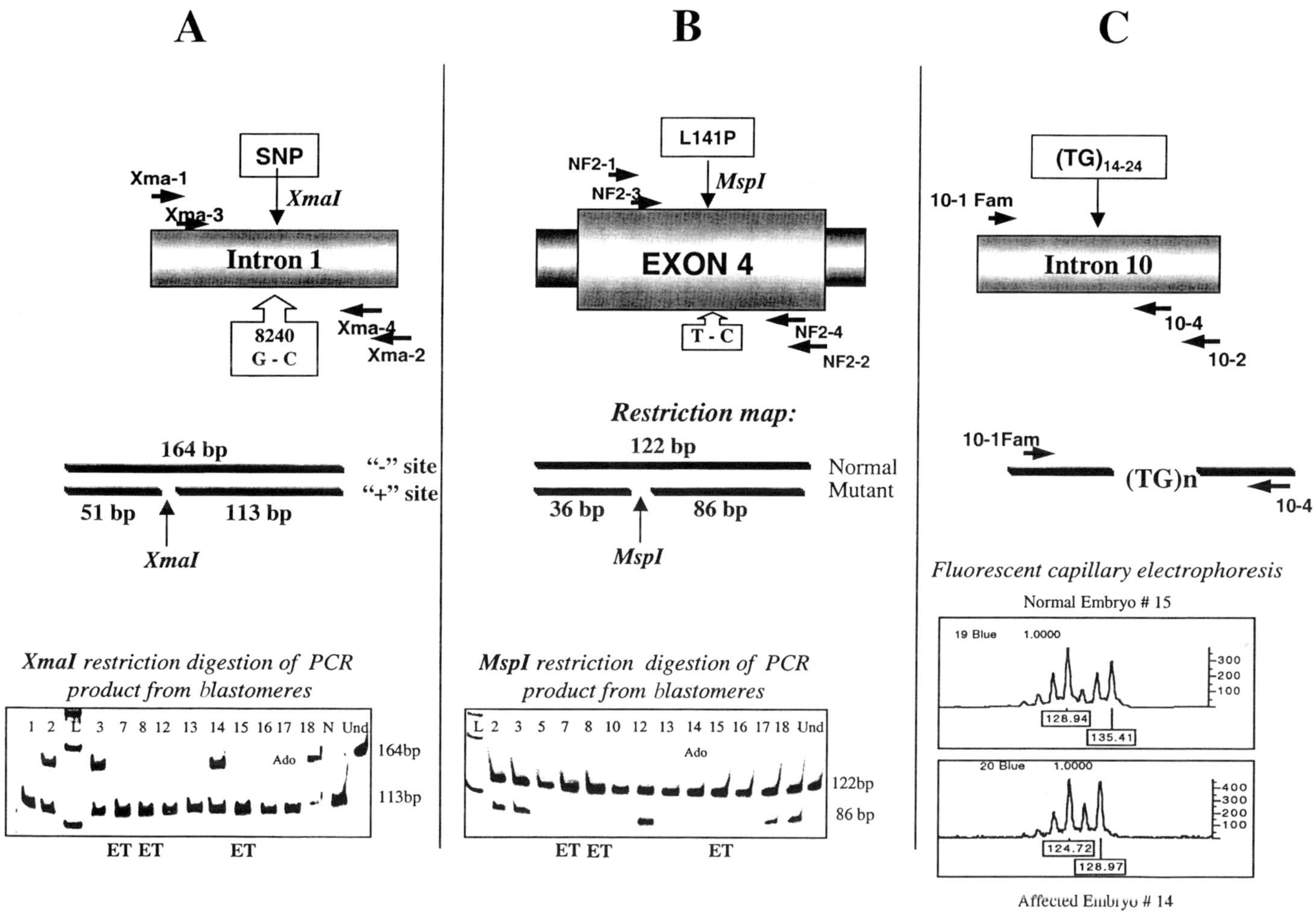

Figure 3.23. PGD for L141P mutation in exon 4 of NF2 gene: mutation and polymorphic marker analysis. (*Top*) Schematic diagram of the mutation (*B*) and linked markers (*A* and *C*) on chromosome 22q12.2. (*Middle*) Restriction maps for *Msp* I (B) and *Xma* I (A) restriction digestion in exon 4 and intron 1 of NF2 gene, and design for the dinucleotide TG repeat testing in intron 10 (*C*). (*Bottom*) Polyacrylamide gel electrophoresis of the restriction digested PCR products of balstomeres, showing seven mutation free-embryos (#5, 7, 8, 10, 13, 15, and 16), evidenced by the presence of a 122 bp undigested fragment, as compared with the presence of a digested 86 bp fragment in addition to a 122 fragment in the rest of embryos (B), in agreement with *Xma* I restriction length polymorphism in intron 1 (*A*), and dinucleotide TG repeats in intron 10 (the latter being detected by fluorescent) shown for normal embryo #15 and abnormal embryo #14, as an example (for the linkages to normal and mutant genes see Figure 3.21).

Table 3.15. Results and outcomes of PGD for cancer predisposition

Disease	# Patients	# Cycles	Cycle by PB	Cycle by PB and BL	Cycle by BL	# ET	# Embryos	Pregnancy	Birth
VHL	3	5	0	2	3	5	10	1	2
FAP	5	13	8	0	5	13	22	2	2
p53	2	4	2	0	2	4	8	1	1
hSNF5	1	1	1	0	0	1	1	0	0
NF2	1	2	0	0	2	2	5	2	2
NF1	2	5	3	0	2	5	11	2	3
RB	3	3	1	0	2	3	5	2	2
HNPCC	1	1	0	0	1	1	2	1	1
Gorlin syndrome	1	1	0	0	1	1	1	1	1
TOTAL	*19*	*35*	*15*	*2*	*18*	*35*	*65*	*12*	*14*

a healthy child free from NF2 mutation, following confirmation of PGD by CVS. Three of five remaining mutation-free embryos that continued development were frozen for future possible use by the couple.

The presented data demonstrate the acceptable diagnostic accuracy of both the PB and blastomere analysis for PGD of NF1 and NF2 [48]. As shown by the follow-up analysis of the mutant embryos or those with insufficient marker information, the PGD results were confirmed in all resulting embryos available for the study. As mentioned, in one of the cycles performed for maternally derived NF1 mutation, the pregnancy resulted in a stillbirth, following confirmation of mutation-free status of the fetus by CVS. Mutation-free status was also confirmed in a baby born following PGD for NF2 (see Figure 3.21).

As seen from the presented cases, together with the other application of PGD to cancer predisposition, the approach seems to be acceptable to the couples at risk, despite important ethical implications. So current genetic counseling services may consider informing patients at risk of having children with a strong genetic predisposition to cancers and late-onset disorders about the availability of PGD, without which these couples may remain childless because of their fear to choose prenatal diagnosis and possible pregnancy termination.

3.6.3 Other Cancers

PGD has also been applied to other cancers, including familial adenomatous polyposis (FAP), von Hippel-Lindau syndrome (VHL), retinoblastoma (RB), and familial posterior fossa brain tumour (hSNF5), hereditary nonpolyposis coli (HNPCC), and nevoid basal cell carcinoma (NBCCS) or Gorlin syndrome [12,13] (Table 3.15). Some of these data were reported previously [49]. All these disorders are relatively rare autosomal dominant conditions, with prevalence of 1 in 5000 in the American populations for FAP, 1 in 15000 for RB, 1 in 36,000 for VHL, and even rarer for others.

With current progress in understanding of the molecular basis of cancers and sequencing of the genes involved in malignancy, the inherited cancer predisposition became one of the emerging indications for PGD. As mentioned, despite extensive discussions of the ethical and legal issues involved in PGD for late-onset disorders with genetic predisposition, an increasing number of patients regard the procedure not only as a favourable option but also the only possible reason for forgoing the pregnancy, which can be established free of mutation from the onset, avoiding their potentially difficult decision to have a pregnancy at high risk of being affected, with the option of prenatal diagnosis and termination of pregnancy if the fetus would be diagnosed to carry a mutant gene.

Patients with FAP usually present with colorectal cancer in early adult life, secondary to extensive adenomatous polyps of the colon, determined by mutation of adenomatous polyposis coli (APC) gene located on chromosome 5 (5q21-q22). Over 826 germ-line mutations have been found in families with FAP, causing a premature truncation of the APC protein, through single amino acid substitutions or frameshifts, with the most common mutation being a 5-bp deletion resulting in a frameshift mutation at codon 1309. The APC mutations lead to a premalignant disease with one or more polyps progressing through dysplasia to malignancy with a median age at diagnosis of 40. Because the mutations in the APC gene are almost completely penetrant, although with striking variation in expression, even presymptomatic diagnosis and treatment of carriers cannot exclude the progression of polyps to malignancy, making PGD an attractive approach for couples carrying APC mutations. However, no successful attempt to apply PGD for FAP has been reported [50].

VHL disease is a cancer syndrome with age-related penetrance, characterized by hemangiomablastomas of the brain, spinal cord, and retina, bilateral renal cysts and renal carcinoma, pheochromocytoma, and pancreatic cysts. Depending on the combination of these clinical features, four different types of the disease have been described. The gene responsible for VHL syndrome consists of three exons and is located on chromosome 3 (3p26-p25), with specific VHL gene mutations correlating with the clinical phenotype. Its normal gene product is a tumour-suppressor protein, which is expressed in most cells and has a variety of functions, including transcriptional and post-trancriptional regulation. More than 300 germ-line mutations have been identified in families with VHL syndrome, consisting of partial or complete gene deletions, and frameshift, nonsense, missense and splice site mutations, most commonly affecting codon 167. Mutations in the VHL gene either prevent its expression completely, or lead to the expression of an abnormal protein. Because only 20% cases of VHL are sporadic, with the remaining 80% being familial, PGD is clearly an attractive option for couples carrying these mutations to avoid the inheritance of these tumour-suppressor gene mutations by their potential children and have a mutation-free child, with no risk of developing cancer.

Predisposition to RB is caused by the germ-line mutations in the RB1 gene, located on chromosome 13 (q14.1–q14.2), and represents a malignant tumour of retina, which occurs in cells with cancer-predisposing mutations, usually before age 5. More than half of patients have the unilateral RB, which may be diagnosed at 24 months, while the bilateral RB is recognizable as early as at 15 months, using direct ophthalmoscopy. The majority of cases are due to a point mutation in the coding regions of the RB1 gene, while partial deletions of the gene were also described. Over 200 distinct mutations have been reported, with the majority resulting in premature termination codon, usually through single base substitutions, frameshift or splice mutations, scattered throughout exon 1 to exon 25 of the RB1 gene and its promoter region. The mutations lead to the loss of the cell cycle regulation function of the RB1 protein, and are nearly completely penetrant in nonsense and frameshift mutations, making PGD and important option for couples at risk.

As mentioned, hSNF5 mutations are very rare and have been found in sporadic rhabdoid tumours of the central nervous system. Rhabdoid tumors are known to be highly malignant neoplasms, usually occurring in children under age 2. Although rhabdoid tumours determined by truncating mutations of the hSNF5 gene are mainly sporadic and have never been previously found in the parents of affected children, a first familial case of posterior fossa brain tumour has recently been described in two generations [51]. The proband presented at the age of 18 months with a cerebellar malignant rhabdoid tumour. Although the parents were healthy, the child's maternal uncle died at age 2 from a posterior fossa choroid plexus carcinoma, and her grandfather's sibling also died as an infant from a brain tumour, suggesting the presence of a germ-line mutation. The couple presented for PGD in order to have a pregnancy free from hSNF5 mutation, also avoiding the birth of a second child with brain tumour.

NBCCS is characrerized by the development of multiple jaw keratocysts, frequently beginning in the second decade of life, and/or basal cell carcinoma from the third decade of life. NBCCS is caused by germ-line mutation of the gene PTCH, localized in chromosome 9q22.3. Prenatal diagnosis of the disease-causing mutation is possible, but because basal cell carcinoma may appear only in adult life, pregnancy termination of the affected fetus may be difficult to accept, so PGD may seem to be a more attractive option for couples at risk.

Nineteen couples, overall, presented for PGD because of cancer predisposition, for whom 35 clinical cycles were performed, including 13 from five couples at risk for FAP, four from two couples with Li-Fraumeni syndrome, five from three couples with VHL, three from three couples with RB, five for two couples with NF1, two for a couple with NF2, and one for hSNF5, NBCCS, and NF2 (see Table 3.15); the data for Li-Frameni syndrome and NF1 and NF2 have been presented above.

For testing the maternal mutations, PB1 and PB2 were removed sequentially following maturation and fertilization of oocytes, while single blastomeres were removed from the eight-cell embryos for testing the paternal mutation. PGD strategy depended on the FAP mutations in the APC gene. One of the examples is presented in Figure 3.24, showing PGD for the maternally derived mutation resulting from the 13-bp deletion in exon 15 of the gene.

The example of PGD for paternally derived mutation causing VHL is shown in Figure 3.25, performed for the father carrying A to T substitution in nucleotide 482 of exon 1. The blastomere biopsy was performed in all three cycles for this couple, as well as for other couples with paternally derived mutations, including two couples with FAP (5-bp deletion in exon 15 and nt2110 insertion "G"), and couples with p53 tumour-suppressor mutation and NF2, described above.

Maternal mutation in a couple at risk for familial porterior fossa brain tumour was due to G to A substitution in a donor splice site of exon 7, which alters the conserved GT sequence at the beginning of the intron violating the GT rule for splice site recognition (Figure 3.26). In this unique case, the mother was unaffected but her daughter, who inherited the mutation, had a brain tumour [51]. Because the mutation was also detected in DNA from her uncle's tumour, suggesting the risk of transmitting the mutation to the next child, PB1 and PB2 were removed in this case to preselect mutation-free oocytes in a standard IVF cycle. This same approach was used in the couple with maternal RB1 mutation, with a bilateral RB found in the mother's childhood and also observed in her father. However, because the mutation in this particular case was unknown, PB1 and PB2 were removed and tested by linked markers, listed in Table 3.16.

DNA testing in all the PGD cycles was performed by multiplex nested PCR analysis, amplifying mutations simultaneously with linked markers both in blastomere and PB1 and PB2, with set of primers listed in Table 3.16. Linkage analysis had been performed for each couple, and the maternal and paternal haplotypes were established to avoid a possible misdiagnosis, which still requires special attention because of the phenomena of allele drop out (ADO) and preferential amplification, known to be frequent in a single cell DNA analysis. For example, in a couple with FAP, two linked markers were found to be informative, including Rsa I restriction site in intron 11, and SspI restriction site in 3′ untranslated area of the APC gene (see Figure 3.24). In a couple with VHL mutation, a single marker was informative, representing a dinucleotide repeat $(CA)_3TG(CA)_3TA(CA)_3TA(CA)_2TAG(CA)n$ (see Figure 3.25), which, however, was absolutely essential to be able to confirm the presence of the normal paternal gene in the embryos selected for transfer.

As seen from Table 3.15, unaffected embryos were preselected in every PGD cycle, with 65 embryos transferred in 35 cycles (1.85 embryos per cycle), resulting in 12 clinical pregnancies (34.3% per transfer), yielding the birth of 14 healthy mutation-free children (there was one stillbirth with confirmed mutation-free status, and one twin pregnancy and birth of healthy twins). PGD cycles from couples with maternal mutations (13-bp deletion and 5-bp deletion in exon 15 of APC gene; RB1; G to A substitution in a donor splice site of exon 7 of hSNF5 gene; TGG → TGA in exon 29 of NF1 gene; and 902insC in p53 tumour-suppressor) were performed by sequential PB1 and PB2 analysis. From 8.5 oocytes per cycle tested by both PB1 and PB2 DNA analysis, 4.5 per cycle were predicted to be free from maternal mutations predisposing to cancer, but only half of them, also having sufficient linked marker information available, resulted in cleaving embryos of acceptable quality for transfer, still yielding clinical pregnancies and birth of unaffected children in one third of them.

One of the PGD cycles for maternally derived mutations, performed for G to A substitution in a donor splice site of exon 7 of hSNF5 gene is presented in Figure 3.26. Of only four oocytes available for testing, three have the results of both PB1 and PB2 analysis, showing that only one oocyte (oocyte # 4) could be predicted to be free of mutation, based on the heterozygous status of PB1 and homozygous mutant PB2. Of the remaining two oocytes, oocyte #2 was clearly mutant as evidenced by the heterozygous PB1 and homozygous normal PB2, suggesting that only the mutant allele was left in the resulting oocyte. The results of mutation analysis in oocytes #2 and 4 were in agreement with both markers in PB1 and PB2 tested.

On the other hand, while based on mutation analysis, oocyte #3 also could have been predicted normal and transferred, as evidenced from the homozygous mutant status of PB1 and homozygous normal status of PB2, the presence of both alleles of D22s1144 marker linked to both the normal and mutant gene in the corresponding PB1 suggested a completely opposite (mutant) genotype

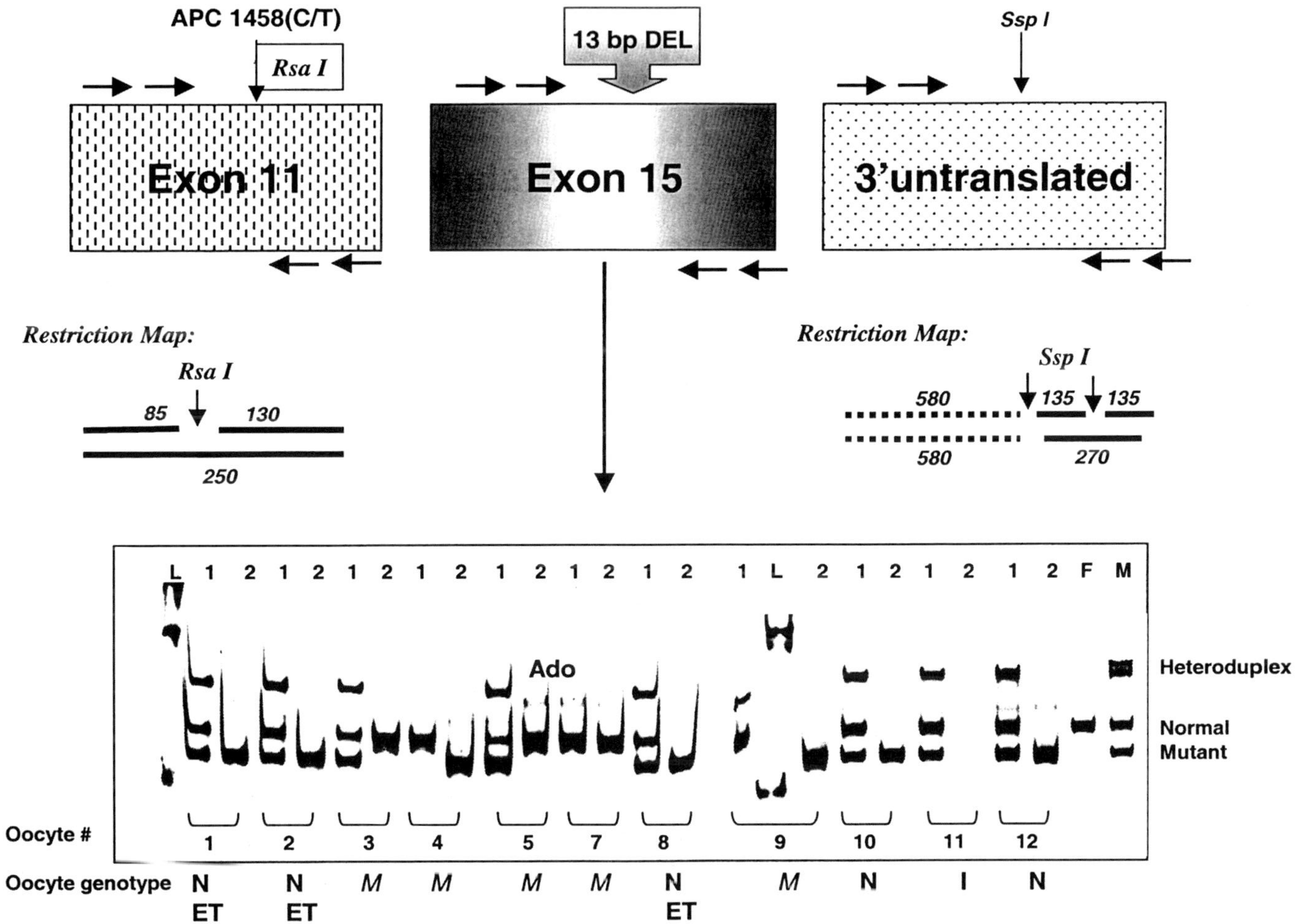

Figure 3.24. PGD for 13-bp deletion in APC gene resulting in birth of a normal child. (*Top*) Schematic diagram of the mutation and linked markers on chromosome 5q21. (*Middle*) Restriction map for *Rsa* I and *Ssp* I restriction digestion in exon 1 and untranslated area of APC gene. (*Bottom*) Polyacrylamide gel electrophoresis of the PCR products of PB1 and PB2 from 11 oocytes in one of the cycles of PGD for FAP, showing mutation-free oocytes #1, 2, 8, 10, and 12, evidenced by normal and mutant fragments in PB1 and a mutant fragment in PB2, in agreement with all two marker analyses.

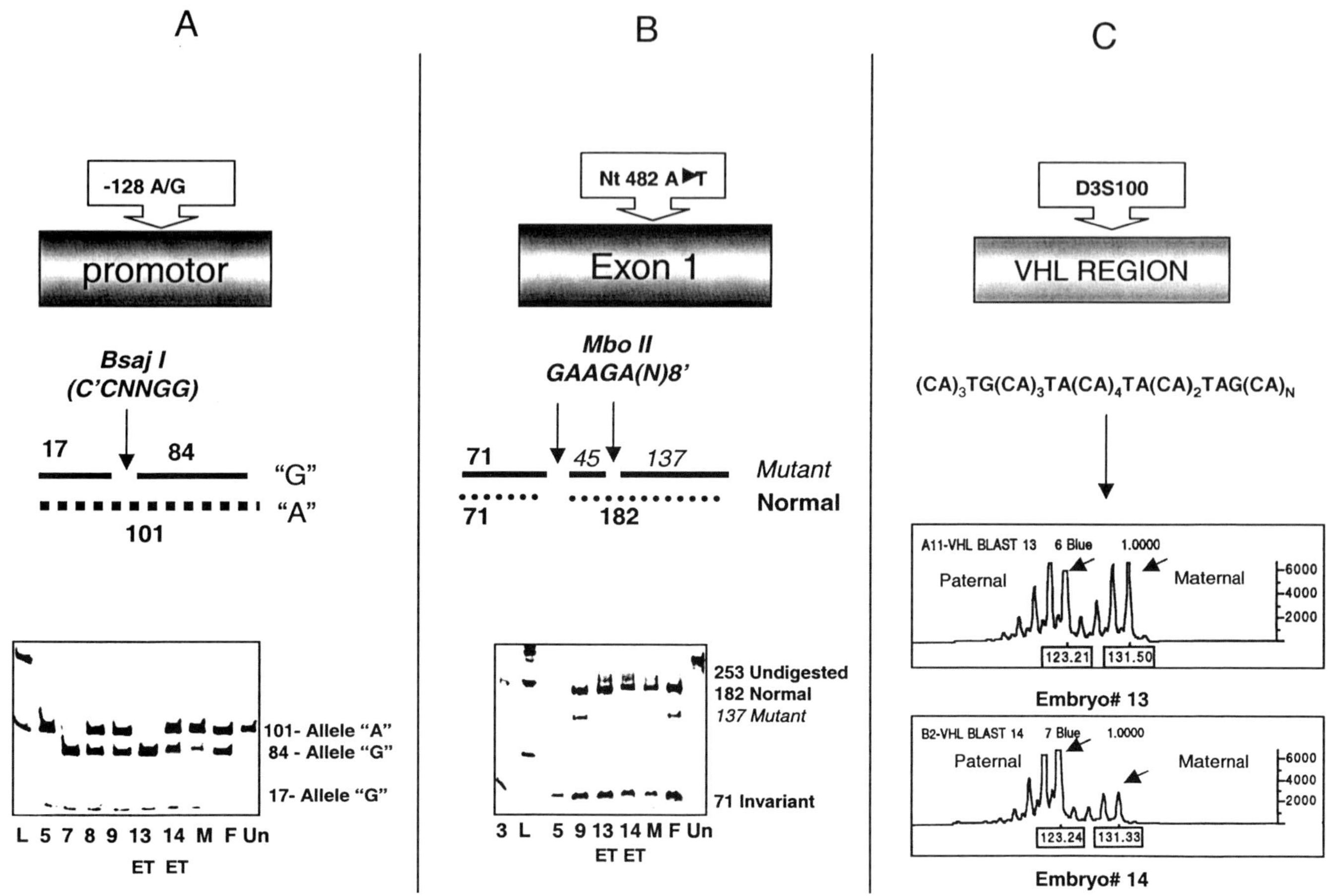

Figure 3.25. PGD for Nt 482 A-T mutation in exon 1 of VHL gene. (*Top*) Schematic diagram of the mutation and linked markers. (*A, top*) Position of G/A SNP in promoter area of the gene (*A, middle*) the restriction map, and the restriction digestion with *Bsaj* I enzyme differentiated "A" and "G" sequences. (*B, top*) Position of nt 482 A-T mutation in exon 1 of VHL gene and the restriction map (*B, middle*). Amplified sequence has invariant restriction fragment (71 bp). Enzyme *Mbo* II digests mutant sequence into two fragments of 45 bp and 137 bp, with the size of the normal allele being 182 bp. (*B, bottom*) Polyacrylamide gel electrophoresis of the *Mbo*II restriction digested PCR products of blastomeres detected mutation-free embryos #13 and 14), evidenced by the presence of a 182 bp undigested fragment, as compared with the presence of a digested 137 bp fragment in the affected embryos. This is in agreement of the results of SNP analysis in promoter (*A, bottom*). (*C, top*) Position and sequence of D3S100 dinucleotide repeat in VHL region. (*C, middle and botttom*) Capillary electrophoregrams of fluorescently labelled dinucleotide repeat PCR products scored by Genotyper. The presence of 123 bp paternal allele in embryos #13 and 14 confirmed the normal status of these embryos predicted by mutation analysis. (ET, shows embryos transferred; M, maternal DNA; P, paternal DNA containing the mutation; L, 100 bp standard.)

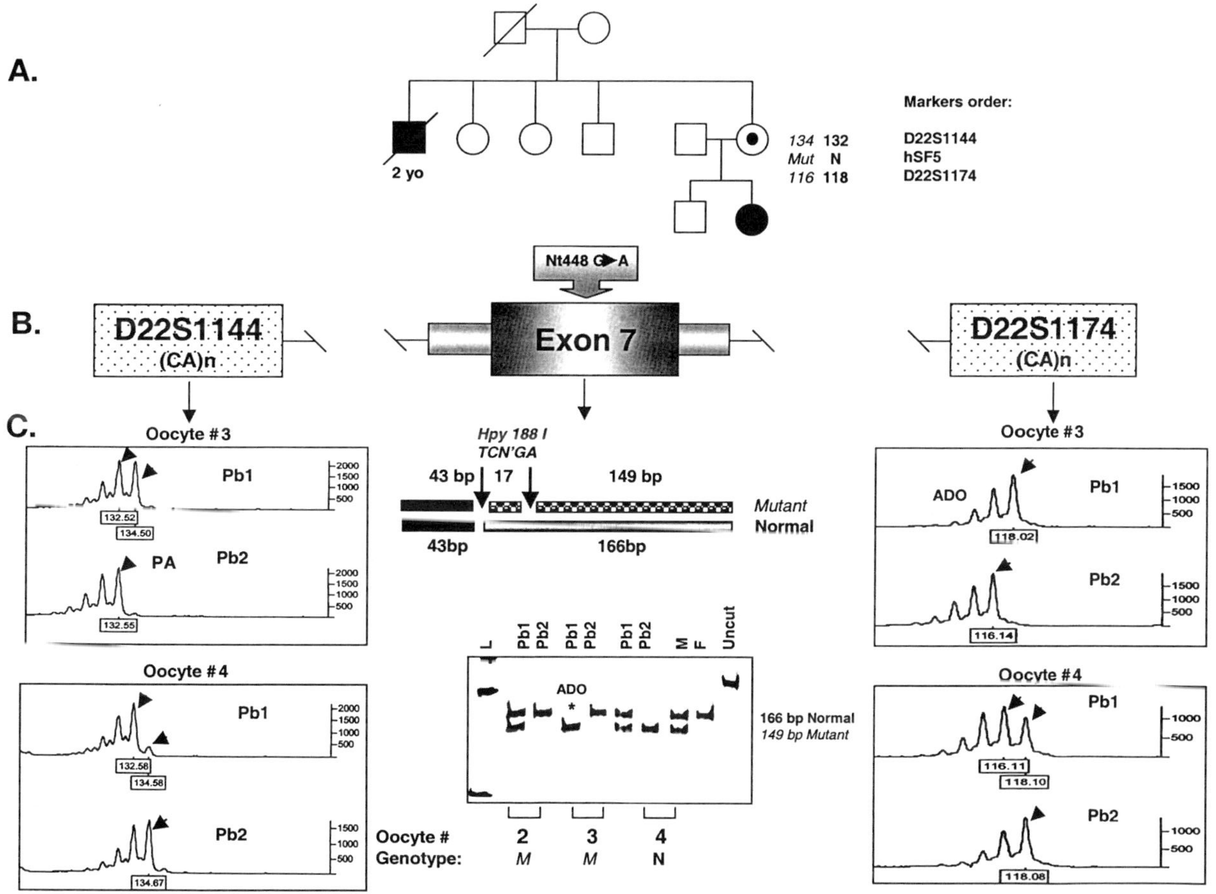

Figure 3.26. PGD for familial porterior fossa brain Tumour (hSNF5). (*A*) Family pedigree with posterior fossa tumours in two generations: an unaffected carrier haplotype was determined by sequential PB1 and PB2 analysis. (*B*) Position of the Nt 443 G → A mutation in exon 7 of hSNF5 gene and tightly linked dinucleotide polymorphic markers *D22s1144* and *D22s1174*. (*C, left*) Fluorescent PCR results of the sequential PB1 and PB2 analysis. One of the second-round PCR primers was labelled with Hex fluorescent dye. Oocyte #3 is predicted to be affected based on heterozygous (132/134) PB1 and 132 bp PCR product linked to normal maternal allele in PB2. Preferential amplification (PA) of 132 bp allele was detected in PB1 from oocyte #4 in agreement with mutation analysis and the second *D22s1174* marker, suggesting that oocyte #4 may be predicted to be normal. (*C, middle*) Restriction map and polyacrylamide gel analysis of *Hpy188 I* restriction digestion of PCR product from PB1 and PB2. Based on this analysis, oocyte #2 is mutant based on heterozygous PB1 and normal PB2. ADO was detected in PB1 from oocyte #3 by polymorphic markers *D22s1144* and *D22s1174*. Without these markers, oocyte #3 genotype would have been predicted normal based on homozygous mutant PB1 and hemizygous normal PB2, which might have lead to misdiagnosis. Oocyte #4 is predicted to be normal based on heterozygous PB1 and mutant PB2, in agreement with marker analysis as mentioned. (*C, right*) Fluorescent PCR results of sequential PB1 and PB2 analysis. One of the second-round PCR primers was labelled with Fam fluorescent dye. ADO of 116 bp allele was detected in PB1 from oocyte #3 by mutation analysis and *D22s1144* marker study. Heterozygous PB1 from oocyte #4 confirmed PA suspected during the *D22s1144* analysis. (N, normal; M, mutant; F, paternal DNA; Mo, maternal DNA; U, uncut PCR product; Nt, nucleotide.)

Table 3.16. Primers and reaction conditions for PGD of cancer predisposition

Gene/ Polymorphism	Upper Primer	Lower Primer	Annealing Tm (°C)
APC	Outside		
13 bp deletion	5′ TCGACATGATGATAATAGGTC3′	5′TTTCTGTTGCTGGATGGTAG3′	62–55
	Inside		
	5′ GACAATTTTAATACTGGCAACA3′	5′TCCAAACTTCTATCTTTTTCAGA3′	58
APC	Outside		
RFLP *RsaI*	5′GGTACCAGTTTGTTTTATTTTAG3′	5′CACAGGTTTTTATCAGTCATTG3′	62–55
	Inside		
	5′GATGATTGTCTTTTTCCTCTTGC3′	5′CTGAGCTATCTTAAGAAATACATG3′	62
APC	Outside		
RFLP *Ssp I*	CTATGCATTAAGAGTAAAATTCC3′	5′GTATTACCCTATCTGAGTGCC3′	62–55
	Inside		
	5′CATTGAAGAAGACTGTTGCCAC3′	5′AAAAGGTTTTCCTCCCCAAATAC3′	62
VHL (Nt 482)	Outside		
Creates *MboII*	5′ACGGCGGGGAGGAGTCG3′	5′CTCAAGGGGCCTCAGTTC3′	62–55
site	Inside		
	5′GCGCCGAGGAGGAGATG3′	5′GGGGCTTCAGACCGTGC3′	58
VHL	Outside		
D3S100	5′GTCTGGTGGCCTGTGAAC3′	5′CTTTCTCTTCGGAATGGGAG3′	62–55
	Inside		
	5′GAAAATGTGTTCATCATCCTC3′	5′TTATCTTATCCCTGCCTCAC3′	58
hSNF5	Outside		
(Nt443)	5′GGGGGAGTTTGTCACCAC3′	5′GGAGGACGGAGCAAACAC3′	62–55
Creates *Hpy188 I*	Inside		
site	5′CCACCATCGCATACAGCA3′	5′GGAGGACGGAGCAAACAC3′	54
D22S1144	Outside		
(Heminested)	5′AAATAGGCAGATGCTGAAA3′	5′ACAGAGCCTCTGGTCCTC3′	62–55
	Inside		
	5′HexGGAAAGCAACTTTGGTAAA3′	5′ACAGAGCCTCTGGTCCTC3′	50
D22S1174	Outside		
(Heminested)	5′GGACATAGCAAACTCTTAGGG3′	5′GAATCTGCTGCTTGCTTTT3′	62–55
	Inside		
	5′FamCACTTCTGAGTTGTTTGAATCTC3′	5′GAATCTGCTGCTTGCTTTT3′	50
RB (GA del)	Outside		
	5′GTAGGCTTGAGTTTGAAGA3′	5′TGAAGTTGTTTTTAAAATGAGA	62–55
	Inside		
	5′FamTGATTTTACTGCATTATGTCAG3′	5′CTTACCAATACTCCATCCAC3′	55
RB	Outside		
Intron 20 STR	5′GACAGGCATTTGGACCAAG3′	5′GCAGTGAGCCGAGATTGC3′	62–55
(CTTT)n	Inside		
	5′HexCCCTACTTACTTGTTAACTG3′	5′GGTAACAGAGTGAGACTCTA3′	56

of oocyte #3, despite the second polymorphic marker D22S1174 showing a correspondence with the mutation analysis. So the availability of the two polymorphic markers made it possible to avoid the transfer of the additional embryo, which may have led to misdiagnosis. Unfortunately, the transfer of only one unaffected embryo back to patient yielded no clinical pregnancy, as there is a reduced chance of establishing pregnancy with a single embryo transfer. The follow-up testing of embryos resulting from oocytes #2 and 3 confirmed the diagnosis, further supporting the need for simultaneous mutation and linked marker analysis for avoiding misdiagnosis in PB-based PGD. Therefore, as previously suggested, the priority in the preselection of mutation-free oocytes should be given to those with heterozygous PB1 and homozygous mutant PB2. If no such oocytes are available, at least three linked polymorphic markers are required to exclude ADO of one of the alleles in apparently homozygous PB1, which is currently the major potential source of misdiagnosis due to ADO and preferential amplification in PB-based PGD.

The results of PB1 and PB2 analysis in another cycle performed for a couple with maternally derived mutation (13 bp deletion in APC gene), resulting in the birth of an unaffected child, was presented in Figure 3.24 above. Of 11 oocytes tested in this case, five were mutant, based on heterozygous PB1 and homozygous normal PB2 in three of them (oocytes #3, 5, and 9), homozygous normal PB1 and homozygous mutant PB2 in one (oocyte #4), and homozygous normal PB1 and PB2 in the other one (oocyte #7). The last pattern of the same homozygous genotype in both PB1 and PB2 may be explained only by ADO of one of the alleles (ADO of the mutant allele in this case) in the corresponding PB1, because PB2 cannot contain the mutant allele without this allele being present in the oocyte following its maturation and the extrusion of PB1, which, therefore, should be heterozygous, corresponding to the oocyte genotype.

As seen from Figure 3.24, one of the oocytes tested had no information about PB2 and could not be genotyped (oocyte #11), while the remaining five oocytes were predicted normal (oocytes #1, 2, 8, 10, and 12), of which three resulted in embryos of acceptable quality and were transferred, yielding a singleton clinical pregnancy and birth of a healthy child following the confirmation by prenatal diagnosis.

The accuracy of PGD was, as usual, confirmed by the follow-up analysis of the embryos resulting from the mutant oocytes, as well as by prenatal diagnosis in the established clinical pregnancies and the blood test after delivery of a child. However, as shown in Figure 3.27, a cord blood sample was not appropriate for confirmation analysis, as it has been contaminated by maternal blood, clearly demonstrated by the presence of the mutant allele, as well as both maternal polymorphic markers in the cord blood sample. The fact that baby has no mutant allele was further confirmed by testing of a hill stick blood sample from the child, which contained no trace of the mutant gene, also in agreement with analysis of polymorphic markers. Therefore, the cord blood sample should not be material for confirmation of PGD.

One of the cycles performed for paternally derived Nt 482 A-G in exon 1 of VHL gene was presented in Figure 3.25 above. As seen from this figure, only one of four embryos in this cycle contained the mutant gene (embryo #9), the remaining three being normal. The haplotype analysis, an obligatory part of PGD for paternally derived dominant mutations, confirmed the presence of a normal paternal gene. As seen from the Figure 3.25, both embryos transferred in this cycle contained the polymorphic 123 bp marker, which is strongly linked to the normal paternal gene, therefore confirming the results of the mutation analysis. The transfer of these two normal embryos back to the patient, however, did not result in a clinical pregnancy. The affected embryo was reanalyzed, confirming the prediction of Nt 482 (A-T) mutation in the embryo.

The presented data demonstrate the acceptable diagnostic accuracy of both the PB and blastomere analysis for PGD of the above conditions. As shown by the follow-up analysis of the mutant embryos or those with insufficient marker information, the PGD results were confirmed in all resulting embryos available for the study, including 14 embryos in blastomere analysis and 48 embryos resulting from mutant ooocytes, detected by PB1 and PB2 analysis. The PGD diagnosis was also confirmed after birth of all 14 children, although as mentioned in the case of FAP, the use of cord blood for PGD confirmation should be avoided to avoid conflicting results due to contamination of cord blood by the maternal cells.

The data also show the need for avoiding misdiagnosis due to ADO and preferential amplification in testing oocytes and blastomeres. The use of the multiplex nested PCR, involving simultaneous amplification of mutation and linked markers, as was discussed above, appeared to be efficient to avoid misdiagnosis [47], with every single marker result reducing the risk of misdiagnosis by half. For example, approximately half of the ADO in PB1, which was estimated to be in the range of 10%, may be detected by applying the first linked marker, further picking up another 2–3% of ADO by the second marker, and reaching a complete detection rate with three markers. The same consideration may be true for blastomere biopsy, however, taking into consideration an approximately double ADO rate in blastomeres compared with PB1 [24, 27]. The other potential approach for the reduction of ADO rates may be the most recent application of a dual-probe real-time PCR, which according to our preliminary results may reduce the ADO rates by half, as discussed previously in Chapter 2.

As also seen from the presented results, indications for PGD are being further extended compared with the practice of prenatal diagnosis, and

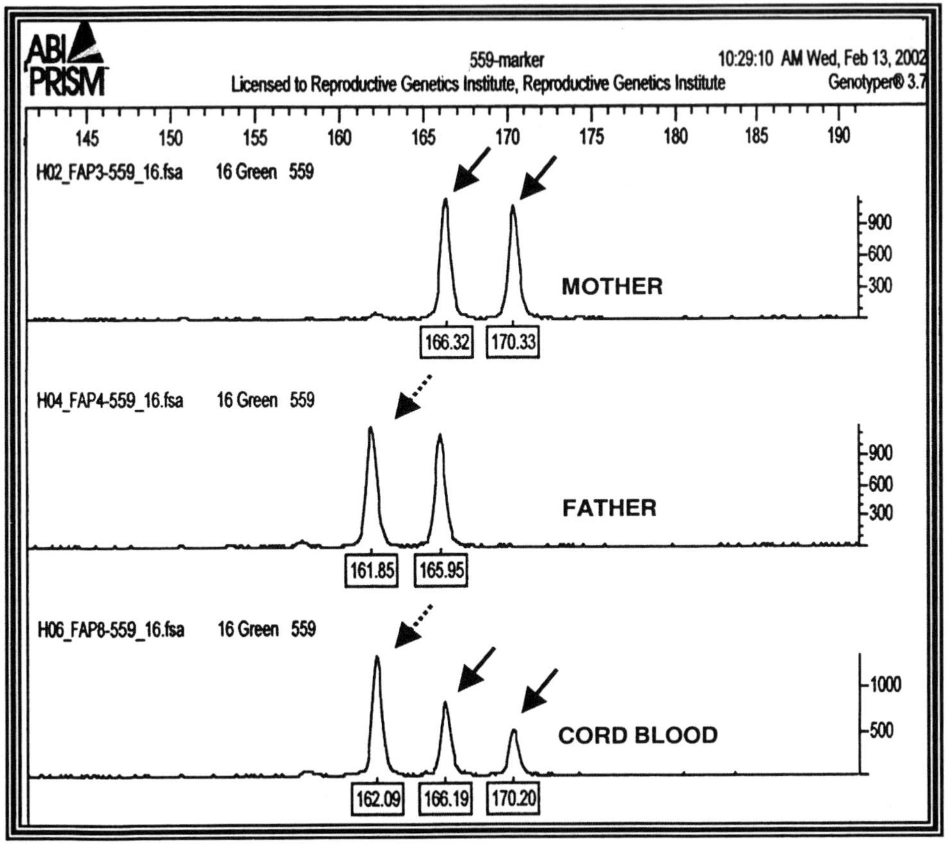

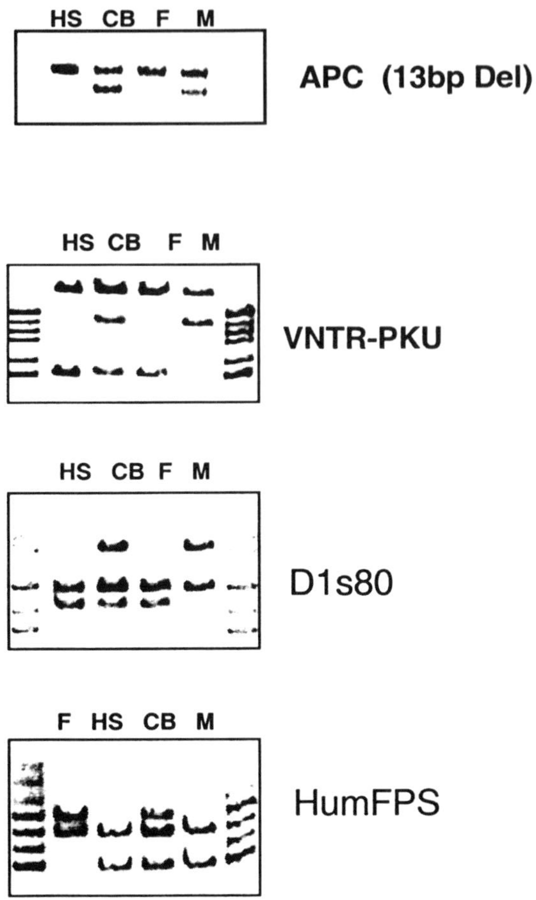

Figure 3.27. Confirmation of PGD for APC mutation by cord and hill stick blood DNA analysis. (*Left*) Fluorescent PCR analysis of D17s559 polymorphic marker: Pick pattern (arrows) evidences the presence of both maternal alleles in cord blood, suggesting the maternal DNA contamination. (*Right*) Only normal allele was detected in hill stick (HS) and paternal (F) samples. Deletion in APC gene was detected in cord blood (CB) and maternal (M) samples. The presence of both maternal and a paternal alleles were confirmed by STR (HumFPS) and VNTR (PKU, D1s80) markers.

at present include PGD for cancer predisposition, which is an acceptable approach for couples at risk, despite important ethical implications. So the genetic counseling services may consider informing patients at risk of having children with a strong genetic predisposition to cancers about the presently available option for PGD, without which these couples may remain childless because of their fear to choose prenatal diagnosis and possible pregnancy termination.

Because such diseases present beyond early childhood, and even later may not be expressed in 100% of the cases, the application of PGD for this group of disorders is still highly controversial. However, initial experience in offering PGD for this indication shows that the availability of PGD allows couples forgoing pregnancy, which otherwise would never be attempted. This may be further demonstrated by the first case of PGD performed for genetic predisposition to Alzheimer disease described below [52].

3.6.4 Alzheimer Disease

Alzheimer disease (AD) is a rare autosomal dominant familial predisposition to presenile form of dementia. Three different genes were found to be involved in this form of AD, including presenilin 1 (PS1) located on chromosome 14 [53], presenilin 2 (PS2) on chromosome 1 [54], and amyloid precursor protein (APP) gene on chromosome 21 [55], which is well known for its role in the formation of amyloid deposits found in the characteristic senile plaques of the patients with AD. The early-onset dementias associated with βAPP mutations are nearly completely penetrant, and, therefore, are potential candidates not only for predictive testing but also for PGD. Of 10 APP mutations presently described, mutations in the exons 16 and 17 were reported in the familial cases with the earliest onset. One such mutation, with as early onset as the middle or late 30s has been reported to be due to a single G to C nucleotide substitution in exon 17, resulting in a valine-to-leucine amino acid change at codon 717 (V717L) [56]. This mutation was identified in three of five family members (siblings) tested, one of whom presented to PGD, described here, which resulted in a pregnancy and birth of a healthy child free from APP mutation.

The patient who presented for PGD was a 30-year-old woman with no signs of AD, carrying V717L mutation, resulting from G to C substitution in exon 17 of APP gene. The predictive testing in the patient was done because of the early-onset AD in her sister, who carried this mutation and developed symptoms of AD at age 38 [56]. This sister is still alive, but her cognitive problems progressed to the point that she was placed in an assisted living facility. Her father had died at age 42 and also had a history of psychological difficulties and marked memory problems. V717L mutation was also detected in one of her brothers, who experienced mild short-term memory problems as early as age 35, with a moderate decline in memory, new learning, and sequential tracking in the next 2–3 years. The other family members, including one brother and two sisters were asymptomatic, although predictive testing was done only in the sisters, who appeared to be free from mutation in APP gene (Figure 3.28).

Two PGD cycles were performed by testing for the maternal mutation using DNA analysis of PB1 and PB2 [57]. A multiplex nested PCR was performed [58] involving the mutation testing simultaneously with the linked polymorphic marker, representing the short tandem repeat (STR) in the intron 1 [(GA)n... (GT)n] [59].

The first-round amplification cocktail for the multiplex nested PCR system contained outer primers for both APP gene and the linked marker, while the second-round PCR used inner primers for each gene. We designed the outer primers APP-1 and APP-102 (see primers in Table 3.17) for performing the first round amplification, and the inner primers APP-101 and APP-103 for the second round of PCR. As shown in Figure 3.29, second-round PCR produces a 115 base pair (bp) product, undigested by *Mnl* I restriction enzyme, corresponding to the normal allele, and two restriction fragments of 72 bp and 43 bp, corresponding to the mutant allele. There was also invariant fragment of 84 bp produced in both normal and mutant alleles, used as a control.

To perform nested PCR for specific amplification of the linked marker (GA)n... (GT)n in intron 1, we designed the outer primers In1-1 and In1-2 for the first round and the inner primers In1-3 and In1-4 for the second round of amplification. The haplotype analysis, based on the polar body genotyping, demonstrated that the affected allele was linked to the 10, and the normal one to the 6 repeats.

A total of 23 oocytes were available for testing in two cycles, of which 15 were tested by both PB1 and PB2 (13 in one cycle and two in the other). The mutation and linked marker analysis in the

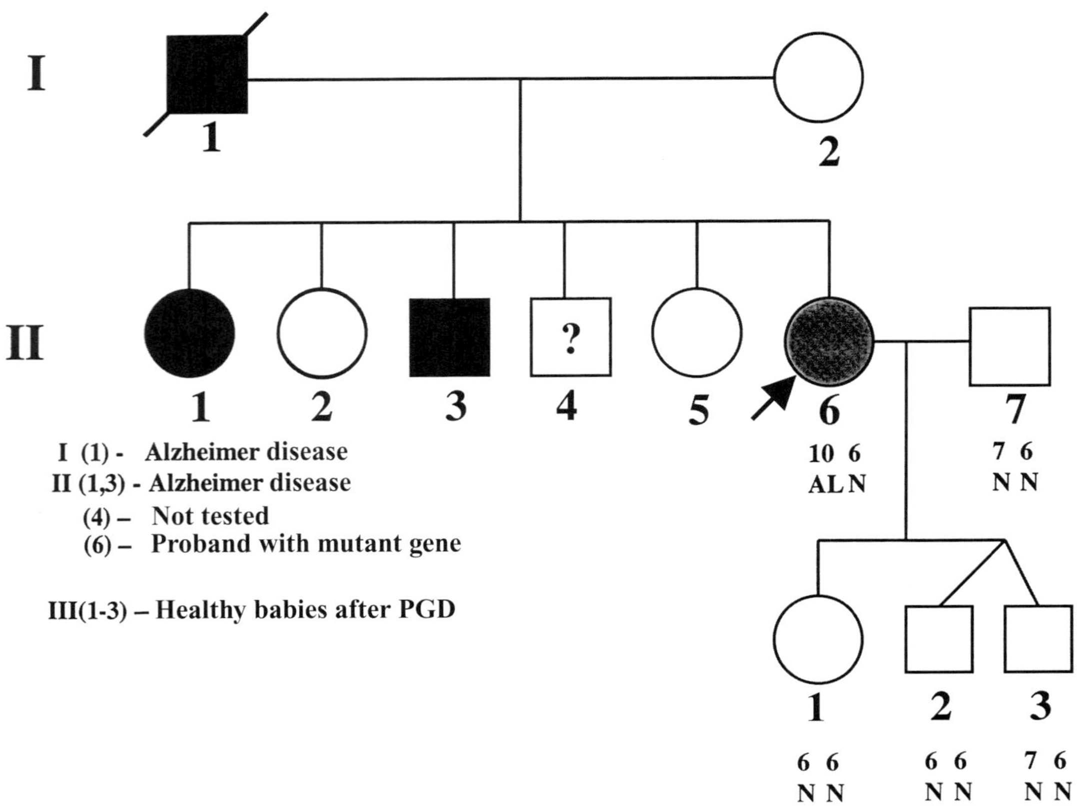

Figure 3.28. PGD for early-onset Alzheimer disease caused by mutation V717L: family pedigree. Two PGD cycles for asymptomatic carrier of the mutant gene (II;6), both resulting in the birth of unaffected children (III). Haplotype analysis shows that these children inherited normal maternal allele linked to the 6 repeats. Her sister (II;1) and brother (II;3) were affected by early-onset AD, as well as her father (I).

intron 1 revealed six normal, all in one cycle, and nine affected oocytes, including two in one cycle and seven in the other. The results of one of the cycles, resulting in the transfer, are presented in Figure 3.29. As seen from this figure, oocytes #4, 9, 14, and 15 were clearly normal, because both mutant and normal genes were present in their PB1, with the mutant gene further being extruded with the corresponding PB2, leaving only normal gene in the resulting oocyte. In addition, oocytes #3 and 13 were also normal, because their corresponding PB1s were homozygous mutant, suggesting, therefore, that the resulting oocytes should have been normal, as further confirmed by the presence

Table 3.17. Primers and reaction conditions for PGD of alzheimer disease

Gene/ Polymorphism	Upper Primer	Lower Primer	Annealing Tm (°C)
APP V717L	Outside APP-1 5′GTGTTCTTTGCAGAAGATG3′	APP-102 5′CATGGAAGCACACTGATTC3′	55
	Inside APP-101 5′GTTCAAACAAAGGTGCAATC3′	APP-103 5′TCTTAGCAAAAAGCTAAGCC	55
Intron 1 (GA)n(GT)n	Outside 5′CCTTATTTCAAATTCCCTAC3′	5′GATTGGAGGTTAAGTTTCTG3′	55
	Inside 5′CAGCATCTGTCACTCAAG3′	5′AATATTTGTTACATTCCTCTC3′	55

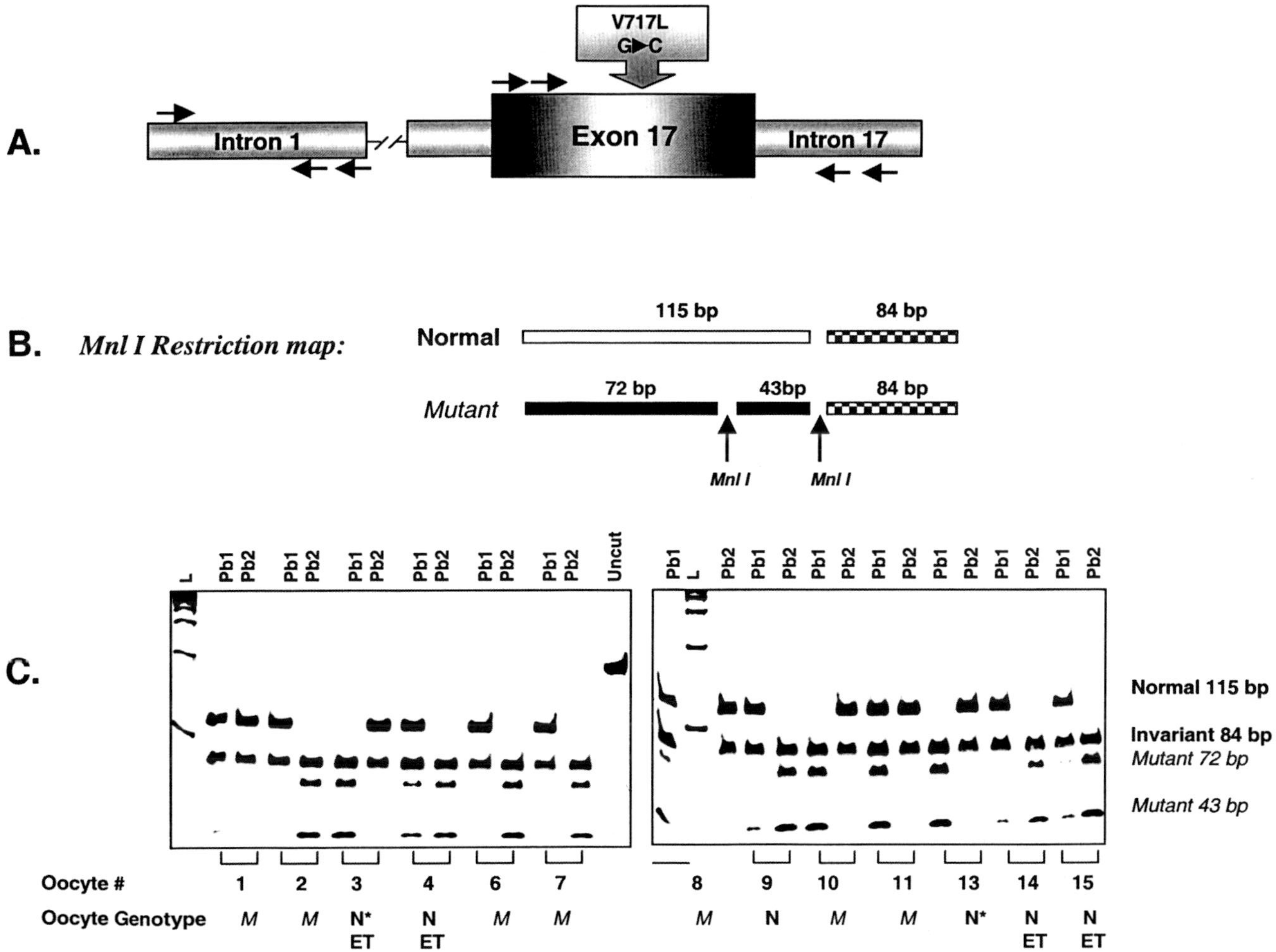

Figure 3.29. Preimplantation diagnosis for V717L mutation in the amyloid precursor protein gene, causing early-onset Alzheimer disease by polar body analysis. (*A*) Map of human APP gene, showing sites and location of V17L G → C mutation and polymorphic markers. (*B*) Restriction map for normal and abnormal alleles. (*C*) Polyacrylamide gel analysis of *Mnl* I restriction digestion, showing six unaffected (N) oocytes (#3, 4, 9, 13, 14, and 15), and seven mutant (M) oocytes (#1, 2, 6, 7, 8, 10, and 11). Four of six embryos resulting from unaffected oocytes (#3, 4, 14, and 15) were transferred back to the patient, resulting in an unaffected pregnancy. Three of these oocytes were with heterozygous PB1 and mutant PB2 (N), and only one (#3) was with homozygous mutant PB1 and normal PB2, leaving 5% probability for misdiagnosis (N*). (ET, embryo transfer; L, ladder (size standard); Uncut, undigested PCR product; *arrows*, fully nested primer sets.)

of the normal gene in the extruded PB2s, also in agreement with the linked markers analysis. However, because only one linked marker was available for testing, a 5% probability for ADO of the normal gene in the corresponding PB1 could not be excluded in this particular PB1.

The remaining oocytes were predicted to be mutant, based on heterozygous PB1 and normal PB2 in five of them (oocytes #1, 2, 8, 10, and 11), and homozygous normal PB1 and mutant PB2 in two (oocytes #6 and 7). The follow-up study of the embryos resulting from these oocytes confirmed their affected status in all but one (oocyte #7). This may be explained by ADO of the mutant allele in the apparently heterozygous PB1, which was undetected because of the amplification failure of the linked marker in this case.

To exclude any probability of misdiagnosis, the priority in the embryo transfer was given to the four of six normal embryos resulting from the oocytes with heterozygous PB2 and mutant PB2. However, only three of these embryos developed into the cleavage stage and could be transferred (#4, 14, and 15), so the additional embryo (#3) was preselected, originating from the oocyte with homozygous mutant PB1 and the normal PB2, as these results were also confirmed by the linked marker analysis. These four embryos were transferred back to the patient, yielding a singleton clinical pregnancy and birth of an unaffected mutation-free child.

The presented results demonstrate feasibility of PGD for early-onset AD, providing a nontraditional option for patients who may wish to avoid the transmission of the mutant gene predisposing to the early AD in their potential children. This may appear for some patients the only reason of undertaking pregnancy, as the pregnancy may be established free from an inherited predisposition to AD from the very onset. Because the disease never presents at birth or early childhood, and even later may not be expressed in 100% of the cases, the application of PGD for AD is still controversial. However, with no current prospect for treatment of AD, which may arise despite presymptomatic diagnosis and follow-up, PGD seems to be the only relief for at-risk couples, such as in the presented case, and the cases of PGD for cancer predisposition described above.

Therefore, prospective parents with risk for AD and other relevant conditions should be informed about this emerging new technology so they can make their reproductive choice based on all available options. This seems to be ethically more acceptable than a denial of the information on the availability of PGD. Presented results of PGD for early-onset AD, together with previously described cases of PGD for the late-onset disorders with genetic predisposition and HLA typing (see below), demonstrate the extended practical implications of PGD, providing prospective couples at genetic risk with wider reproductive options for having unaffected children of their own.

So, for the diseases with genetic predisposition and with no current prospect for treatment, arising despite presymptomatic diagnosis and follow-up, PGD may be offered as the only relief for at-risk couples.

3.7 Blood Group Incompatibility

Although the at-risk pregnancies for blood group incompatibility, including that caused by Kell (K1) genotype or RhD, may be detected by prenatal diagnosis, in order to be treated by an intrauterine transfusion, the potential complication for the fetus cannot be completely excluded even after the procedure. Pregnancy termination in such cases will also be unacceptable, as the antibodies to K1, for example, are developed only in 5% of persons obtaining incompatible blood. On the other hand, some of the at-risk couples have had such an unfortunate experience of haemolytic disease of the newborn (HDN), resulting in neonatal death, that they regard PGD as their only option to plan another pregnancy. This makes PGD attractive for patients at risk for alloimmunization, although such conditions have rarely been an indication for prenatal diagnosis. We performed the first PGD for maternal-fetal incompatibility caused by K1 genotype, which is presented below.

The K1 system is one the major antigenic systems in human red blood cells, comparable in importance to RhD, as it may cause maternal-fetal incompatibility leading to severe HDN in sensitised mothers. K1 allele is present in 9% of the population, in contrast to its highly prevalent allelic variant K2. The gene is located on chromosome 7 (7q33), consisting of 19 exons, with the only C to T base substitution in exon 6 in K1 compared with K2 antigen, that leads to a threonine to methionine change at amino acid residue 193, preventing N-glucosylation [60, 61]. C to T base substitution also creates a *Bsm* I restriction enzyme

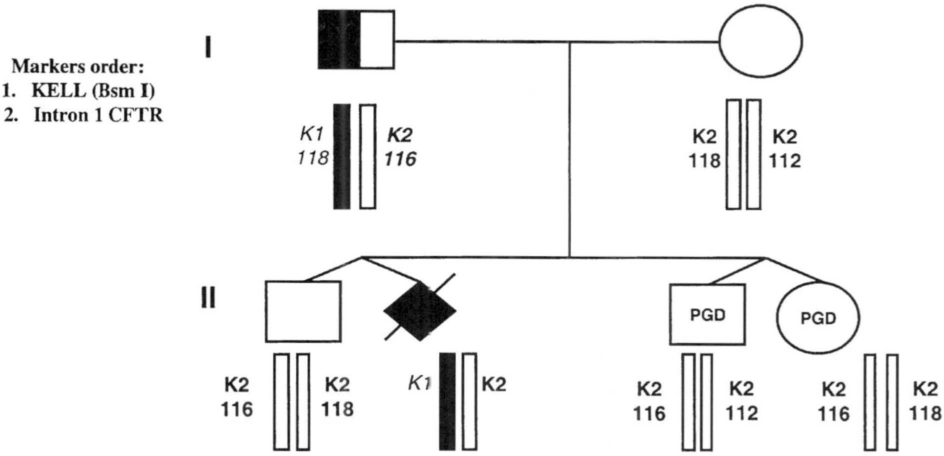

Figure 3.30. Preimplantation genetic diagnosis for Kell genotype: family pedigree. (*Top*) The father has *K1/K2* genotype, *K1* allele linked to 118 bp repeats and *K2* allele to 116 bp repeats of intron 1 of *CFTR* polymorphic marker, while the mother has *K2/K2* genotype, one allele linked to 118 bp repeats and the other to 112 bp repeats of intron 1 of *CFTR* polymorphic marker. (*Bottom*) Reproductive outcomes of this couple, including previous twin pregnancy resulting in death of one of the twins near birth due to HDN. Two healthy twins with *K2/K2* genotype resulting from PGD were born, confirmed also by linked polymorphic markers.

site, providing a reliable DNA test for diagnosis of K1 genotype.

In the case of pregnancy with a K1 fetus in a K2 mother, antibodies to K1 may be developed, leading to maternal-ofetal incompatibility causing severe HDN. As mentioned, because prenatal diagnosis may not always prevent the potential complications for the fetus, i.e., stillbirth or neonatal death, PGD is clearly an attractive option for preventing both Kell and rhesus haemolytic diseases.

Two couples presented for PGD, both with paternal K1/K2 genotype, i.e., heterozygous for C to T base substitution in exon 6, creating a *Bsm* I restriction enzyme site. In one couple, a 36-year-old mother had a previous dizygotic twin pregnancy, resulting in death of one of the twins carrying K1 allele near the birth, due to HDN (Figure 3.30).

In the second couple, a 37-year-old mother had three previous pregnancies, of which the first resulted in the birth of a healthy boy carrying the K1 allele, the second in the birth of a normal K2/K2 boy, and the third in a premature delivery of a 32-week female carrying the K1 allele, who died the day after birth with the clinical features of severe HDN.

To establish paternal haplotypes, a single sperm analysis was performed, to be able to undertake the linked marker analysis, in addition to Kell genotyping. Short tandem repeats (STR) associated with the cystic fibrosis (CFTR) gene were used, which are known to be located close to K1 and K2 alleles, with extremely rare recombination rates [62]. This analysis showed the presence of one informative linked marker (CFTR Intron 1) in the first and three (D7S550, CFTR intron 6 and CFTR intron 8) in the second couple. K1 allele was linked to the 118 (CFTR intron 1) repeat in the first (see Figure 3.30), and to the 158 (D7S550), 7 (CFTR intron 6), and 124 (CFTR intron 8) repeats in the second couple. Thus, in PGD cycles for both couples, multiplex nested PCR analysis was performed [46,57] by testing the *Bsm* I restriction site simultaneously with the linked polymorphic markers, including the CFTR intron 1 in the first, and the D7S550, CFTR intron 6, and CFTR intron 8 in the second couple. Outside and inside primer sequences and primer melting temperatures for DNA analysis in both couples are shown in Table 3.18 [63–65]. PCR products were identified by restriction digestion using *Bsm* I for Kell gene (Figure 3.31), and by capillary electrophoresis and scoring by Gynotyper for STRs (results not shown).

Overall, five PGD cycles were performed, including one for the first and four for the second couple, using blastomere biopsy. Based on both *Bsm* I restriction digestion and STR analysis, K1-allele-free embryos were preselected for transfer

Table 3.18. Primers and reaction condition for PGD of Kell genotype

Gene/Polymorphism	Upper Primer	Lower Primer	Annealing Tm (°C)
Kell Gene (Nested PCR)	Outside 5′TCAGCCCCCTCTCTCTCCTT3′	5′GTGTCTTCGCCAGTGCATCC3′	62–55
	Inside 5′AAGCTTGGAGGCTGGCGCAT3′	5′CCTCACCTGGATGACTGGTG3′	50
D7S550 (Heminested PCR)	Outside 5′ACTATCATCCACAATCCACTCC3′	5′GCAGTTGGGTTATTTCAAGTCT3′	62–55
	Inside 5′ACTATCATCCACAATCCACTCC3′	5′HexGATGTTGTGATTAGAGTTGCTGTA3′	56

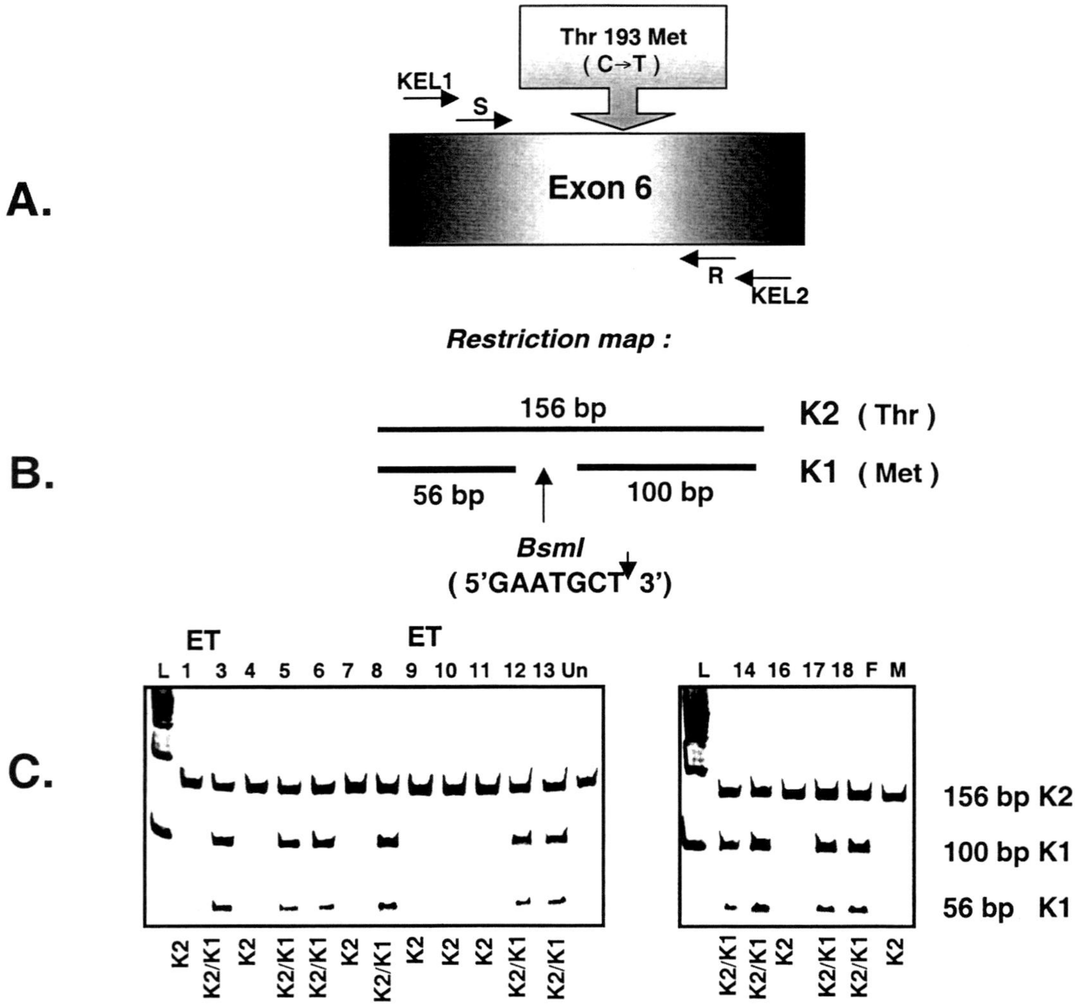

Figure 3.31. Preimplantation genetic diagnosis for Kell genotype. (*A*) Schematic diagram showing C ⟶ T substitution in exon 6 of *KEL* gene on chromosome 7; *arrows* demonstrate the positions of nested primers. (*B*) Restriction map for *Bsm*I digestion, showing the gain of *Bsm*I site by the *K1* allele (*lower line*). (*C*) Polyacrylamide gel electrophoresis of the *Bsm*I digested PCR products of 16 blastomeres from the first PGD couple, demonstrating *K1*-allele-free genotype in embryos #1, 4, 7, 9, 10, 11, and 17, from which embryos #1 and 9 were transferred, resulting in a twin pregnancy and birth of healthy *K1*-allele-free children. The remaining nine embryos have *K1/K2* genotype. (L, standard; F, paternal DNA amplified from sperm; M, maternal normal amplified DNA; Un, undigested PCR product; *K1/K2*, affected blastomere; *K2/K2*, normal blastomere.)

back to patients, while those predicted to contain K1 allele were exposed to confirmatory analysis using genomic DNA from these embryos in order to evaluate the PGD accuracy.

In a single PGD cycle for the first couple, 16 embryos at day 3 were tested, of which nine were excluded from transfer due to the presence of the K1 allele (see Figure 3.31). The remaining seven embryos did not contain the K1 allele (embryos #1, 4, 7, 9, 10, 11, and 17), also in agreement with the absence of the linked intron 1 CFTR (118 repeats) marker, but showed the presence of the paternal K2 allele, evidenced also by the presence of the intron 1 CFTR (116 repeats) marker, linked to the paternal K2 allele. One of the K1-allele-free embryos (embryo #11) did not develop, four were frozen (embryos #4, 7, 10, and 17), while the remaining two K1-allele-free embryos (embryos #1 and 9) were transferred back to the patient, resulting in a clinical twin pregnancy and birth of a healthy boy and girl homozygous for the K2 genotype (see Figure 3.30).

In the second couple, a total of 20 embryos at day 3 were tested for the K1 allele in four PGD cycles (eight in the first, five in the second, three in the third, and four in the fourth cycle). Nine of these embryos were heterozygous for the K1 allele, i.e., unsuitable for transfer, while the remaining 11 embryos were predicted to be free from this allele. Seven of these embryos were transferred, including two in the first, one in the second, one in the third, and three in the fourth cycle; however, yielding no clinical pregnancy.

Follow-up analysis of the K1-allele-containing embryos excluded from transfer was possible in a total of nine cases from both couples. Although the predicted genotypes were confirmed overall, six cases of recombination between K alleles and two linked markers (CFTR intron 6 and CFTR intron 8) were observed, suggesting a limited value of these two markers on their own. However, no recombination was observed between the gene and one linked marker in both cases (intron 1 CFTR in the first and D7S550 in the second couple), which allowed verifying the absence of the K1 allele and detecting both maternal and paternal K2 alleles in the embryo, so improving considerably the reliability of the diagnosis. For example, the preselection of the K1-free embryos shown in Figures 3.30 and 3.31 was based not only on the absence of the K1 allele, which may be also explained by ADO, but also on the absence of the linked intron 1 CFTR (118 repeats) marker and the presence of both the paternally and maternally derived K2 alleles, evidenced by the presence of polymorphic markers linked to the paternal and maternal K2 alleles. In other words, the absence of the K1 allele together with the presence of both paternal and maternal K2 alleles allowed reliably preselecting the K1-allele-free embryos for transfer.

The presented cases are the only PGD cycles resulting in the birth of unaffected K1-free children in the worldwide PGD experience. A number of attempts have been undertaken also to perform PGD for rhesus disease, which initially did not result in a clinical pregnancy [66]. The most recent attempt, however, yielded a clinical pregnancy and birth of a healthy girl confirmed to be blood type Rh-negative [67]. A couple with mother Rh-negative and father RhD-positive had two children one of whom was affected by HDN, with hyperbilirubinemia and neonatal jaundice, as well as significant haemolytic anemia. Because of RhD alloimmunization, the couple was presented with the dilemma of whether to attempt further pregnancy, there is a tendency for rhesus disease to worsen with each subsequent rhesus-incompatible pregnancy in sensitised women. Using blastomere biopsy and direct PCR amplification with analysis by capillary electrophoresis of fluorescently labelled amplicons, RhD-negative embryos were preselected for transfer, yielding an unaffected pregnancy and birth of a healthy child.

Both Kell and rhesus disease are quite prevalent in the populations, taking into consideration approximately 15% frequency for RhD and 9% for K1 antigen, presenting the risk for alloimmunization that may lead to HDN in some of the at-risk couples. Therefore, PGD may be a practically useful option for these couples to avoid the establishment of the RhD or K1 pregnancy in the sensitised mothers.

Thus, PGD for Kell genotype and other red blood group systems is feasible, providing a novel approach for sensitised mothers to avoid the risk of having children with HDN [68].

3.8 Congenital Malformations

Congenital malformations are highly prevalent (29.3/1000 live births) and are usually sporadic.

As described in Chapter 1, the major reduction of congenital malformations may be expected from population-based preventive measures, such as folic acid fortification of major foodstuffs, which may prevent the birth of tens of thousands of children with congenital malformations in North America. However, this will not have sufficient impact on the prevention of inherited forms, for which PGD might be an important option. In fact, with progress of the human genome project, an increasing number of inherited forms are being described, which, therefore, may be avoided through PGD. An example of one of the first applications of PGD for congenital diseases was PGD for sonic hedgehog (SHH) gene mutation, which is presented below.

3.8.1 Sonic Hedgehog Mutation

SHH gene is a human homologue of the *Drosophila* gene encoding inductive signals involved in patterning the early embryo, known to be (functionally) highly conserved in many species. The gene was mapped to chromosome 7 (7q36), previously designated as the locus for the gene involved in holoprosencephaly (HPE3) [69].

The available data provide the evidence that SHH mutations may cause the failure of cerebral hemispheres to separate into distinct left and right halves, leading to HPE, which is one of the most common developmental anomalies of forebrain and midface [70]. Although the majority of HPE are sporadic, familial cases are not rare, with clear autosomal dominant inheritance.

In our first case, the couple presented for PGD with two children showing the clinical signs of HPE [71] (Figure 3.32). One of them, a female with severe HPE and cleft lip and palate died shortly after birth. The chromosomal analysis performed using peripheral blood lymphocytes of both this child and the parents was normal, but DNA analysis in the child's autopsy material demonstrated the presence of SHH nonsense mutation due to GAG>TAG sequence change leading to premature termination of the protein at position 256 (Glu256→stop) [70] (Figure 3.33). SHH protein is an intercellular signalling molecule, which is synthesised as a precursor undergoing autocatalytic internal cleavage into a highly conserved domain (SHH-N) with signalling activity, and a more divergent domain (SHH-C), which, in addition to precursor processing, acts as an intramolecular cholesterol transferase crucial for proper patterning activity in animal development. Although the effect of the above nonsense mutation on SHH function is unknown, the resulting protein may fail to fulfill the expected signalling function in early morphogenesis [12,70].

The same mutation was found in their 5-years-old son, who was born after a full-term, normal pregnancy weighing 6 pounds, with a birth length of $18\frac{1}{3}$ inches. This child has less severe facial dysmorphisms, which included microcephaly, Rathke's pouch cyst, single central incisor, and choanal stenosis (the last was treated surgically after birth with dilatation). There was also clinodactily of the fifth fingers and incurved fourth toes bilaterally. The child's growth was slow in the first two years, but since then he has been maintaining a reasonably good growth and presently has normal social and cognitive development.

The couple had another pregnancy, which ended in spontaneous abortion due to Turner syndrome (45, X), showing no inheritance of the SHH mutation. The mutation was not found in either parent's genomic DNA, although the paternity testing showed that the father was in fact a biological father of both affected children. This clearly suggested a de novo gonadal mutation in one of the parents, which has been identified by a single sperm genotyping in the present study (see below).

Two PGD cycles were performed based on single blastomere analysis removed from the eight-cell embryos and tested by multiplex nested PCR analysis, involving specific mutation testing simultaneously with linked marker analysis [47, 57]. Of 15 embryos in the first cycle, 12 were available for blastomere biopsy at the eight-cell stage. Blastomeres from four embryos failed to amplify, leaving eight with available data for mutation analysis. Seven of these eight embryos appeared to contain the mutant allele, while only one embryo was mutation-free and transferred, yielding no clinical pregnancy.

The second PGD cycle was performed one year later, in which 19 embryos were available, of which 10 were acceptable for blastomere biopsy and DNA analysis. Of these 10 biopsied single blastomeres, only one failed to amplify, leaving nine with available data for SHH gene and the marker, to identify the mutation free-embryos for transfer

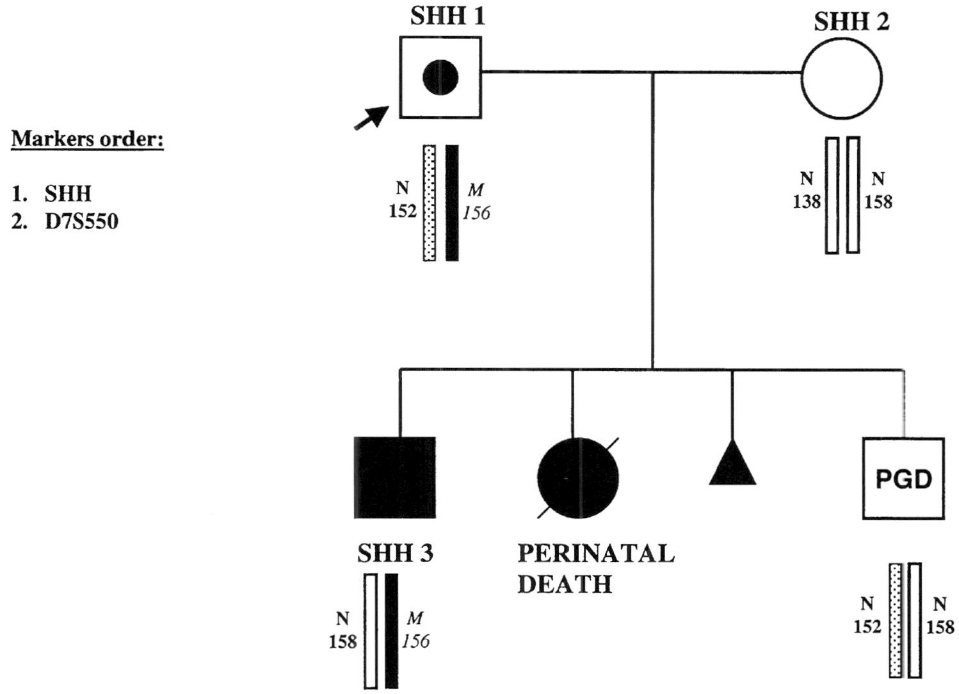

Figure 3.32. Preimplantation diagnosis for SHH mutation: family pedigree. (*Top*) The father has a gonadal mosaicism for SHH mutation, which is linked to 156 bp dinucleotide C-A repeat allele of D7S550 polymorphic marker, while the mother is normal, with one normal allele linked to 158 bp repeat and the other to 138 bp repeat alleles. (*Bottom*) Reproductive outcomes of this couple, including three previous pregnancies, one resulting in the birth of affected child with holoprosencephaly, carrying the mutant gene (*left*), one in perinatal death, also carrying the mutant gene (*middle, circle*), and one in a spontaneously aborted fetus with Turner syndrome, free from SHH mutation (*middle, triangle*). PGD (right) shows the outcome of preimplantation diagnosis, resulting in an unaffected clinical pregnancy and birth of a healthy child, following confirmation of the mutation-free status by amniocentesis.

(see Figure 3.33). Prior to PGD cycles, a single sperm testing was performed, which identified mosaicism for SHH mutation. As the mutation was shown to lead to the gain of an *Xba* I restriction site [70], the normal allele was identified as undigested PCR product, the mutant allele being represented by two fragments, as a result of *Xba* I digestion (see Figure 3.33).

To avoid misdiagnosis in mutation analysis due to ADO, which exceeds 10% in single blastomere DNA analysis, a closely linked microsatellite DNA marker D7S550 was tested in the same reaction as internal control. The list of primers used in the first and second round PCR for mutation and linked marker analysis and reaction conditions are presented in Table 3.19. A haplotype analysis showed that the mutant allele was linked to 156 bp dinucleotide CA repeat, while the normal gene was linked to 152 bp repeat allele in 7q36 (see Figure 3.32). Although other linked markers have been also described [69], they were not informative in the present couple.

As seen from the Figure 3.33, four ADOs were observed in mutation analysis, including ADO of the mutant allele in embryos #2, 9, and 17, and ADO of the normal allele in embryo #10. This was based on the marker analysis, showing that in all four cases the embryos were heterozygous. In other words, three of these four embryos (#2, 9, and 17) could have been misdiagnosed as normal without linked marker analysis. In addition to these three embryos, embryo #19 also contained the mutant gene.

The remaining four embryos were free of mutant gene, as confirmed by marker analysis, showing that all these embryos contained two normal

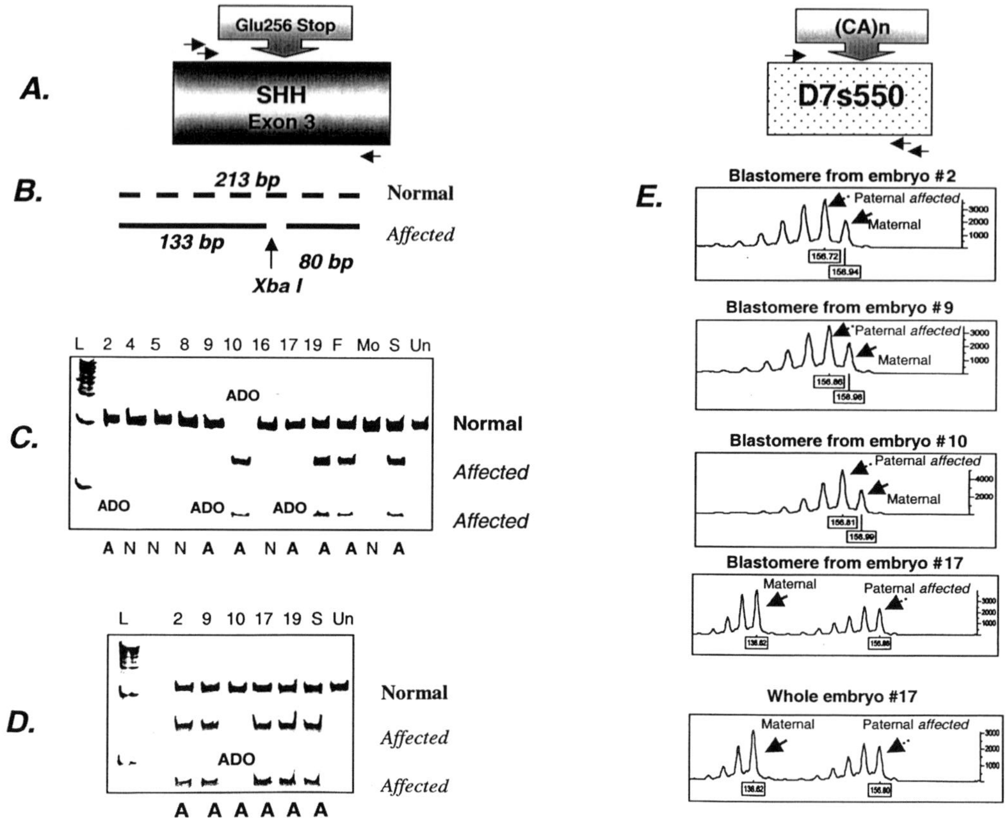

Figure 3.33. PGD for SHH mutation involving confirmation of the presence of both maternal and paternal normal genes in preselected mutation-free embryo. (*A*) Schematic diagram of the mutation and D7S550 linked marker on chromosome 7; *arrows* demonstrate the positions of heminested primers. (*B*) Restriction map for *Xba*I digestion, showing the gain of *Xba*I site by the mutant allele (*lower line*). (*C*) Polyacrylamide gel electrophoresis of the *Xba*I digested PCR products of nine blastomeres from PGD cycle. (*D*) Follow-up DNA analysis of genomic DNA from five embryos predicted affected by blastomere testing. (*E*) Capillary electrophoregrams of fluorescently labelled PCR product of tightly linked marker D7S550. Paternally derived 156 bp dinucleotide C-A repeat linked to SHH mutation is shown by *arrows* (paternal affected) in blastomeres of embryos #2, 9, 10, 17, and the genomic DNA of the whole embryo #17, in which ADO of mutant gene was seen (see panel C, embryo #17). Maternally derived 158 bp dinucleotide C-A repeats of D7S550 polymorphic marker are shown by *arrows* (maternal) in blastemeres of embryos #2, 9, and 10, while the other maternally derived 138 bp repeats of D7S550 are shown in blastomeres of embryo #17. (L, 100 bp standard; ADO, allele drop out; F, paternal DNA amplified from sperm; Mo, maternal normal amplified DNA; S, amplified DNA from affected baby; Un, undigested PCR product; A, affected blastomere; N, normal blastomere.)

Table 3.19. Primers and reaction conditions for the detection of Glu256Stop mutation in sonic hedgehog gene and linked marker D7S550

Gene/Polymorphism	Upper Primer	Lower Primer	Annealing Tm (°C)
SSH (Heminested)	Outside GAGCAGGGCGGCACCAA	GGCCGAGTCGTTGTGC	62–55
	Inside GGCACCAAGCTGGTGAAG	GGCCGAGTCGTTGTGC	56
D7S550 (Heminested)	Outside ACTATCATCCACAATCCACTCC	GCAGTTGGGTTATTTCAAGTCT	62–55
	Inside ACTATCATCCACAATCCACTCC	GATGTTGTGATTAGAGTTGCTGTA	56

alleles, including the paternal one linked to 152 repeat and either the normal maternal allele linked to 138 repeat (embryos #4 and 5) or the other normal maternal allele linked to 158 bp repeat (embryos #8 and 16). Two of these embryos (embryos #4 and 5) were transferred back to the patient, resulting in a singleton pregnancy and birth of a healthy child following confirmation of the mutation-free status by amniocentesis. The other two mutation-free embryos (embryos #8 and 16) were frozen for future use by the couple.

The presented data demonstrate a diagnostic accuracy of the multiplex-PCR-based blastomere analysis, despite the well-known high ADO rate in this type of single cells. As mentioned, ADO is an important limitation of PCR analysis in single cells, due to allele-specific amplification failure, which is particularly high in single blastomere analysis, shown to be at least two times higher than in single fibroblasts and polar bodies. As shown by the follow-up analysis of the preselected mutant embryos, PGD results were confirmed in all resulting embryos available for study, which is in accordance with extensive data, described above, based on testing of hundreds of oocytes and embryos for different single-gene disorders. Although, ideally, three linked markers are needed to completely exclude the risk for misdiagnosis due to ADO, the use of only one linked marker in the present study was quite reliable, because the preselection of mutation-free embryos was based not only on the presence of the normal paternally derived normal allele, but also on the presence of the second normal allele linked to the maternal linked marker. In other words, the absence of the mutant gene together with the presence of the two normal alleles, identified by different linked makers, led in this case to the correct identification of embryos as normal or carrying the SHH mutation.

The presented case demonstrates the clinical relevance of PGD for familial HPE [71]. Because of the high prevalence of congenital craniofacial anomalies, this approach may have practical implications for at-risk couples. A great intrafamilial clinical variability of HPE from alobar HPE and cyclopia, to cleft lip and palate, microcephaly, ocular hypertelorism, and even normal phenotype, suggests the interaction of the SHH gene with other genes expressed during craniofacial development and the possible involvement of environmental factors. This may explain the fact that almost one third of carriers of SHH mutations may be clinically unaffected. Therefore, even in familial cases, the detection of SHH mutations in prenatal diagnosis might not justify pregnancy termination, making PGD a more attractive option for couples at risk for producing a progeny with HPE, as demonstrated by the first PGD for this mutation mentioned, which resulted in birth of a healthy mutation free baby.

Of other congenital malformations, PGD was performed for Currarino, Crouzon, and Holt-Oram syndromes [72, 73], all resulting in birth of mutation-free children. In these cases, the application of prenatal diagnosis may be limited by the factors that modify clinical manifestations and confound prediction of an individual's phenotype making PGD an attractive choice, as shown in the examples of PGD for Currarino syndrome [74] and Crouzon syndrome, presented below.

3.8.2 Currarino Syndrome

Currarino syndrome (CS) is a severe autosomal dominant disorder caused by homeobox gene HLXB9 mutation, involving partial sacral agenesis, presacral mass, and anorectal malformations, which are among the most common digestive anomalies requiring neonatal surgery

[12, 13]. This homeobox gene is located on chromosome 7q36, between microsatellite DNA markers D7S559-D7S2423 [75]. The abnormalities observed in CS are caused by the disturbances in early embryonic development of the human tail bud, leading to the formation of positioning defects of the neural tube, notochord, somites, and hindgut. More than two dozen different HLXB9 intragenic mutations and microdeletions were detected in patients with CS, including frameshift and nonsense mutations in intron 1 and missense mutations in homeodomain, resulting in a nonconservative substitution of a highly conserved amino acid [12, 76–78].

Although a carrier screening and prenatal diagnosis of CS is currently available, the decision about termination of pregnancy will be controversial, as approximately half of the carriers of the mutation are asymptomatic [78]. Therefore, the absence of the genotype–phenotype correlation and an extremely high variability of phenotype in the carriers, from the above severe triad to minor sacral abnormalities undetectable without x-ray or even a completely asymptomatic carrier status, make PGD an important alternative for at-risk couples wishing to avoid CS in their offspring.

The couple presented for PGD with a previous child diagnosed to have CS. The child was born with imperforate anus, an anterior meningocoele, and a typical sickle-shaped hemisacrum revealed by x-ray of the sacral region (Figure 3.34, IV 1).

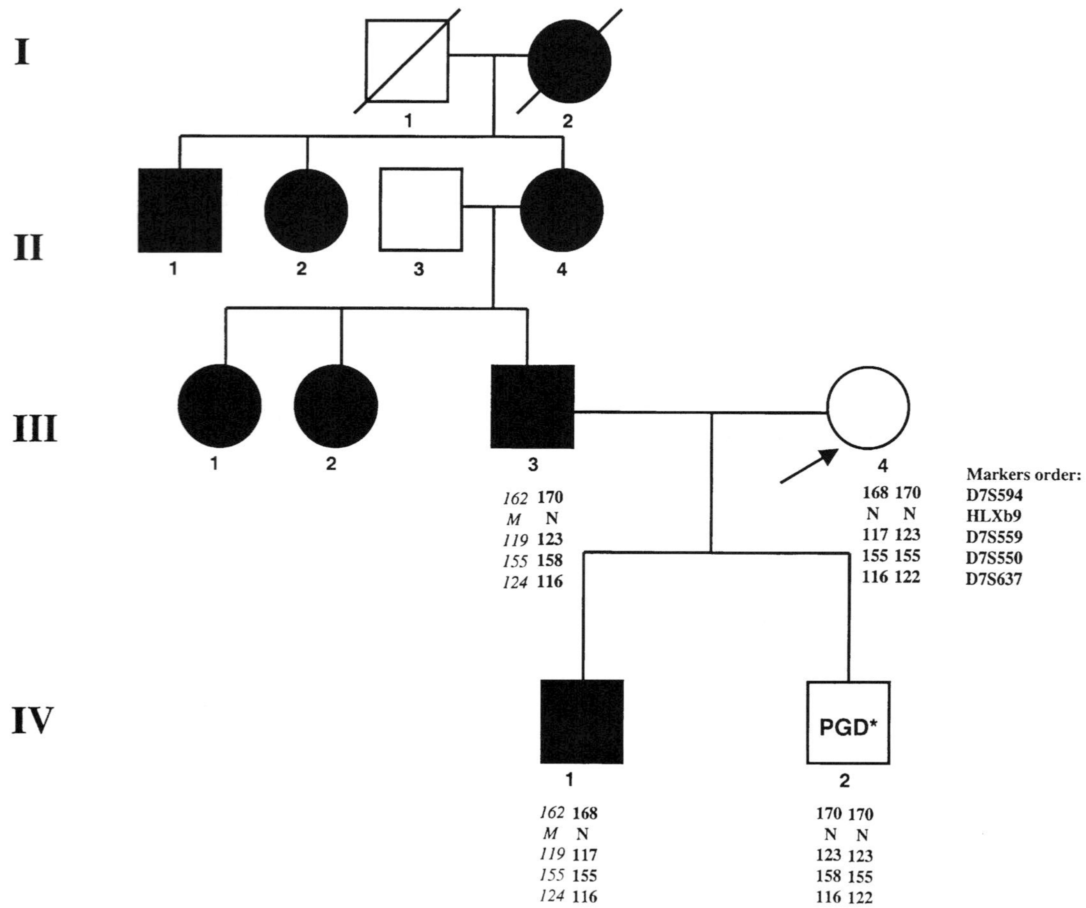

Figure 3.34. Preimplantation diagnosis for Currarino syndrome: family pedigree. Affected father (III, 3) has two affected sisters; homeobox gene HLXB 9 mutation was inherited from their mother (II, 4) who also had two affected siblings—a brother (II, 1) and sister (II, 2). Reproductive outcome is shown in *lower panel*, showing the previous affected child (IV, 1) and the healthy baby boy born after PGD (IV, 2).

The father was also born with an anal stricture (see Figure 3.34, III 3) requiring anal dilatation, and also had sacral defect detected by x-ray, involving a central anomaly from S2 downward. One of his two sisters (see Figure 3.34, III 1) was born with imperforate anus and anterior meningocoele and also had rectovaginal fistula, and a vesicoureteric reflux resulted in the need for renal transplantation; a sacral x-ray showed the same central defect with absence of the distal one third of the sacrum. The other sister (see Figure 3.34, III 2) was clinically asymptomatic, but sacral x-ray revealed no coccyx and a MRI scan disclosed an anterior meningocoele. His mother had undeveloped coccyx and urinary tract bilateral ureteropelvic junction obstruction (see Figure 3.34, II 4). Of her three siblings, only one of two brothers had anal stricture (see Figure 3.34, II 1), while the asymptomatic sister (see Figure 3.34, II 2) was identified as a carrier of the mutation because of the finding of imperforate anus in her grandson. As seen from the pedigree, the father inherited the mutation from his grandmother (see Figure 3.34, I 2), who was probably the first affected member of the family, known to have constipation but normal sacral x-ray.

DNA analysis in this family demonstrated the presence of homeobox HLXB9 mutation due to frameshift insertion of a cytosine into a stretch of six cytosines at positions 125–130 in exon 1 of the gene, leading to the introduction of premature termination codon [74]. The primer sequences for mutation testing and their positions are presented in Figure 3.35 and listed in Table 3.20.

PGD cycle was performed using single blastomeres, removed from the eight-cell embryos and tested by multiplex nested PCR analysis, involving specific mutation testing simultaneously with linked marker analysis. PCR product was identified by fragment length analysis using capillary elecrophoresis (see Figure 3.35).

Prior to initiating PGD cycle, a single sperm testing was performed to establish paternal haplotypes. To be able to identify a possible ADO in the mutation analysis, four closely linked microsatellite dinucleotide DNA markers, D7S559, D7S550, D7S637, and D7S594, were tested in the same reaction with the HLXB9 gene [75]. As seen from Figure 3.34 and Table 3.21, the mutant allele was linked to 119, 155, 124, and 162 bp, and the normal allele to 123, 159, 116, and 168 bp repeats of D7S559, D7S550, D7S637, and D7S594 markers, respectively. Testing for these markers in the affected child and the mother showed that the parents shared the same site of two of four markers linked to normal allele making theses markers of limited value for preselection of mutation-free embryos for transfer.

A total of 17 embryos were available for testing in a single PGD cycle. Single blastomeres were removed from these embryos, of which three failed to amplify either HLXB9 alleles or polymorphic markers (embryos #1, 9, and 14), suggesting the lack of a nucleus in these blastomeres, and two showed amplification of polymorphic markers only with no signal detected for the HLXB9 gene (embryos #8 and 12) (see Table 3.21). Of a total of 12 embryos with results for the mutation and linked marker analysis, 11 were with conclusive results, of which 5 were predicted to contain the mutant allele in agreement with the presence of repeat markers (embryos #3, 5, 6, 11, and 13). One of these embryos contained only paternal alleles (embryo #13), which may be explained by the absence of the maternal chromosome 7, which may be due to mosaicism, known to be very frequent at the cleavage stage.

The remaining six embryos were predicted to be free of mutation, but only in three of them (embryos #2, 4, and 16) could ADO of the mutant allele be excluded based on the linked marker analysis (see Table 3.21 and Figure 3.35). Two of these embryos were transferred, yielding a singleton pregnancy and birth of a mutation-free child, following confirmation of diagnosis by amniocentesis (Figure 3.36). In the other three embryos predicted to contain the normal allele (embryos #7, 10, and 15), ADO of the mutant gene could not be excluded, because of the parents' sharing the same site of polymorphic markers linked to the normal allele. For example, one of these embryos (embryo #15) was informative only for one linked marker, D7S637 (116/122 bp), which may have suggested the presence of both paternal and maternal normal alleles, assuming that both of these alleles could not be of maternal origin. However, this may have been also due to the uniparental disomy 7 of maternal origin, which was not supported by the other linked markers.

The probability of ADO of the mutant gene in the other two embryos (embryos #7 and 10) could have not been excluded either, despite the fact that these embryos were heterozygous for the two linked markers, D7S637 (116/122) and D7S550 (159/155), because the detected alleles (116 and 159 bp), linked to paternal normal gene, may have

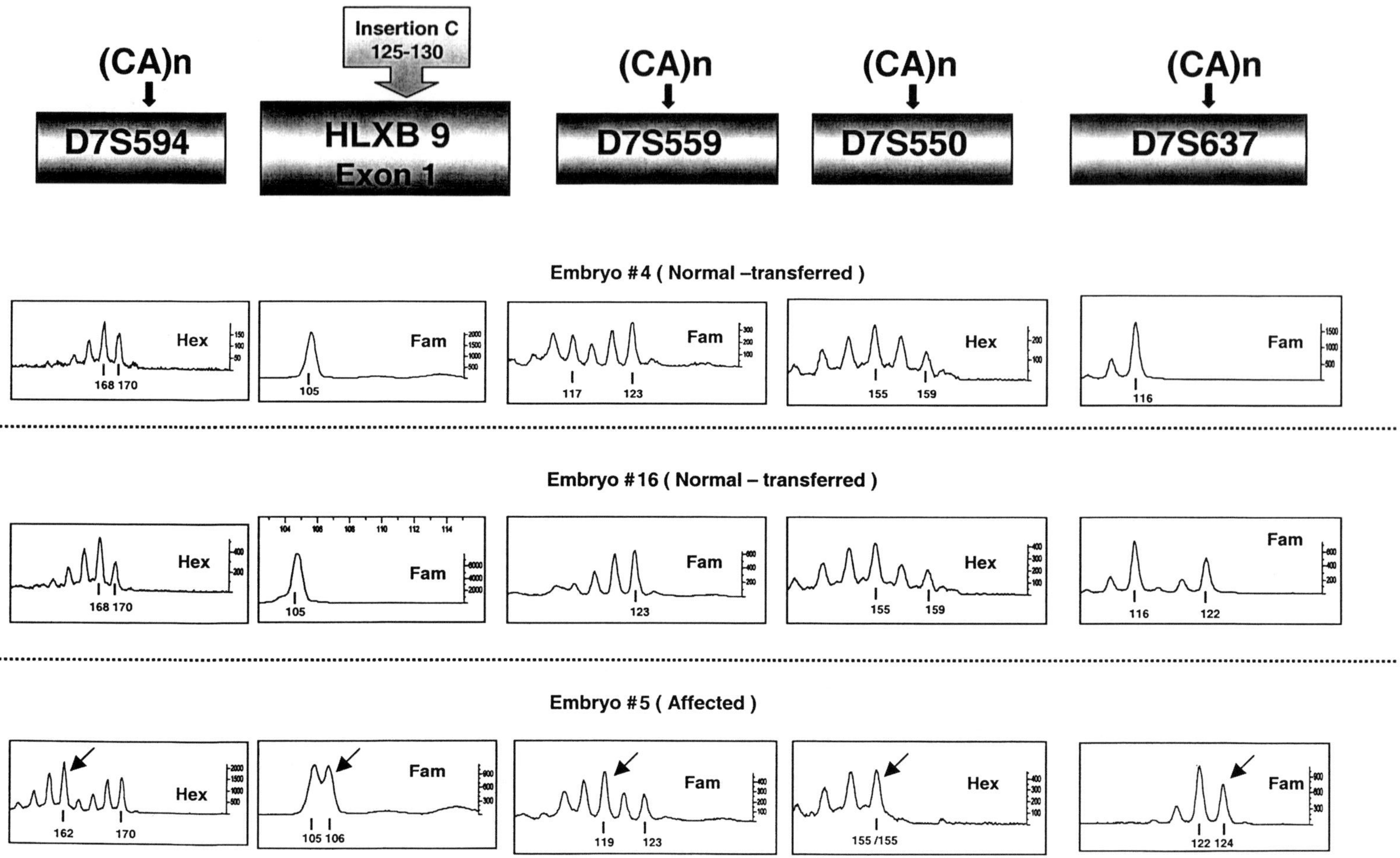

Figure 3.35. Preimplantation diagnosis for homeobox gene HLXB 9 mutation in exon 1 causing Currarino syndrome. (*Top*) location of the mutation in HLXB9 gene and linked markers on chromosome 7. The other three panels show capillary elecrophoregarms of fluorescently labelled PCR products of HLXB9 alleles and each of the four linked markers. Paternally derived mutant allele is shown by *arrow* in embryo #5 (*bottom*), in agreement with paternally derived markers (CA repeats) linked to the mutant gene. The mutant allele is absent in embryos #4 and 16 (*middle*), also in agreement with all four markers. These embryos have been transferred back to the patient resulting in the birth of mutation-free baby. According to marker analysis, the baby originates from the transfer of embryo #16.

Table 3.20. Primers and reaction conditions for PGD of Currarino syndrome

Gene/ Polymorphism	Upper Primer	Lower Primer	Annealing Tm (°C)
HLXb9	Outside		
Insertion C 125-130	5′TGGCAAATAACCACCATAACAA3′	5′CGACCGACGTGACCAAGG3′	62–55
(Heminested PCR)	Inside		
	5′FamCGAGCCGATGGAAAAATCC3′	5′CGACCGACGTGACCAAGG3′	55
D7S559	Outside		
(Heminested PCR)	5′GACCACAGTAAGAAATAACCCATTA3′	5′AAAAATAATAAAACAGAAGATGCCA3′	62–55
	Inside		
	5′GACCACAGTAAGAAATAACCCATTA3′	5′FamAAGTACTGTCTTGTTAAATTTTTGC3′	55
D7S637	Outside		
(Heminested PCR)	5′GCACAGAACTAAATACACCCCAT3′	5′CTTCCCCTAAAGTTAACATCTTG3′	62–55
	Inside		
	5′GCACAGAACTAAATACACCCCAT3′	5′FamAACTATAGTACAATATCAAAACCAAGA3′	55
D7S594	Outside		
(Heminested PCR)	5′CAGCCTGTGTATGGTTACTTTTTA3′	5′GAAGTTCTACAGGCTGAAATCAA3′	62–55
	Inside		
	5′HexAAAAGTATGCATTTATGTATTTTTT3′	5′GAAGTTCTACAGGCTGAAATCAA3′	55
D7S550	Outside		
(Heminested PCR)	5′ACTATCATCCACAATCCACTCC3′	5′GCAGTTGGGTTATTTCAAGTCT3′	62–55
	Inside		
	5′ACTATCATCCACAATCCACTCC3′	5′HexGATGTTGTGATTAGAGTTGCTGTA3′	55

Table 3.21. Summary of results of PGD for Currarino syndrome

Blastomere #	D7S594	HLXb9	D7S559	D7S550	D7S637	Predicted Genotype	ET
1	**FA**	**FA**	**FA**	**FA**	**FA**	Embryo : INCONCLUSIVE	NO
2	170	N	123	159/155	116/122	Embryo : NORMAL	YES
3	**162**/170	**M**/N	**119**/123	155/155	**124**/122	Embryo : **AFFECTED**	NO
4	170/168	N	123/117	159/155	116/116	Embryo : NORMAL	YES**
5	**162**/170	**M**/N	**119**/123	155/155	**124**/122	Embryo : **AFFECTED**	NO
6	**162**/168	**M**/N	**119**/117	155/155	ADO/116	Embryo : **AFFECTED**	NO
7	170	N	123	159/155	116/122	Embryo : NORMAL	YES*
8	170	FA	123	155/155	**124**/122	Embryo : INCONCLUSIVE	NO
9	FA	FA	FA	FA	FA	Embryo : INCONCLUSIVE	NO
10	170	N	123	159/155	116/122	Embryo : NORMAL	YES*
11	**162**/168	**M/N**	**119**/117	155/155	**124**/116	Embryo : AFFECTED	NO
12	170/168	FA	**119**/117	155/155	**124**/116	Embryo : INCONCLUSIVE	NO
13	**162/**	**M/**	**119/**	155/	**124/**	Embryo : AFFECTED	NO
14	FA	FA	FA	FA	FA	Embryo : INCONCLUSIVE	NO
15	170	N	123	155	116/122	Embryo : NORMAL	YES*
16	**170**/168	FA	123/ADO	159/155	116/122	Embryo : NORMAL	YES**
17	170	FA	FA	159/155	116/122	Embryo : INCONCLUSIVE	NO
Father	**162**/170	**M**/N	119/123	**155**/159	**124**/116		
Mother	168/170	N/N	117/123	155/155	116/122		
Affected child	**162**/168	**M**/N	119/117	**155**/155	**124**/116		

ADO, allele drop out; FA, failed amplification; ET, embryo transfer; M, mutant allele; N, normal; Affected haplotype is **bold**.
*Chance of misdiagnosis due to potential ADO or recombination.
** Transferred embryos.

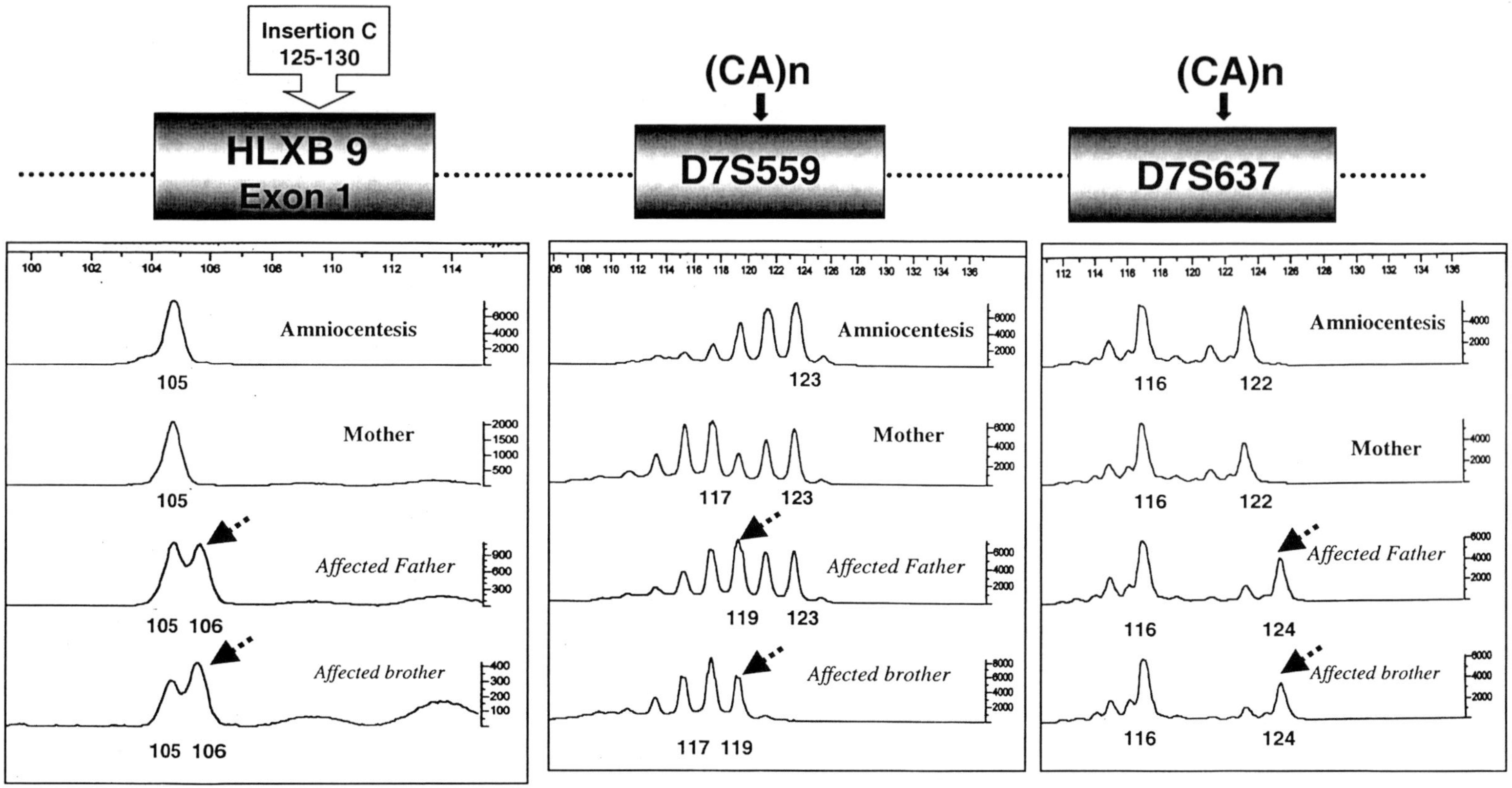

Figure 3.36. Confirmation of Currarino syndrome–free pregnancy by aminocentesis. (*Top*) Location of the mutation in HLXB9 and linked markers D7S559 and D7S637 on chromosome 7. *Upper row* capillary elecrophoregarms of fluorescently labelled PCR products of HLXB9 gene and two linked markers shows the results obtained from amniotic fluid, evidencing the mutation-free status of the fetus, confirming the results of PGD. *Two lower panels from the bottom* show patterns for affected father and the first affected child, in which paternal mutant allele is shown by arrow, in agreement with paternally derived markers (CA repeats) linked to the mutant gene. *Second panel from the top* shows the normal pattern for the mother.

Table 3.22. Results and outcomes of PGD for congenital malformations

Disease	# Patients	# Cycles	Cycle by PB	Cycle by PB and BL	Cycle by BL	# ET	# Embryos	Pregnancy	Birth
Craniofacial dysostosis, type 1 (CFD1)	2	3	0	2	1	3	8	2(1)*	1
Currarino syndrome	1	1	0	0	1	1	2	1	1
Sonic hedgehog (SHH)	1	2	0	0	2	2	3	1	1
Treacher Collins-Franceschetti syndrome (TCOF)	1	1	1	0	0	1	2	1(1)*	0
TOTAL	*5*	*7*	*1*	*2*	*4*	*7*	*15*	*5(2)**	*3*

*Ongoing pregnancies; number of fetuses is shown in parentheses.

also derived from the mother who shares the same site of linked marker, linked to the normal allele. Finally, one of the embryos (embryo #12) showed the failure of amplification of HLXB9 alleles, but may have been predicted to be mutant, based on the presence three of four polymorphic markers linked to the mutant gene. However, this was not in agreement with the presence of 170 bp repeat of D7S594 marker, linked to the normal paternal allele, which may probably be due to recombination of the paternal alleles. Unfortunately, due to parents sharing two of four markers linked to the normal allele, only three of six potentially normal embryos could be preselected for transfer, because of the inability to completely exclude the risk for misdiagnosis due to ADO of the mutant paternal allele.

The presented case is the first PGD for homeobox-containing gene mutations, demonstrating the clinical relevance of PGD for this group of genes involved in transcriptional regulation during early embryonic development. Because of high prevalence of congenital anomalies determined by the mutations causing the formation and positioning defects, this approach may have practical implications for at-risk couples carrying such mutations, and also mutations causing other groups of familial dysmorphologies, as a principally new option for a large group of couples to avoid the risk of having children with a wide range of formation and positioning abnormalities at different stages of embryogenesis. The data on the stage-specific expression of homeobox genes observed in CS will be also of relevance for the development of PGD for these transcriptionally relevant anomalies to offer different options for couples at risk for producing progeny with inherited predisposition to congenital malformations.

3.8.3 Crouzon Syndrome

A total experience of PGD for congenital malformations performed in our series is presented in Table 3.22, including Crouzon syndrome (CFD1), which is described below. It is a dominantly inherited craniosynostosis caused by mutations in the fibroblast growth factor receptor 2 gene (*FGFR2*) on chromosome 10q [12, 79]. It is quite rare, with birth prevalence of 15–16 per million [80]. PGD was previously applied to testing of G/A base substitution at codon 568 in a couple with one affected child who died at age 18 months during corrective surgery. Of two PGD cycles performed, one resulted in birth of an unaffected child [73]. We performed three PGD cycles for two couples at risk of producing offspring with Crouzon syndrome, presented in Table 3.22, of which two resulted in unaffected clinical pregnancies (primers and reaction conditions are presented in Table 3.23). In one of these cases, the mother was a carrier of C3422Y mutation in exon B of FGFR2 gene and also had one previous child affected with Crouzon syndrome (Figure 3.37). Two clinical cycles were performed for this patient, using PB1 and PB2 testing, based on restriction digestion with *HpyCH4* V, cutting the normal allele into 74 bp and 20 bp fragments and leaving the mutant allele intact. Two informative linked markers, D10S190 and *Msp* I polymorphism, were amplified simultaneously with the causative gene for testing PB1 and PB2, allowing the preselection

Table 3.23. Primer sequences and reaction conditions for PGD of Crouzon syndrome

Gene/Polymorphism	Upper Primer	Lower Primer	Annealing Tm (°C)
FGFR2	Outside		
Cys 342 Tyr	5′TAACACCACGGACAAAGAGA3′	5′ACCCAGAGAGAAAGAACAGTAT3′	62–45
(G—A)	Inside		
(Heminested; Mismatched inside primer; Hpy CH4V cuts normal allele)	5′TAACACCACGGACAAAGAGA3′	5′AATAGAATTACCCGCCA*T*G3′	50
D10S190	Outside		
(Heminested)	5′AGTGTATGGTTCCTTCCCCTG3′	5′CAGGCATGTAACACAGCAAAAG3′	62–45
	Inside		
	5′HexCTGGTTGTTTAAGTGTATGGTTCC3′	5′CAGGCATGTAACACAGCAAAAG3′	55
D10S 1483	Outside		
(Heminested)	5′CCAATGCTATCCCGGCTAT3′	5′AAGCGTGTAACATTTGGTATGC3′	62–45
	Inside		
	5′CCAATGCTATCCCGGCTAT3′	5′HexCAAATGCAATAACATTAAACCATT3′	55
D10S 209	Outside		
(Heminested)	5′TCATATCCTGCTAACATTACCAACA 3′	5′CCCACAGGTCACTATGCTTACTT3′	62–45
	Inside		
	5′FamCTGAGCCAGTGGGATGAGAG3′	5′CCCACAGGTCACTATGCTTACTT3′	55
D10S 1230	Outside		
(Nested)	5′GCTCTGGTTATTTGCTGCCT3′	5′GCAGCAGCTTTGTTTTCCA3′	62–45
	Inside		
	5′HexTTTCACCTGACTTTCCTAATAATCC3′	5′AGAGCAAGCAACTAAATTATTTTTC3′	55
D10S 1757	Outside		
(Heminested)	5′CAAAGACAATTTAGGAAATTCAAAA3′	5′CATTCCCTCGTGTATCAGCC3′	62–45
	Inside		
	5′CAAAGACAATTTAGGAAATTCAAAA3′	5′FamCTTTCTGCCTAAAACATATCCCT3′	55
D10S 1679	Outside		
(Heminested)	5′GCAGTGCCTGAGGCTTGTG3′	5′CCATGAGGGTACTATAGAAAGTTG3′	62–45
	Inside		
	5′HexCCATGAGGGTACTATAGAAAGTTG3′	5′CCATGAGGGTACTATAGAAAGTTG3′	55

of mutation-free oocytes for transfer. Of nine oocytes tested in the first cycle, five appeared to be mutation-free, but the transfer did not result in clinical pregnancy. In the other cycle, 22 oocytes were available for testing, of which nine were mutation-free (Figure 3.38). The transfer of two of these embryos resulted in a singleton pregnancy and birth of an unaffected child, confirmed to be free of the causative gene.

As can be seen from overall data on PGD for congenital malformations (see Table 3.22), seven PGD cycles were performed for five couples, resulting in transfer of embryos in every cycle. Two embryos on average were transferred in each of seven cycles, yielding unaffected pregnancies in all but two cycles, of which three resulted in birth of unaffected children and two are still ongoing following confirmation of PGD diagnosis by prenatal diagnosis.

3.9 Dynamic Mutations

Dynamic mutations, which represent trinucleotide repeat expansion, are currently among most frequent indications for PGD, following X-linked disorders, CF, and hemoglobin disorders, although PGD for this group of diseases was introduced only in 1995, initially offered to couples at risk for producing offspring with myotonic dystrophy (DM) [81, 82]. PGD for XRM1, another important example of the dynamic mutations, was described in Section 3.3, demonstrating

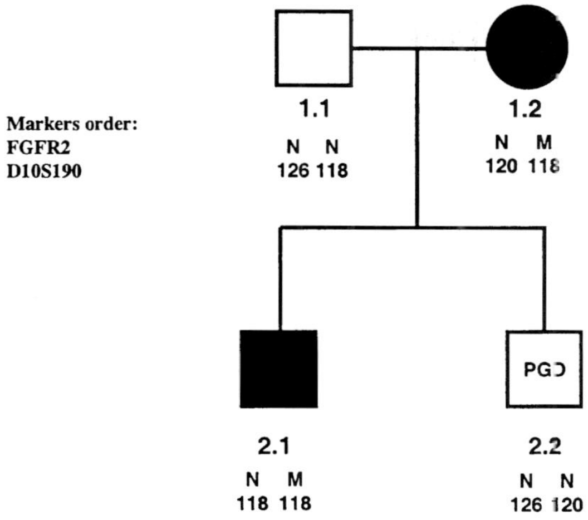

Figure 3.37. PGD for Crouzon syndrome: family pedigree. The mother (1.2) is the carrier of mutation in FGFR2 gene and had previous child with Crouzon syndrom (2.1). PGD resulted in the birth of a healthy unaffected boy (2.2).

the complexity of performing PGD for this group of diseases [83, 84].

In addition to DM and FRAXA, current PGD practices include PGD for Huntington's disease (HD), spinal and bulbar muscular atrophy (SBMA), and spinocerebellar ataxia (SCA) types 2, 3, 6, and 7. Overall, approximately 300 PGD cycles were performed for more than 150 patients worldwide, resulting in the transfer of unaffected embryos in approximately 85% of cycles, and yielding 20% clinical pregnancies per transfer [5]. The majority of these cases were done for DM, XRM1, and HD. The accuracy of PGD for dynamic mutations has been reported to be improved with the application of fluorescent PCR with the expanded long template (ELT) kit, which enabled reducing the ADO rate from 30–35% in both conventional and fluorescent PCR to as low as 5% in the testing for DM [83]. The other attractive approach for improving the accuracy of PCR analysis for this group of diseases, similar to those mentioned in Chapter 2, involved the application of real-time PCR, which was found to reduce the ADO rate by half, in comparison to conventional or fluorescent PCR. The application of these approaches together with the simultaneous testing of sufficient number of linked markers may allow avoiding the risk for misdiagnosis completely in PGD for dynamic mutations.

Table 3.24 summarizes our experience of 95 PGD cycles for 60 couples at risk for producing offspring with DM, FRAXA, Huntington disease (HD), SCA2, SCA3, SCA6, and SCA7. Among these conditions, almost half (43 cycles) were performed for FRAXA, also described previously in section 3.3, which resulted in 37 embryo transfers yielding 16 unaffected pregnancies and 15 clinical children born. The second largest group was DM, involving 33 PGD cycles performed for 19 at-risk couples, of which 22 resulted in the transfer of mutation-free embryos, yielding 11 clinical pregnancies and 8 births of healthy children. Overall, 168 (2.3 per transfer on the average) unaffected embryos were selected for transfer in 74 (77.8%) of 95 cycles, resulting in 33(44.6%) pregnancies and birth of 28 children.

As the expanded alleles are usually not amplified in oocyte cell PCR, PGD for dynamic mutation is mainly based on the identification of normal alleles based on the testing of sufficient number of linked markers. To avoid misdiagnosis, at least three closely linked markers should be present confirming the inheritance of the normal allele from the affected parents. In practice, the presence of normal alleles from both parents should be confirmed by the analysis of the maternal and paternal haplotypes, as demonstrated by the example of PGD for DM, presented in Figures 3.39 and 3.40.

As seen from Figure 3.39, the affected mother with MD (*DMPK*) has an expanded allele linked to 159 and 151 repeat markers. Of six embryos tested, using three closely linked markers (see Figure 3.39), three were affected (embryos #2, 9, and 15), including one with trisomy 19 (embryo #2), evidenced by the presence of both mutant and normal maternal alleles, and one set of paternally derived markers. The fact that this embryo is trisomic for chromosome 19, to which the gene for *DMPK* is assigned (19q13.2–q13.3), was also confirmed by FISH analysis using specific fluorescent probe for chromosome 19 and 13, showing three signals for this chromosome as well as three signals for chromosome 13, suggesting double trisomy 13 and 19. The remaining three embryos (embryos #1, 8, and 11) were normal, in which all three markers were in agreement with the presence of the maternal normal allele, together with the presence of one of the paternal alleles. So two of these embryos (embryos #1

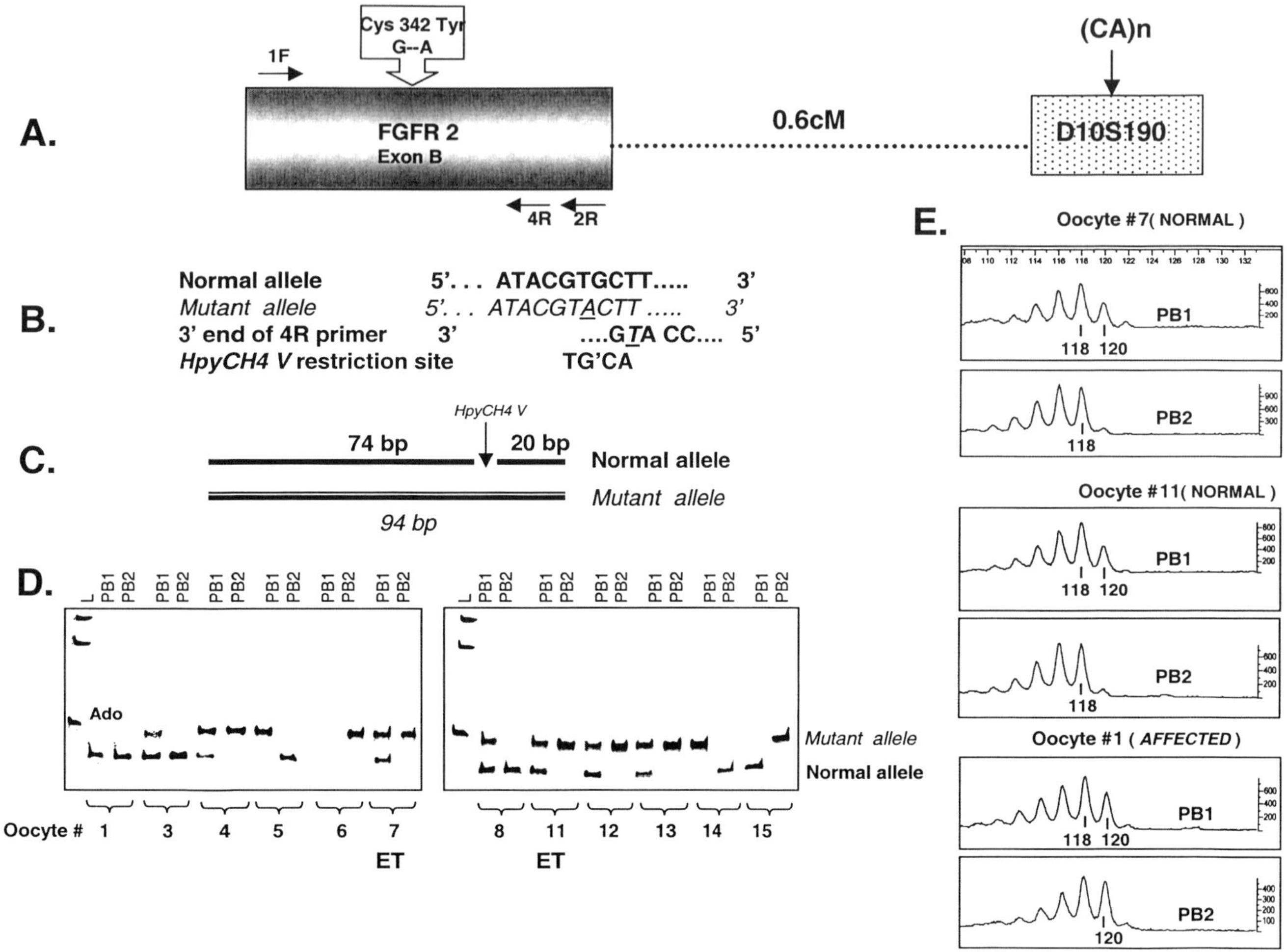

Figure 3.38. PGD for mutation in FGFR2 gene causing Crouson syndrome. (*A*) Map of human FGFR2 gene, showing sites and location of C342Ymutation and position of linked D10S190 dinucleotide STR. Horizontal arrows show primer sets for heminested PCR. (*B*) Primer design and (*C*) restriction map for normal and abnormal alleles. (*D*) Mutation analysis of 12 oocytes by sequential PB1 and PB2. ADO of the mutant allele was detected in oocyte #1 both by sequential PB1 and PB2 (identical genotype of both) and linked marker analysis (presence of 118 bp and 120 bp bands). STR profile for the PB1 and PB2 from oocytes #1, 7, and 11 confirmed the results of mutation analysis (*E*), suggesting that the embryos #7 and 11 are normal and may be transferred. (L, size standard; bp, base pair; ET, embryo transfer.)

Table 3.24. Results and outcomes of PGD for dynamic mutations

Disease	# Patients	# Cycles	# Cycles performed by PBs	# Cycles performed by PB and Blast	# Cycles performed by Blast	# ET	# Embryos	Pregnancy	Birth
DM	19	33	6	13	14	22	48	11	8
FRA X	29	43	16	24	3	37	86	16	15
HD	6	10	0	1	9	8	17	3	1
SCA2	3	4	1	3	0	4	11	1	2
SCA3	1	2	1	0	1	1	2	1	2
SCA6	1	2	1	1	0	1	2	0	0
SCA7	1	1	1	0	0	1	2	1	0
TOTAL	*60*	*95*	*26*	*42*	*27*	*74*	*168*	*33*	*28*

and 11) were transferred, resulting in an unaffected clinical pregnancy.

Machado-Joseph disease (SCA3), for which PGD is shown on Figure 3.40, is a neurodegenerative disorder characterized by cerebellar ataxia 3, pyramidal and extapiramidal signs, peripheral nerve palsy, external ophthalmoplegia, and facial and lingual fasciculation and bulging. The mother was a carrier of the expanded allele, inherited from her mother, closely linked to 166, 137, 109, and 126 repeat markers of D14S617, D14S1015, D14S1016, and D14S1050, respectively. PGD was performed by sequential PB1 and PB2 analysis (primer sequences are listed in Table 3.25) in seven oocytes, five of which appeared to be affected and only two without expansion, which were transferred back to the patient, yielding an unaffected twin pregnancy, and birth of healthy children, following confirmation of PGD by amniocentesis.

Finally, a nondisclosure PGD has been considered for couples at risk for HD, involving the transfer of the disease-free embryos while the

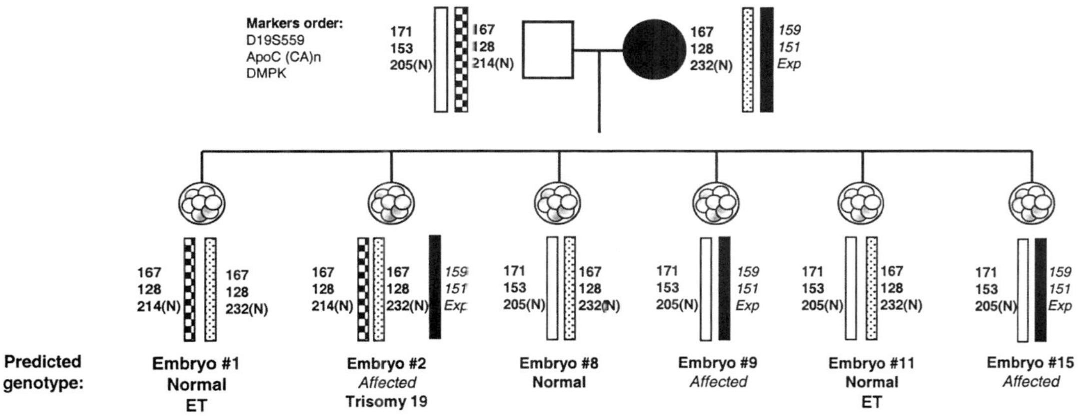

Figure 3.39. PGD for myotonic dystrophy combined with chromosome 19 aneuploidy testing. (*Top*) Maternal haplotype based on PB1 and PB2 multiplex DNA amplification of the normal allele of DMPK gene (19q13.2–13.3) and linked markers (*right*). Paternal haplotype based on blastomere analysis (open and darker). As the expansion of $(CTG)^n$ repeat in DMPK gene is not detectable at single cell level (darker bar in maternal haplotype corresponds to the affected allele), linked polymorphic markers are used for PGD; clearer bar represents the normal allele and tightly linked markers. (*Botttom*) Six embryos tested for the presence of the normal number of $(CTG)^n$ repeat in DMPK gene using polymorphic marker pattern. Embryos #1, 8, and 11 were predicted normal based on the presence of the normal maternal chromosome (haplotype) (clearer bar) and one paternal chromosome 19. Embryos #2, 9, and 15 were affected evidenced by the presence of the maternal affected haplotype (darker bar). Embryo #2 contains two sets of maternal and one set of paternal polymorphic markers, suggesting trisomy 19 of maternal origin. Therefore, amplification of only $(CTG)^n$ repeat in DMPK gene would have revealed the normal status based on the presence of only normal maternal and paternal alleles, which would have led to misdiagnosis. The follow-up FISH analysis confirmed trisomy 19, as predicted, and also incidental trisomy 13 (data not shown). (DMPK, dystrophia myotonica protein kinase gene; ET, embryo transfer.)

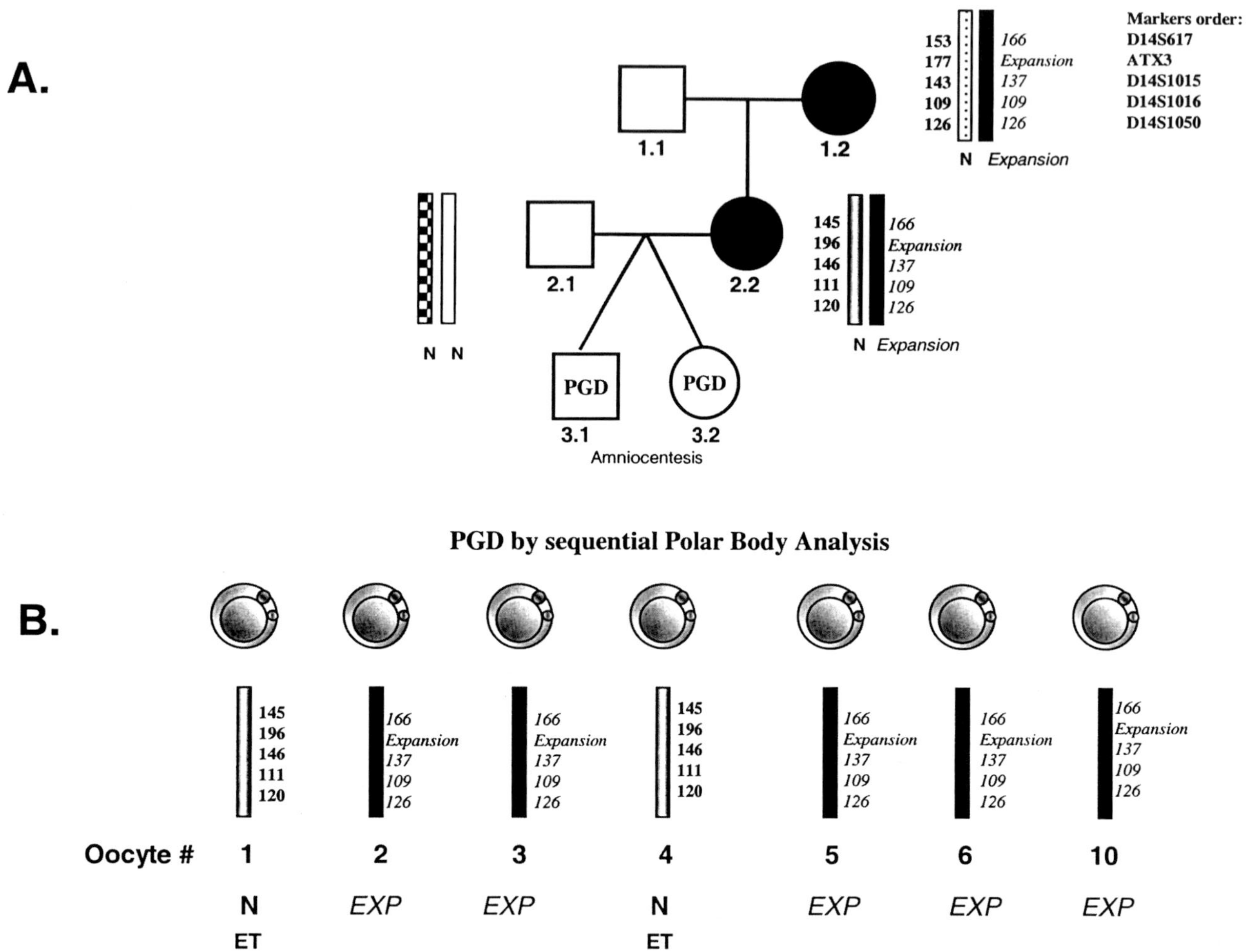

Figure 3.40. PGD for Machado-Joseph disease (SCA3) resulted in birth of healthy twins. (*A*) Family pedigree with haplotype analysis, showing that the mother (2.2) inherited the expanded allele (darker bar) from her mother (1.2). PGD resulted in birth of two unaffected twins (3.1 and 3.2) following amniocentesis. (*B*) Seven oocytes were tested for the presence of the expanded allele by sequential PB1 and PB2 analysis, showing that only two oocytes (oocytes #1 and 4) were free of the expansion. Both embryos resulting from these mutation-free oocytes were transferred, yielding birth of unaffected twins (3.1 and 3.2).

Table 3.25. Primers and reaction conditions for PGD of Machado-Joseph disease

Gene/Polymorphism	Upper Primer	Lower Primer	Annealing Tm (°C)
Mutation in ATX3 gene	Outside 5'TACCAGTGACTACTTTGATTCGTGA3'	5'GCTCCTGAACTGGTGGCTG3'	62–45
	Inside 5'TACCAGTGACTACTTTGATTCGTGA3'	5'FamTCTGTCCTGATAGGTCCCCC3'	55
D14S995 (Heminested)	Outside 5'TCTGGAAGGCAGAGAGGTTG3'	5'GTAGATAATATGTTTGGCGAGGG3'	62–45
	Inside 5'HexCCACTGCACTCCAGCCTG3'	5'GTAGATAATATGTTTGGCGAGGG3'	58
D14S1050 (Heminested)	Outside 5'AGACAGGAGTGGGCAGAAAT3'	5'ATCTCCCATACTGACTCTCCCT3'	62–45
	Inside 5'AGACAGGAGTGGGCAGAAAT3'	5'HexGTGACAATTTAGAGGGGGGAC3'	58
D14S1015 (Heminested)	Outside 5'TATTTTGTCCTTTTTTATGGCTG3'	5'AACTACAGAAAAAATTGAAAGTAGAAC3'	62–45
	Inside 5'TATTTTGTCCTTTTTTATGGCTG3'	5'HexCTATTTGATTCAGAAATCCCATTAC3'	58
D14S1016 (Heminested)	Outside 5'TAAAAAGGGAAAGTAGTTAGAATGG3'	5'GAAGATGAACCTGAACATGGC3'	62–45
	Inside 5'HexGTTATGACATATATTTTTTCCCCAC3'	5'GAAGATGAACCTGAACATGGC3'	58
D14S617 (Heminested)	Outside 5'ATTCTCCTCTTATTATAGGACTTTA3'	5'GCAACAGAACAAGATCTGTCTC3'	62–45
	Inside 5'FamTGTGGAGAAATTAAGTTTTAGGTG3'	5'GCAACAGAACAAGATCTGTCTC3'	58

prospective parents do not learn their own status. Parents receive no information about number of oocytes obtained after hormonal stimulation, number of embryos formed, and the number of embryos available for transfer. However, there might be no unaffected embryos for transfer, or no affected embryos can be found, suggesting the parent is genetically normal. The other alternative in HD is an exclusion-PGD testing, but embryos with detected grandparental allele would be excluded from transfer notwithstanding that only half of these embryos would contain the affected allele. However, in our experience, a direct testing of embryos from known disease gene carriers was performed, currently including 10 PGD cycles for six couples at risk, resulting in transfer of 17 mutation-free embryos in eight cycles, yielding three unaffected clinical pregnancies.

In conclusion, our present experience comprises 562 PGD cycles for Mendelian conditions, which is presently the largest series in the world. The number of conditions for which PGD is being performed is expanding gradually, with the present number being close to 100. As can be seen from Table 3.26, the outcome of PGD for single-gene disorders was more favourable than in a routine IVF, resulting in 194 pregnancies (38.8% pregnancy rate), despite transferring of only 2.2 embryos per cycle on the average. It is of note that unaffected embryos for transfer were available in 89% PGD cycles, and only 1.6 repeated cycles were required for couples to achieve the transfer of unaffected embryos. Overall, 187 children were born following the procedure, with only two misdiagnoses (1.1%), one in PGD for CFTR, involving misdiagnosis of a compound heterozygote embryo as heterozygous at the very beginning of the introduction of PGD, when the phenomenon of ADO was not yet appreciated, and another in PGD for FRM1, involving the transfer of the embryo with the predicted 5% risk of misdiagnosis, that the couple decided to transfer in addition to other two with 100% accuracy, to improve their chances to become pregnant. Assuming that the overall experience has involved the genetic testing of more than 5000 oocytes and embryos, which resulted in preselection and transfer of 1094 unaffected embryos, the applied technique may be considered

Table 3.26. Overall results and clinical outcome of PGD for single gene disorders

Disease	# Patients	# Cycles	# ET	# Embryos	Pregnancy	Birth
Adenomatous polyposis of the colon	5	13	13	22	2	1(1)*
Adrenoleukodystrophy	2	3	3	7	1	1
Alpha-1 antitrypsin deficiency	1	2	1	2	0	0
Alport syndrome	1	1	1	2	0	0
Basal cell nevus syndrome (BCNS) (GORLIN)	1	1	1	1	1	1
Blood group—Kell-Cellano system	4	7	6	11	2	3
Brain tumour, posterior fossa of infancy, familial	1	1	1	1	0	0
Ceroid lipofuscinosis, neuronal 2, late infantile (CLN2)	1	2	1	1	0	0
Charcot-Marie-Tooth disease, axonal, type 2E	1	1	1	2	1	2
Charcot-Marie-Tooth disease, demyelinating, type 1A (CMT1A)	5	6	5	8	2	3
Charcot-Marie-Tooth disease, demyelinating, type 1B (CMT1B)	1	1	1	3	0	0
Charcot-Marie-Tooth disease, X-linked, 1 (CMTX1)	2	2	2	4	0	0
Choroideremia (CHM)	1	1	1	2	1	3
Citrullinemia, classic	1	1	1	3	0	0
Colon cancer, familial nonpolyposis, type 1	1	1	1	2	1	1
Congenital adrenal hyperplasia (CAH)	2	4	3	6	2	3
Craniofacial dysostosis, type 1 (CFD1) (Crouzon)	2	3	3	8	2	1(1)*
Currarino syndrome	1	1	1	2	1	1
Cystic fibrosis (CF)	63	106	91	212	40	41(4)*
Darier-White disease (DAR)	1	1	1	1		
Dystrophia myotonica 1	19	33	22	48	11	8
Early-onset familial Alzheimer disease	1	2	2	3	2	3
Epidermolysis bullosa dystrophica, Pasini type	1	1	0	0	0	0
Epiphyseal dysplasia, multiple, 1 (EDM1)	1	2	1	3	0	0
Exostoses, multiple, type 1	1	1	1	2	0	0
Fabry disease	1	1	1	2	1	1
Fanconi anaemia, complementation group C (FANCC)	2	2	1	2	1	2
Fragile site mental retardation 1	29	43	37	86	16	15
Gaucher disease, type 1	2	5	3	6	0	0
Glycogen storage disease type VI	1	1	1	2	0	0
Haemoglobin—alpha locus 1 (HBA1)	3	5	4	7	1	1
**Haemoglobin—beta locus (HBB)	70	108	100	242	4	27
Haemophilia A; hemophilia B	4	5	5	11	2	2
HLA + ectodermal dysplasia, hypohidrotic, with immune deficiency	1	1	1	2	1	1
HLA genotyping	12	16	16	33	8	8
HLA + adrenoleukodystrophy	1	1	1	0	0	0
HLA + Fanconi anaemia, complementation group A (FANCA)	4	9	7	8	2	1
HLA + Fanconi anaemia, complementation group C (FANCC)	1	2	1	2	1	1
HLA + immunodeficiency with hyper-IgM, type 1 (HIGM1)	1	2	2	2	1	1
HLA + Wiskott-Aldrich syndrome (WAS)	1	1	0	0	0	0
Huntington disease (HD)	6	10	8	17	3	1
Hurler syndrome	2	3	3	6	0	0
Hydrocephalus, X-linked (L1CAM)	3	8	8	18	0	0
Hypophosphatasia, infantile	2	4	4	8	0	0
Incontinentia pigmenti (IP)	3	4	4	8	1	1
Li-Fraumeni syndrome (LFS)	2	5	3	5	1	1
Long-chain 3-hydroxyacyl-CoA dehydrogenase deficiency (HADHA)	1	3	4	13	3	5
Machado-Joseph disease (MJD)	1	2	1	2	1	2
Marfan syndrome (MFS)	4	8	8	16	3	3
Metachromatic luekodystropy	1	2	1	1	0	0
Mucopolysaccharidosis type II (Hunter) Hunter-McAlpine craniosynostosis syndrome	1	2	2	3	1	2

Table 3.26. *Continued*

Disease	# Patients	# Cycles	# ET	# Embryos	Pregnancy	Birth
Muscular dystrophy, Duchenne type (DMD); Becker type (BMD)	8	14	13	36	5	6
Myotubular myopathy 1 (MTM1)	1	2	1	3	0	0
Neurofibromatosis, type 1 (NF1)	2	5	8	11	2	3
Neurofibromatosis, type 2 (NF2)	1	2	2	5	2	2
Neuropathy, hereditary sensory and autonomic, type 3 (HSAN3)	2	2	2	5	2	4
Norrie disease (NDP)	2	4	4	7	0	0
Oculocutaneous albinism, TYPE 1 (OCA1) type 2 (OCA 2)	2	4	4	8	1	1
Optic atrophy 1 (OPA1)	1	1	1	2		
Ornithine transcarbamylase deficiency, hyperammonemia due to	4	5	5	8	4	4
Osteogenesis imperfecta congenita (OIC)	2	4	3	7	1	2
Osteopetrosis, autosomal recessive	1	1	1	2	0	0
Pelizaeus-Merzbacher-like disease (PMLD)	2	2	2	5	1	2
Phenylketonuria	1	1	1	4	1	2
Polycystic kidney disease 1 (PKD1)						
Polycystic kidney disease 2 (PKD2)	7	8	8	16	2	4(1)*
Polycystic kidney disease, autosomal recessive (ARPKD)	1	1	1	3	0	0
Retinoblastoma (RB1)	3	3	3	5	2	2
Sickle cell anaemia	12	17	17	38	6	2
Sonic hedgehog (SHH)	1	2	2	3	1	1
Spinal muscular atrophy, type 1 (SMa1)	6	7	6	11	4	8
Spinocerebellar ataxia 2 (SCA2)	3	4	4	11	1	2
Spinocerebellar ataxia 6 (SCA6)	1	2	1	1	0	0
Spinocerebellar ataxia 7 (SCA7)	1	1	1	2	1	0
Symphalangism, proximal (SYM1)	1	1	1	3	1	1
Tay-Sachs disease (TSD)	3	4	4	8	3	2
Torsion dystonia 1, autosomal dominant (DYT1)	2	6	6	16	2	1
Treacher Collins-Franceschetti syndrome (TCOF)	1	1	1	2	1	1
Tuberous sclerosis type 1	2	3	3	4	0	0
Tuberous sclerosis type 2	1	1	1	2	0	0
von Hippel-Lindau syndrome (VHL)	3	5	4	10	1	2
TOTAL	357	562	500 (89%)	1094 (2.2)	194 (33.8%)	187(7)

*Ongoing pregnancy; number of fetuses is shown in parentheses.
**This includes 38 PGD cycles combined with HLA typing presented in Chapter 4 (see Table 4.1).

to be highly accurate and reliable. The distribution of different conditions for which PGD was performed has been changing gradually, with the increase of the proportion of common diseases with genetic predisposition and nondisease testing, including preimplantation HLA matching, described next in Chapter 4.

References

1. Handyside AH, Kontogiani EH, Hardy K, Winston RML. Pregnancies from biopsied human preimplantation embryos sexed by Y-specific DNA amplification. Nature1990;344:768–770.
2. Verlinsky Y, Ginsberg N, Lifchez A, Valle J, MoiseJ, Strom CM. Analysis of the first polar body: Preconception genetic diagnosis. Hum Reprod 1990;5:826–829.
3. Verlinsky Y, Cohen J, Munné S, et al. Over a decade of preimplantation genetic diagnosis experience: A multicenter report. Fertil Steril 2004;82:292–294.
4. Kuliev A, Verlinsky Y. Thirteen years' experience of preimplantation diagnosis: Report of the Fifth International Symposium on Preimplantation Genetics. Reprod BioMed Online 2004;8:229–235.
5. Kuliev A, Verlinsky Y. Current features of preimplantation genetic disgnosis. Reprod BioMed Online 2002;5:296–301.
6. Kuliev A, Rechitsky S, Verlinsky O, et al. Preimplantation diagnosis of thalassemia. J Assist Reprod Genet 1998;15:219–25.
7. Kuliev A, Rechitsky S, Verlinsky O, et al. Birth of healthy children following preimplantation diagnosis for thalassemias. J Assist Reprod Genet 1999;16:219–225.

8. Verlinsky Y, Milayeva S, Evsikov S, et al. Preconception and preimplantation diagnosis for cystic fibrosis. Prenat Diagn 1992;12:103–10.
9. Handyside AH, Lesko JG, Tarin JJ, Winston RM, Hughes MR. Birth of a normal girl after in vitro fertilization and preimplantation diagnosis testing for cystic fibrosis. N Engl J Med 1992;327:905–909.
10. Goossens V, Sermon K, Lissens W, et al. Clinical application of preimplantation genetic diagnosis for cystic fibrosis. Prenat Diagn 2000;20:571–581.
11. ESHRE Preimplantation Genetic Diagnosis (PGD) Consortium. Data Collecttion II (May 2002). Hum Reprod 2002;15:2673–2683.
12. Online Mendelian Inheritance in Man (OMIM). 2001. John Hopkins University http://www.ncbi.nlm.nih.gov/Omim
13. Online Human Gene Mutation Database (HGMD). 2004. CELERA http://archive.uwcm.ac.uk/uwcm/mg/search/1119297.html
14. Slaugenhaupt SA, Blumenfekl A, Gill SP et al. Tissue-specific expression of a splicing mutation in the IKBKAP gene causes familial dysautonomia. Am J Hum Genet 2001;68:598–605.
15. Blumenfeld A, Slaugenhaupt SA, Liebert CB, et al. Precise genetic mapping and haplotype analysis of the familial dysautonomia gene on human chromosome 9q31. Am J Hum Genet 1999;64:1110–1118.
16. Rechitsky S, Verlinsky O, Kuliev A, Ozen S, Masciangelo C, Lifchez A, Verlinsky Y. Preimplantation genetic diagnosis for familial dysautonomia. Reprod BioMed Online 2003;6:488–493.
17. Fallon MS, Harton BS, Sisson BS, et al. Preimplantation genetic diagnosis for spinal muscular atrophy type 1. Neurology 1999;53:1087–1090.
18. Daniels G, Pettigrew R, Thornhill A, et al. Six unaffected livebirths following preimplantation diagnosis for spinal muscular atrophy. Mol Hum Reprod. 2001;7:995–1000.
19. Ozelius L, Hewett J, Page CE, et al. The early-onset torsion dystonia gene (DYT1) encodes an AATP-binding protein. Nat Genet 1997;17:40–48.
20. Ozelius L, Klein C, et al. The TOR1A (DYT1) Gene family and its role in early onset torsion dystonia. Genomics 1999;62:377–384.
21. Valente EM, Povey S, Waener TT, Wood NW, Davis MB. Detailed haplotype analysis in Ashkenazi Jewish and non-Jewish British dystonic patients carrying the GAG deletion in the DYT1 gene: Evidence for a limited number of founder mutations. Ann Hum Genet 1999;63:1–8.
22. Kwiatkovski DJ, Nygaard TG, Schuback DE, et al. Identification of a highly polymprphic microsatellite VNTR within the argininosuccinate synthetase locuc: Exclusion of the dystonia gene on 9q32-34 as the cause of dopa-responsive dystonia in a large kindred. Am J Hum Genet 1991;48:121–128.
23. Risch NJ, De Leon D, Ozelius L, et al. Genetic analysis of idiopathic torsion dystonia in Ashkenazi Jews and their recent descent from a small founder population. Nat Genet 1995;9:152–159.
24. Rechitsky S, Strom C, Verlinsky O, et al. Allele drop out in polar bodies and blastemeres. J Assist Reprod Genet 1998;15:253–257.
25. Rechitsky S, Verlinsky O, Kuliev A, et al. Preimplantation genetic diagnosis for early onset torsion dystonia. Reprod BioMed Online, 2004;8:224–228.
26. De Vos A, Sermon K, De Rijcke M, et al. Preimplantation genetic diagnosis for Charcot-Marie-Tooth disease type 1A. Mol Hum Reprod 2003;9:429–435.
27. Verlinsky Y, Rechitsky S, Verlinsky O, Strom C, Kuliev A PGD for ornithine transcarbamilase deficiency. Reprod BioMed Online 2000;1:45–47.
28. Platteau P, Sermon K, Seneca S, et al. Preimplantation genetic diagnosis for fragile Xa syndrome: Difficult but possible. Hum Reprod 2002;17:2807–2812.
29. Ray PF, Gigarel N, Bonnefont JP, Attie T, Hamamah S, Frydman N, Vekemans M, Frydman R, Munnich A. First specific preimplantation genetic diagnosis for ornithine transcarbamilase deficiency. Prenat Diagn 2000;20:1048–1054.
30. Verlinsky Y, Rechitsky S, Verlinsky O, Strom C,Kuliev A. Preimplantation testing for pnenylketonuria. Fertil Steril 2001;76:346–349.
31. Harris PC, Thomas S, Ratcliffe PJ, Breuning MH, Coto E, Lopez-Larrea C. Rapid genetic analysis of families with polycystic kidney disease 1 by means of a microsatellite marker. Lancet 1991;338:1484–1487.
32. Peral B, Ward CJ, San Milan JL, Thomas S, Stalings RL, Moreno F, Harris P. Evidence of linkage desequilibrium in the Spanish polycystic kidney disease I population. Am J Hum Genet 1994;54:899–908.
33. Snarey A, Thomas S, Shneider MC, Pound SE, Barton N, Wright AF, et al. Linkage desequilibrium in the region of the autosomal dominant polycystic kidney disease gene (PKDI). Am J Hum Genet 1994;55:365–371.
34. Verlinsky Y, Rechitsky S, Verlinsky O, Ozen S, Beck R, Kuliev A. Preimplantation genetic diagnosis for polycystic kidney disease. Fertil Steril 2004;82:926–929.
35. Verlinsky Y, Rechitsky S, Verlinsky O, et al. Preimplantation diagnosis for p53 tumor suppressor gene mutations. Reprod BioMed Online 2001;2:102–105.
36. Holstein M. p53 Mutations in human cancers. Science 1991;253:49–53.
37. Huson SM, Compston DAS, Clark P, Harper PS. A genetic study of von Recklinghausen neurofibromatosis in south-east Wales. I. Prevalence, fitness, mutation rate, and the effect of parental transmission on severity. J. Med Genet 1989;26:704–711.
38. Clementi M, Barbujani G, Trolla L, Tenconi R. Neurofibromatosis-1: Maximum likelihood estimation of mutation rate. Hum Genet 1990;84:116–118.
39. Simpson JL. Celebrating preimplantation genetic diagnosis of p53 mutations in Li-Fraumeni syndrome. Reprod BioMed Online 2001;3:2–3.
40. Cram D. Preimplantation genetic diagnosis for familial cancer. Reprod BioMed Online 2001;3:3–4.
41. Klose A, Robins PN, Gewies A, et al. Two novel mutations in exon 19a and 20 and BsaI polymorphism in a newly characterized intron of the neurofibromatosis type 1 gene. Hum Genet 1998;102:367–371.
42. Xu G, Nelson L, O'Connell P, White R. An Alu polymorphism intragenic to the neurofobromatosis type 1 gene. Nucleic Acids Res 1992;19:3764.
43. Hoffmeyer S, Assum G. An Rsa I polymorphism in the transcribed region of the neurofibromatosis (NF1) gene. Hum Genet 1994;93:481–482.
44. Weissenbach J, Gyapay G, Dib C, Vignal A, Morisetti J, Milasseau P, Vaysseix G, Lathrop M. A second-generation linkage map of the human genome. Nature 1992;359:794–801.

45. Legoix P, Legrand M-F, Ollagnon E, Lenoir G, Thomas G, Zucman-Rossi J. Characterization of 16 polymorphic markers in the NF2 gene: Application to hemizygosity detection. 1999;13:290–293.
46. Rechitsky S, Strom C, Verlinsky O, et al. Accuracy of preimplantation diagnosis of single-gene disorders by polar body analysis of oocytes. J Assist Reprod Genet 1999;16:192–198.
47. Rechitsky S, Verlinsky O, Strom C, et al. Experience with single-cell PCR in preimplantation genetic diagnosis: How to avoid pitfalls. In: Hahn S, Holzgreve W. (eds.). Fetal Cells in Maternal Blood. New Developments for a New Millennium. 11th Fetal Cell Workshop. Karger, Basel, 2000, pp. 8–15.
48. Verlinsky Y, Rechitsky S, Verlinsky O, et al. Preimplantation diagnosis for newrofibromatosis. Reprod BioMed Online 2002;4:102–105.
49. Rechitsky S, Verlinsky O, Chistokhina A, et al. Preimplantation genetic diagnosis for cancer predisposition. Reprod BioMed Online 2002;4:148–155.
50. Ao A, Wells D, Handyside A, Winston R, Delhanty J. Preimplantation genetic diagnosis of inherited cancer: Familial adenomatous polyposis coli. J. Assist Reprod Genet 1998;15:140–144.
51. Taylor MD, Gokgoz N, Andrulis IL, et al. Familial posterior fossa brain tumors of infancy secondaty to germline mutation of the hSNF5 gene. Am J Hum Genet 2000;66:1403–1406.
52. Verlinsky Y, Rechitsky S, Verlinsky O, et al. Preimplantation diagnosis for early onset Alzheimer disease caused by V717L mutation. JAMA 2002;287:1018–1021.
53. Sherington R, Rogaev EI, Liang Y, et al. Cloning of the gene bearing missense mutations in early onset familial Alzheimer disease. Nature 1995;375:754–760.
54. Levy-Lehad E, Wasco W, Poorkaj P, et al. Candidate gene for the chromosome 1 familial Alzheimer's disease locus. Science 1995;269:973–977.
55. Goate AM, Chantier-Harlin MC, Mullan M, et al. Segregation of missense mutation in the amyloid precursor protein gene with familial Alzheimer disease. Nature 1991;349:704–706.
56. Murrel J, Hake AM, Quaid KA, Farlow MR, Ghetti B. Early-onset Alzheimer disease caused by a new mutation (V717L) in the amyloid precursor Protein gene. Ach Neurol 2000;57:885–887.
57. Verlinsky Y, Kuliev A. An Atlas of Preimplantation Genetic Diagnosis Parthenon, New York, 2000.
58. Rechitsky S, Verlinsky O, Amet T, et al. Reliability of preimplantation diagnosis for single gene disorders. Mol Cell Endocrinol 2001;183:565–568.
59. Tupler R, Rogaeva E, Vaula G. et al. A highly informative microsatellite repeat polymorphism in intron 1 of the human amyloid precursor protein (APP) gene. Hum Mol Genet 1993;2:620–621.
60. Lee S, Zambas E, Wu X, Reid M, Zelinsky T, Redman C. Molecular basis of the Kell (K1) phenotype. Blood 1995;85:912–916.
61. Lee S, Zambas E, Green ED, Redman C. Organization of the gene for encoding the human Kell group protein. Blood 1995;85:1364–1370.
62. Purohit KR, Weber JL, Ward LJ, Keats JB. The Kell blood group locus is close to the cyctic fibrosis locus on chromosome 7. Hum Genet 1992;89:457–458.
63. Reid ME, Rios M, Powell VI, Charles-Pierre D, Malavade V. DNA from blood samples can be used to genotype patients who have recently received a transfusion. Transfusion 2000;40:48–53.
64. Zielenski J, Rozmahel R, Bozon D, Kerem BS, Grzelczak Z, Riordan J, Rommens J, Tsui L-H. Genomic DNA sequence of the cystic fibrosis transmembrane conductance regulator (CFTR) gene. Genomics 1991;10:214–228.
65. Chehab FF, Johnson J, Louie E, Goossens M, Kawasaki E, Elrich H. A dimorphic 4-bp repeat in the cystic fibrosis gene is in absolute linkage disequilibrium with the delta F508 mutation: Implications for prenatal diagnosis and mutation origin. Am J Hum Genet 1991;48:223–226.
66. Van Den Veyver IB, Chong SS, Cota J, et al. Fetus-placenta-newborn: Single cell analysis of the RhD blood type for use in preimplantation diagnosis in the prevention of severe haemolitic disease of the newborn. Am J Obstet Gynecol 1995;172:533–540.
67. Seeho SKM, Burton G, Leigh D, Marshall JT, Persson JW, Morris JM. The role of preimplantation genetic diagnosis in the management of severe rhesus alloimmunization: First unaffected pregnancy: Case report. Hum Reprod 2005;20:697–701.
68. Verlinsky Y, Rechitsky S, Seckin O, Masciangelo C, Ayers J, Kuliev A. Preimplantation diagnosis for Kell genotype. Fertil Steril 2003;80:1047–1051.
69. Muenke M, Gurrieri F, Bay C, et al. Linkage of a human brain malformation, familial holoprosencephaly, to chromosome 7 and evidence for genetic heterogeneity. Proc. Natl Acad Sci USA 1994;91:8102–8106.
70. Nanni L, Ming JE, Bocian M, et al. The mutational spectrum of the sonic hedgehog gene in holoprosencephaly: SHH mutations cause a significant proportion of autosomal dominant holoprosencephaly. Hum Mol Genet 1999;8:2479–2488.
71. Verlinsky Y, Rechitsky S, Verlinsky O, et al Preimplantation diagnosis for sonic hedgehog mutation causing familial holoprosencephaly. N Engl J Med 2003;348:1449–1454.
72. Abou-Sleiman PM, Apessos A, Harper JC, Serhal P, Delhanty JDA. Pregnancy following preimplantation genetic diagnosis for Crouson syndrome. Mol Hum Reprod 2002;8:304–309.
73. He J, McDermont DA, Song Y, Gilbert F, Kligman I, Basson C. Preimplantation genetic diagnosis of human congenital heart disease and Holt-Oram syndrome. Am J Med Genet 2004;1264:93–98.
74. Verlinsky Y, Rechitsky S, Schoolcraft, Kuliev A. Preimplantation diagnosis for homeobox gene HLXB9 mutation causing Currarino syndrome. Am J Med Genet 2005;134A:103–104.
75. Ross AJ, Rui-Perez V, Wang Y, Hagan DM, Scherer SW, Lynch SA, Lindssy S, Custard E, Belloni E, Wilson DI, Wadey R, Goodman F, Orstavic KH, Monclair T, Robson S, Reardon W, Burn J, Scambler P, Strachan T. A homeobox gene, HLXB9, is the major locus for dominantly inherited sacral agenesis. Nat Genet 1998;20:358–361.
76. Belloni E, Martucciello G, Verderio D, Ponti E, Seri M, Jasonni V, Torre M, Ferrari M, Tsui L-C, Scherer SW., et al. Involvement of the HLXB9 homeobox gene in Currarino syndrome. Am J Hum Genet 2000;66:312–319.
77. Hagan DM, Ross AJ, Strachan T, Lynch SA, Ruiz-Perez V, Wang YM, Scambler P, et al. Mutation analysis and embryonic expression of the HLXB9 Currarino syndrome gene. Am J Hum Genet 2000;66:1504–1515.

78. Kochling J, Karbasiyan M, Reis A. Spectrum of mutations and genotype-phenotype analysis in Currarino syndrome. Eur J Hum Genet 2001;9:599–605.
79. Reardon W, Winter RM, Rutland P, et al. Mutations in the fibroblast growth factor gene cause Crouzon syndrome. Nat Genet 1994;8:98–103.
80. Cohen MM, Kreiborg S. Birth prevalence studies of the Crouzon syndrome: Comparison of direct and indirect methods. Clin Genet 1992;41:12–15.
81. Sermon K, Lissens W, Joris H. et al. Clinical application of preimplantation diagnosis for myotonic dystrophy. Prenat Diagn 1997;17:925–932.
82. Sermon K, Seneca S, Vanderfaeillie A, et al. Preimplantation diagnosis for fragile X syndrome based on the detection of the non-expanded paternal and maternal CGG. Prenat Diagn 1999;19:1223–1230.
83. Sermon K, Seneca S, De Rycke M, et al. PGD in the lab for triplet diseases—myotonic dystrophy, Huntington's disease and fragile-X syndrome. Mol Cell Endocrinol 2001;183:S77–S85.
84. Verlinsky Y, Rechitsky S, Verlinsky O, Strom C, Kuliev A. Polar body based preimplantation diagnosis for X-linked genetic disorders. Reprod BioMed Online 2002;4:38–42.

4

Preimplantation HLA Typing

Preimplantation HLA matching has never been an indication for prenatal diagnosis because of a possible clinical pregnancy termination after finding the fetus HLA unmatched. However, PGD for such purpose should be acceptable because only a limited number of the embryos are usually preselected for transfer anyway, which in this case will represent unaffected embryos with a perfect match for affected siblings at need for a transplant. The first case of preimplantation HLA typing was introduced in combination with mutation analysis for Fanconi anemia (FA), with the objective of establishing an unaffected pregnancy yielding a potential donor progeny for transplantation in an affected sibling [1, 2].

4.1 PGD for Fanconi Anemia with Preimplantation HLA Matching

One of the severe congenital disorders requiring stem cell transplantation from a family member is FA, which was the first disease for which cord blood transplantation was introduced [3]. FA is an autosomal recessive disorder, characterised by inherited bone marrow failure, congenital malformations, and an increased predisposition to the development of leukemia. It is genetically heterogeneous, involving different complementation groups (FANCA, FANCB, FANCC, FANCD, and FANCE), one of the most severe being FANCC mutation leading to aberrantly spliced transcripts (IVS4+4A-T), which result in inactivating FANCC protein [4–6]. Bone marrow transplantation is the only treatment, which restores definitively hematopoiesis in FA patients. However, because any modification of the conditioning is too toxic for these patients leading to a high rate of transplant-related mortality, the HLA-identical cord blood transplantation from a sibling is particularly valuable for FA, to avoid late complications due to severe GVH [7, 8].

A couple presented for PGD with both parents being unaffected carriers of IVS 4+4 A-T mutation in FANCC gene. Their affected 6-year-old daughter had two copies of this mutation, requiring an HLA-compatible donor for bone marrow transplantation. The couple requested PGD for FANCC, together with HLA testing of embryos, so to have an unaffected child who may also be a compatible cord blood donor for their affected daughter.

PGD was performed using a standard IVF protocol combined with micromanipulation procedure to biopsy single blastomeres from the day 3 cleaving embryos, as described in Chapter 2. Blastomeres were tested for IVS 4+4 A-T mutation in FANCC gene using polyacrylamide gel analysis of PCR product digested with *Sca* I restriction enzyme, according to the method of single-cell PCR analysis, also described in Chapter 2.

The outer primers IVS4-1 (5′-GTCATAAAAGGCACTTGCAT-3′) and IVS4-2 (5′-GGCACATTCAGCATTAAACA-3′) were designed for performing the first round of amplification, while using the previously described inner primers [9] for the second round of PCR. The second-round PCR produces a 131-bp product, undigested by *Sca* I restriction enzyme, corresponding to the mutant allele, and two restriction fragments of 108 bp and 23 bp, corresponding to the normal allele, due to introduction of the restriction site by application of 4R inner primer with single base modification of T to G.

Nested PCR for specific amplification of HLA A gene exons 2 and 3 was performed using gene-specific outer primers Asp5 (5′GCCCCGAACCC-TC(CT)TCCTGCTA 3′) and Asp3 (5′CCGTGG-CCCCTGGTACCCGT 3′) [10], followed by four separate second-round PCR with allele-specific inner primers 085 (5′TCCTCGTCCCCAGGCT-CT 3′) and 98 (5′GCAGGGTCCCCAGGTCCA 3′) for A2 allele [11], 140 (5′GGTTCTCACACCAT-CCAGATA 3′) and 142 (5′ CAGGTATCTGCGGA-GCCCG 3′) for A1 allele, 140 (5′ GGTTCTCAC-ACCATCCAGATA 3′) and 126 (5′CCACTCCAC-GCACGTGCCA 3′) for A3 allele [11], and 118 (5′TCCATGAGGTATTTCTACACC 3′) and 145 (5′GCAGGGTCCCCAGGTTCG 3′) for allele A26 [11]. As haplotype analysis for father, mother, and the affected child showed different polymorphic short tandem repeat (STR) marker (GAAA)n (C2–4–4) located in between HLA-A - HLA-B (in HLA-E - HLA-C region), a heminested PCR system was designed to study the number of repeats in blastomeres from different embryos [12]. The first-round amplification cocktail for this system contained outer primers P1-1 (5′-GGCTTGACTTGAAACTCAGAG-3′) and P1-3 (5-TATCTACTTATAGTCTATCACG-3′), while the second-round PCR used, in addition to P1-1, the inner primer P1-2 (5′-CTTCAAACAATACG-CAATGACA-3′). The nested PCR system for HLA B allele discrimination included outer primers Bout 1 (5′-GAGGGTCGGGCGGGTC-TCAG-3′) and Bout 2 (5′-TGGGGGATGGGGA-GTCGTGAC-3′) for the first round of amplification. The second round of amplification for HLA B35 was performed using inner PCR primers CG4 (5′-GACGACACCCAGTTCGTGA-3′) and 35in (5′-GAAGATCTGTGTGTTCCGG-3′). Accordingly, the second round of amplification of HLA B41 was performed with primers CG3 (5′-CTCTGGTTGTAGTAGCCGC-3′)and 41up (5′-CCACGAGTCCGAGGAAGG-3), and HLA B44 with primers 41up and GC2 (5′-GCTCTGG-TTGTAGTAGCGGA-3′) [13].

Blastomere genotyping for IVS 4+4 A-T mutation in FANCC gene was performed in four clinical cycles, involving the mutation analysis in 33 embryos, including seven in the first, four in second, eight in third and 14 in fourth cycle. Of 30 embryos with results, 19 were heterozygous carriers, six were homozygous affected, and five were homozygous normal. Of 14 embryos tested for mutation in the last cycle, only one was homozygous affected, three were homozygous normal, two did not amplify, with the remaining being heterozygous unaffected.

Testing for HLA A (A2, A26) and HLA B (B35, B44) in these 24 unaffected embryos, including 19 heterozygous and five homozygous normal embryos, revealed five heterozygous unaffected embryos for transfer with HLA match for the affected sibling requiring transplantation of stem cells. The results of HLA typing in 14 embryos in the last cycle revealed only one unaffected heterozygous embryo being HLA-identical to the affected child, which therefore was transferred back to the patient. Similarly, two unaffected HLA-matched embryos were available for transfer in the first, one in second, and one in third cycle. However, only the transfer in the last cycle resulted in a clinical pregnancy and birth of a healthy carrier of FANCC gene, following confirmation of the results of both mutation analysis and HLA matching by CVS. Umbilical cord blood of the baby was collected at birth and transplanted to the affected sibling, resulting in a successful hematopoietic reconstitution. Nine embryos predicted normal but carrying HLA genes different from the sibling were developed to the blastocyst stage and frozen, while 5 affected embryos were exposed to PCR analysis, confirming the blastomere diagnosis.

As described in Chapter 3, the practical application of PGD has recently been extended for new indications, which appeared to be different from those used in prenatal diagnosis, such as the late-onset disorders with genetic predisposition, which could have hardly been considered candidates for prenatal genetic diagnosis. HLA testing is the most recent and most unexpected addition to the indications for PGD. Presented results demonstrate feasibility of preimplantation HLA matching as part of PGD, with a prospect for the application of this approach to the other inherited conditions, such as thalassemias, also requiring an HLA-compatible donor for bone marrow transplantation. Although initially this was the first and only experience of PGD for HLA testing, it provided a realistic option for couples desiring to avoid the birth of an affected child, together with the establishment of a healthy pregnancy, potentially providing an HLA match for an affected sibling. The data showed that the HLA testing in single blastomeres was accurate, and may also be applied as primary indication, i.e., in cases not requiring mutation testing, such as for couples having affected children with leukemia or other cancers, awaiting an HLA-compatible donor with no success for years. These new indications make PGD a genuine alternative to conventional prenatal diagnosis, providing patients with important prospect not only to avoid an inherited risk without facing termination of pregnancy, but also to establish a pregnancy with particular genetic parameters to benefit the affected member of the family.

4.2 Current Experience of PGD with Preimplantation HLA Matching

In addition to FANCC, we have now undertaken HLA genotyping in more than two dozen cycles in combination with PGD for thalassemia, FANCA, hyperimmunoglobulin M syndrome (HIGM), X-linked adrenoleukodystrophy (X-ALD) and Wiscott-Aldrich syndrome (WAS), and X-linked hypohidrotic ectodermal displasia with immune deficiency (HED-ID) [14], confirming the usefulness of preimplantation HLA matching as part of PGD, which potentially provides an HLA-matched progeny for treatment of an affected sibling.

Among these conditions, the most prevalent ones are thalassemias, which are the most common autosomal recessive diseases in the Mediterranean region, Middle East, and Southeast Asia, with heterozygous frequency of thalassemia mutations reaching 14% in Greece and Cyprus. The prevalence of the other conditions is much lower, 1 in 17,000 for X-ALD, 1 in 100,000 for FA, and even rarer for WA, HIGM and HED-ID.

FANCA, similar to FANCC, is an autosomal recessive disorder causing an inherited bone marrow failure with increased predisposition to leukemia. As mentioned, bone marrow transplantation is the only treatment for FA, as it restores hematopoiesis in FANCA patients. However, because any modification of the conditioning is too toxic for these patients, as in FANCC, leading to a high rate of transplant-related mortality, the HLA-identical cord blood transplantation from a sibling is particularly valuable, to avoid late complications due to severe GVH, as mentioned above.

Beta-thalassemia is an autosomal recessive disease affecting the production of β-globin chains resulting in a severe anaemia, which makes the patients transfusion-dependent starting from six months after birth, so bone marrow transplantation is the only option for radical treatment. At present, 371 different mutations have been described in the β-globin gene, located in chromosome 11 (11p15.5), causing congenital anaemia of variable severity [15]. Prenatal diagnosis has been applied widely for more than two decades resulting in considerable reduction of new cases of thalassemia in many populations [16]. Considerable progress also has been achieved in treatment of the disease by bone marrow transplantation [17], the application of which is still limited to the availability of HLA-matched stem cells, making PGD an attractive option for couples with thalassemic children. PGD for thalassemia has already been provided for a few years [18], so HLA typing is presently offered in the same framework, allowing couples not only to avoid the birth of another child with thalassemia, but also to produce an unaffected child who may be an HLA match to the thalassemic sibling, and a potential stem cell donor.

X-ALD affects the nervous system and the adrenal cortex, with three main phenotypes. One of them manifests between ages 4 and 8 as attention deficit disorder, followed by progressive impairment of cognition and behaviour, vision, hearing, and motor function, leading to total disability within two years. Another phenotype, called adrenomyeloneuropathy, manifests in the late 20s as progressive paraparesis, sphincter disturbances, and hearing loss, while the third presents with primary adrenocortical insufficiency by age 7–8. Regardless of the presence of symptoms, 99% of patients have an elevated concentration of very-long-chain fatty acids (VLCFA). The disease is caused by mutations of the ABCD1 gene, with more than 200 different mutations reported at the present time, which may be detected by PCR and direct sequencing, except for large deletions identified by Southern blot analysis. Carrier screening and prenatal diagnosis are available and the same method may be applied for PGD with simultaneous HLA typing.

HIGM is a rare immunodeficiency characterised by normal or elevated serum IgM levels, with absence of IgG, IgA, and IgE, which results in an increased susceptibility to infections, manifested in the first few years of life, and a high frequency of autoimmune haematological disorders, accompanied by gingivitis, ulcerative stomatitis, fever, and weight loss. HIGM is caused by

Table 4.1. Results and outcomes of PGD with preimplantation HLA typing

Disease	Maternal Age	Patient/ Cycle	# Embryos Total/Amplified	# Normal Embryos Nonmatch	# Normal Embryos Match	# Abnormal Embryos Nonmatch	# Abnormal Embryos Match	# Transfer/ # Embryos	Pregnancy/ Birth
Thalassemia	36.1*	18 / 31	311 / 289	163	41	75	10	22 / 38	3 / 2
FANCA+C	34.2*	5 / 11	73 / 68	39	9	18	2	8 / 10	3 / 2**
WAS	37	1 / 1	2 / 2	0	1	1	0	1 / 1	0
X-ALD	39	1 / 1	4 / 4	2	0	0	2	0	0
HIGM	33	1 / 2	24 / 21	14	2	3	2	2 / 2	1 / 1
HED-ID	30	1/1	16 / 16	9	2	4	1	1 / 2	1 / 1
TOTAL		*27/47*	*430/400*	*227*	*55*	*101*	*17*	*34/53*	*8/6*

*mean maternal age.
**One couple decided to transfer partially matched embryo carrier of the paternal mutation in FANCA gene.

mutation in the CD40 ligand gene (CD40LG) located in chromosome Xq26, which leads to a defective CD40 ligand expression, resulting in the failure of T cells to induce IgE synthesis in interleukin-4-treated B cells. Although a regular administration of intravenous immunoglobulins may be used for treatment, the best results were obtained by HLA-matched bone marrow transplantation, which makes PGD the method of choice for those who cannot find an HLA match among their relatives.

WAS is a lethal X-linked immune deficiency, in which lymphocyte dysfunction and thrombocytopenia result in severe infections, bleeding episodes, and increased risk of lymphoproliferative malignancies. While supportive therapy may increase survival rate, the only hope for avoiding early mortality is bone marrow transplantation. WAS is caused by a mutation in the WAS gene mapped to the Xp11.22-11.23 region, which results in actin polymerization, with T lymphocytes of males exhibiting a severe disturbance of the actin cytoskeleton. The gene has 12 exons that encode a 502 amino acid cytosolic protein, expressed exclusively in hematopoietic cells.

HED-ID is a congenital disorder of teeth, hair, and eccrine sweat glands, inherited as an X-linked recessive condition, caused by approximately two dozens different mutations in the IKK-gamma gene (IKBKG, or NEMO) located in Xq28. The gene consists of 10 exons and codes for a scaffold protein that binds IKK-alpha and IKK-beta, essential for forming a functional IKK complex. The disease is characterised by susceptibility to microbial and streptococcal infections, dysgammaglobulinemia, poor polysaccharide-specific antibody responses, and depressed antigen-specific lymphocyte proliferation. Intravenous immunoglobulins and prophylactic antibiotics may be useful in improving clinical status, but bone marrow transplantation is required to prevent early mortality.

Twenty-seven couples, overall, presented for PGD combined with HLA typing, for whom 47 clinical cycles were performed, including 11 from five couples with FANCA and FANCC, 31 from 18 couples with thalassemia, two from a couple with HIGM, and one from a couple with WAS-, X-ALD, and HED-ID-affected siblings in a family (Table 4.1).

PGD cycles were performed using a standard IVF protocol, coupled with micromanipulation procedures described in Chapter 2 [19]. For testing of the maternal mutations and the maternal HLA match, the first and second polar body (PB1 and PB2) were removed sequentially following maturation and fertilization of oocytes, while single blastomeres were removed from the eight-cell embryos for testing of both the paternal and maternal mutations and the paternal and maternal HLA match.

Two of five couples at risk for producing a progeny with FA carried IVS 4+4 A-T mutation in FANCC gene, the strategy for testing of which was described above. Three couples were carriers of FANCA gene mutations, including one with different maternal and paternal mutations, the maternal one involving A*T*G to A*A*G substitution in exon 1, resulting in methionine to lysine aminoacid substitution, and the paternal 14-bp deletion in exon 2, representing a frameshift mutation. The paternal mutation was detected as the size difference in capillary electrophoresis of PCR product, and the maternal by *Nla* II restriction digestion, which cuts the normal sequence, leaving the mutant sequence uncut. In the other couple, only paternal mutation was known, representing T1131A mutation, due to *A*CT to *G*CT substitution in exon 34, which creates restriction site for *Fsp*4H I. In addition, another restriction enzyme, *TspR* I, was used, which cuts the normal sequence on the opposite. No mutation was identified in the third couple, leaving the only possibility of choosing unaffected embryos by linkage analysis, using five closely linked polymorphic markers.

A couple at risk for producing a progeny with X-ALD had an affected son with G343D mutation, representing a sequence change from aspartic acid to glycine at amino acid position 343, caused by a single (G to A) sequence change in the nucleotide 1414 (G1414A) of the ABCD gene. PGD was based on *Fok* I restriction digestion, which creates two fragments in the PCR product of normal gene, leaving the mutant one uncut.

The other couple at risk for producing a progeny with WAS had two affected sons carrying the missence Leu39Pro mutation in exon 1 of the WAS gene, due to a single nucleotide (CTT to CCT) substitution at position 150, which leads to substitution of leucine by proline at position 39. The mutation testing was done using *Scr* FI restriction digestion, cutting the mutant and leaving the normal gene product intact.

Seventeen couples had thalassemia mutations, including 619-bp deletion, -28, codon 39, IVS I-5, R30T, IVS I-110, IVSI-1, IVS 2-1, codon 41/42, codon 8/9 (+G), and cap1. The maternal mutations were analyzed by PB1 and PB2, as described elsewhere [18, 19], followed by paternal mutation testing and HLA typing in blastomeres, particularly when parents were carriers of different mutations, as shown in Figure 4.1.

A single couple with a HIGM-affected child had CYS218STOP mutation in exon 5, which eliminates restriction site for *Cac*8 I, so this enzyme creates two fragments in PCR product from the normal gene, leaving the mutant gene product uncut. For higher accuracy, another restriction enzyme (*Mnl* I) was applied, which creates three fragments in the mutant PCR product, compared with two fragments in the normal gene (Figure 4.2).

Finally, in one couple, the mother was a carrier of L153R mutation, resulting from T to G change (CTG→CGG) in exon 4 of the NEMO (IKBKG) gene, replacing leucine with arginine at position 153 of the resultant protein. Because of the presence of closely linked pseudogene with a normal sequence at the position of the mutation, which is co-amplified with the transcribed gene, a special design was developed to avoid misdiagnosis (Figure 4.3).

Approaches for PGD and combined HLA typing have been described above and reported previously [2, 19, 20]. In brief, DNA testing cycles were performed by the multiplex nested PCR analysis, amplifying mutations simultaneously with linked markers, HLA alleles, and short tandem repeats in the HLA region. Linkage analysis has been performed for each couple to avoid a possible misdiagnosis, which requires special attention because of the phenomena of allele drop out (ADO) and preferential amplification, which are frequent in a single-cell DNA analysis. For example, six informative linked markers are available for testing (see Figure 3.1) together with mutation analysis for thalassemias (three of them are shown in Figure 4.1). In couples with FANCA mutations, six markers were used, representing dinucleotide repeats essential in confirming the presence of the normal gene in the embryos selected for transfer. In the couple with HIGM, four closely linked dinucleotide markers were applied, together with CYS218STOP mutation in exon 5 and HLA alleles, allowing improved accuracy of PGD (see Figure 4.2), and eight closely linked STRs were used in the couple with HED-ID (six of eight markers are shown in Figure 4.3).

Because patients referring for combined PGD and HLA typing are usually of advanced reproductive age, the copy number of the chromosome, in which the causative gene tested is localised, was analysed using corresponding satellite markers, as well as the copy number of chromosome 6, where HLA genes are localized. A set of STRs throughout the HLA region [20, 21] and chromosome-specific satellite makers allowed identifying monosomy, trisomy, or uniparental disomy of chromosome 6, as well as the possibility of recombination, which may strongly affect the accuracy of HLA matching.

Primers for multiplex PCR analysis for thalassemias and FANCC were described above and reported elsewhere [2, 19, 20]; others, including those for FANCA, X-ALD, HIGM, WAS, and HED-ID, are listed in Table 4.2.

As usual, the patients gave consent, which was approved by Institutional Review Board, that, based on the multiplex mutation and marker analysis, unaffected HLA embryos would be preselected for transferring back to the patients, unaffected but non-HLA-matched embroys would be frozen for possible future use by the couple, while those predicted mutant would be exposed

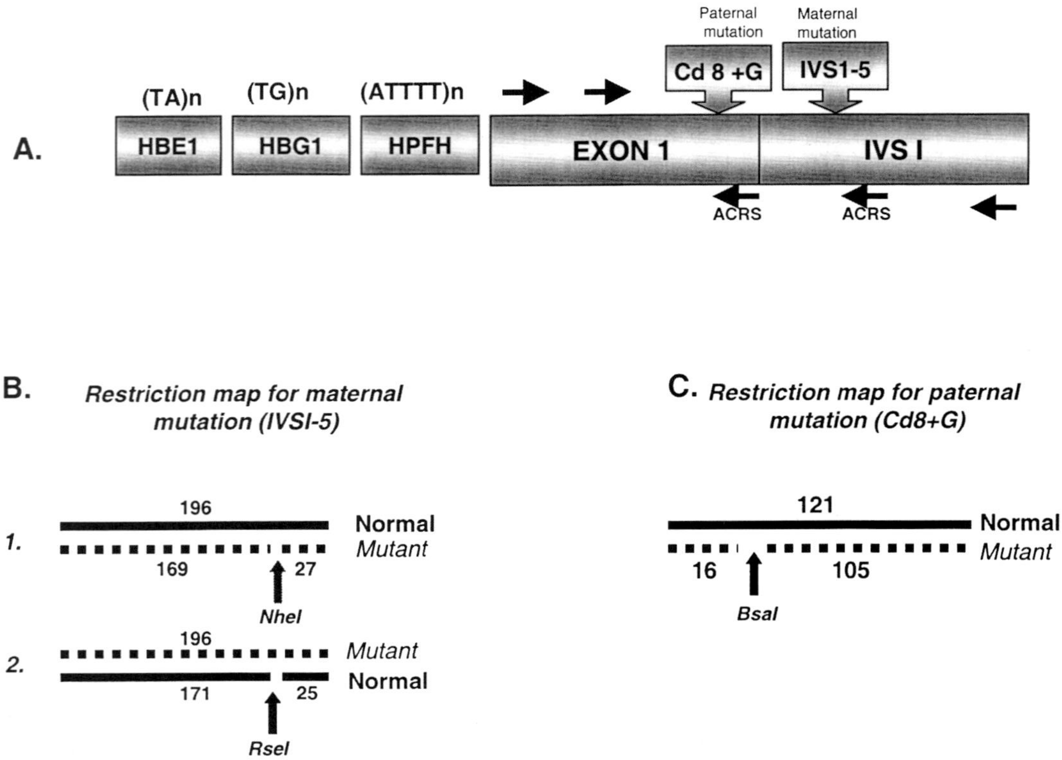

Figure 4.1. PGD Design for two thalassemia mutations in combination with HLA typing (see also Figure 4.4). (*A*) Position of two different mutations and informative linked polymorphic markers used in PGD. *Arrows* represent heminested primers designed for the mutations analysis. Application of outside primers for testing both maternal and paternal mutation improves the ADO detection rate. Two different artificially created restriction site primers (ACRS) were used in the second round of heminested PCR to detect the maternal and paternal mutations. (*B*) Restriction map for the maternal mutation and (*C*) restriction map for the paternal mutation. The mentioned artificially created restriction site primers (ACRS) introduces *Bsa* I restriction site to the mutant paternal sequence. Two different ACRS primers were designed to detect maternal mutation, one introduced *Nhe* I restriction site to the mutant sequence, and the other introducing *Rsa* I restriction site to the normal beta globin sequence, both increasing the reliability of the mutation analysis to overcome potential problems due to incomplete digestion of the PCR product.

to the confirmatory analysis using the genomic DNA from these embryos to evaluate the accuracy of the single-cell-based PGD.

Overall, of 430 oocytes or embryos tested in 47 PGD cycles performed for 27 couples carrying the above-mentioned mutations (approximately 9.2 per cycle), 400 were with results (8.5 per cycle), from which 282 unaffected embryos (6 per cycle) were detected. Although the overall number of HLA-matched embryos was 72 (18%), only 55 were also unaffected (13.7%), of which 53 developed appropriately to be acceptable for transfer in 34 of 47 PGD cycles (1.6 embryos per transfer on the average). This yielded eight unaffected HLA-matched pregnancies (23.5% per transfer), and birth of six healthy children HLA matched to affected siblings (see Table 4.1), including two free from FA, two free from thalassemia, one free from HIGM, and one free from HED-ID, following confirmation of PGD by amniocentesis (one pregnancy is still ongoing).

The largest group of PGD cycles (31 cycles for 18 couples) was performed for thalassemias. Of 311 oocytes and embryos tested, 289 were with conclusive results, of which 85 were predicted to be homozygous mutant and 204 unaffected carriers or normal. The HLA typing revealed 51 (17.6%) embryos to be matched to the affected siblings, but only 41 were also unaffected (14.2%), of which 38 developed appropriately to be acceptable for transfer in 22 of 31 PGD cycles, resulting in three unaffected HLA-matched pregnancies. One of these

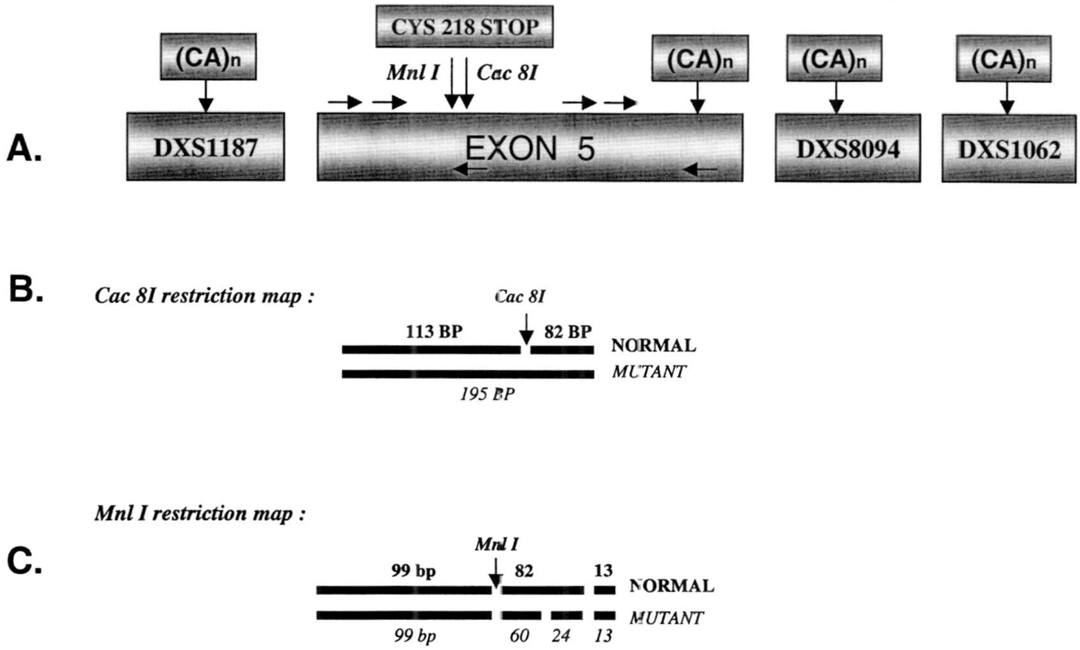

Figure 4.2. PGD design for HIGM in combination with HLA typing (see also Figure 4.5). (*A*) Position of the C218X mutation in exon 5 of CD40 ligand gene (Xq26.3) and tightly linked dinucleotide polymorphic markers inside the gene (exon 5) and outside the gene (DXS1187, DXS8094, DXS1062). Horizontal arrows represent primers positions. Vertical arrows indicate the location of *MnlI* and *Cac8I* restriction sites, and the positions of the dinucleotide polymorphic markers. (*B*) Restriction map of the *CacI* restriction digestion. (*C*) Restriction map for the MnI I digestion.

pregnancies resulted in a stillbirth, and the other two in birth of two unaffected HLA-matched children, one of which is presented in Figure 4.4. As parents were carriers of different mutations (IVSI-5 and Cd8 + G) in this case, both PB and blastomere analysis were performed so that mutation-free oocytes may be selected (oocytes #1, 2, 3, 6, 8, and 9) prior to blastomere analysis, used

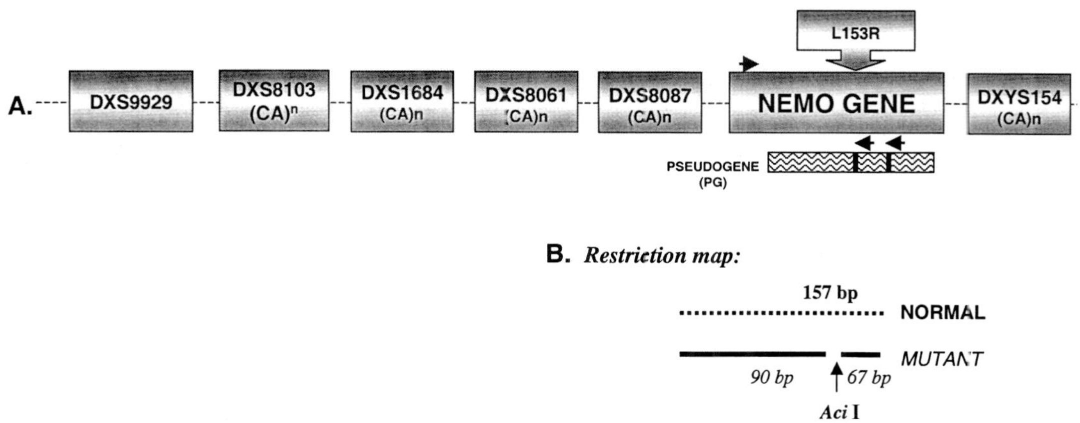

Figure 4.3. PGD design for the mutation in NEMO gene in combination with HLA typing (see also Figure 4.6). (*A*) Position of the L153R mutation in exon 5 of CD40 ligand gene (Xq26.3) and tightly linked dinucleotide polymorphic markers inside the gene (exon 4). (*B*) Restriction maps for *AciI* restriction enzyme.

Table 4.2. List of mutations and primers used for PGD of FAA, WAS, XALD, HYGM and HED-ID combined with HLA typing

Gene/Polymorphism	Upper Primer	Lower Primer	Annealing Temp.
WAS(WASP gene) Leu39Pro	5′ TCAGCAGAACATACCCTCC 3′	5′ AGAGAGAGAAGGAGGAGAGG 3′	62-45°C
Heminested PCR ScrF I cuts mutant allele	5′ TCAGCAGAACATACCCTCC 3′	5′ GAAGAAACGGTGGGGAC 3′	53°C
WASP-DXS6940	5′ TCAAATTGTGGGACGTACAC 3′	5′ GGTCCCCGTTCTTTTTG 3′	62-45°C
Heminested FL-PCR	5′ FAM CAACAAACAGACCAGGGAC 3′	5′ GGTCCCCGTTCTTTTTG 3′	53°C
HED-ID (NEMO-IKBKG gene) Leu153Arg (T—G)	5′ GCTGACAGGAAGTGGCTTTTTA 3′	5′ CTCATTGCTTTTGGAAACCCT 3′	62-45°C
Aci I cuts mutant allele Heminested PCR	5′ GCTGACAGGAAGTGGCTTTTTA 3′	5′ CGACTCACCGACCCTCCA 3′	60°C
HED-ID – DXYS154	5′ ACTCTCACCTATCCTATTCAACTTA 3′	5′ AAGTGATCCTCCCGCTTC 3′	62-45°C
Heminested FL-PCR	5′ HEX ACATGATATTATATGTAGAAAATCC 3′	5′ AAGTGATCCTCCCGCTTC 3′	50°C
HED-ID – DXS1684	5′ AGCACCCAGTAAGAGACTG 3′	5′ TGAATCAATCTATCCATCTCTC 3′	62-45°C
Nested FL-PCR	5′ FAM CAGGCCACTACCACTTATG 3′	5′ TACTGTTTTCCACTCTAATGC 3′	55°C
HED-ID – DXS8103	5′ GTGAAGCCAAGGTGGGAGGAT 3′	5′ GCCCTGGGGTACACAAGCC 3′	62-45°C
Heminested FL-PCR	5′ HEX CACAGGCGTTCAAAACCAGC 3′	5′ GCCCTGGGGTACACAAGCC 3′	55°C
HED-ID – DXS8087	5′ TGAGGCAGGGCGCACTTG 3′	5′ CAGGAGGCCGTGTGAGAGC 3′	62-45°C
Heminested FL-PCR	5′ TGAGGCAGGGCGCACTTG 3′	5′ FAM GGCTGCGCCAGTGAACAA 3′	55°C
HED-ID – DXS8061	5′ GTCCATGAGGTATCCAAACAGG 3′	5′ GTCACAGTGACTCCATCCCAT 3′	62-45°C
Heminested FL-PCR	5′ GTCCATGAGGTATCCAAACAGG 3′	5′ HEX GCTTGTAATCTCCAACCCCTA 3′	55°C
HED-ID – DXS9929	5′ AGAGGGAGGCCAGCCATC 3′	5′ CCTTGGTGACATCGCTGTA 3′	62-45°C
Nested FL-PCR	5′ CACAGGCCTCCTGCATGATG 3′	5′ HEX TTCCAGGCTGGGGCTGCAC 3′	55°C
X-ALD (ABCD1 gene) G343D (G—A) Heminested PCR	5′ TTCTGGAACGCCTGTGGT 3′	5′ CTGGGTCTCACCTGACTCTG 3′	62-45°C
Hae III cuts normal allele	5′ TTCTGGAACGCCTGTGGT 3′	5′ TGGCAGTGATGATGGGGA 3′	55°C
ABCD1-DXS1073	5′ GAAACTTAGAGGGTTGGCTT 3′	5′ CCCCAAAGAATGCCCT 3′	62-45°C
Heminested PCR	5′ ACACTGCTCCCCTTGCC 3′	5′ HEX CCGAGTTATTACAAAGAAGCAC	55°C
HIGM (CD40 gene) Cys218Stop (C—A) Heminested PCR	5′ TTCGAGTCAAGCTCCATTTATAG 3′	5′ AAAGGACGTGAAGCCAGTG 3′	62-45°C
Cac 8I cuts Normal allele, Mnl I cuts mutant allele	5′ AGCCTCTGCCTAAAGTCCC 3′	5′ AAAGGACGTGAAGCCAGTG 3′	55°C
CD40-DXS8094	5′ AATGTCACCATTAACAGTTTGG 3′	5′ GTAGCAACCCCATCAAACTAAA 3′	62-45°C
Heminested FL-PCR	5′ AATGTCACCATTAACAGTTTGG 3′	5′ HEX CTTGTCCAGGACAGTTACATGA 3′	55°C
CD40-DXS1187	5′ TTAGAGCAGAGGTTTCTAGTCTTTC 3′	5′ GAGAAAGTCACTGAACAGAGGAGTT 3′	62-45°C
Heminested FL-PCR	5′HEX TCCTTTCTACCCACTTTTTTCAA 3′	5′ GAGAAAGTCACTGAACAGAGGAGTT 3′	55°C
CD40-DXS1062	5′ CCCACCTAATAGGATTTTATGAA 3′	5′ CTAGTATTTCAAGAGCCAATGATTA 3′	62-45°C
Heminested FL-PCR	5′ CCCACCTAATAGGATTTTATGAA 3′	5′ FAM CCTCAGTCAGTTGCCTGTTAAG 3′	55°C
3′ CD40-(CA)n	5′ TTCCAATCTCTCTTTCTCAATCTCT 3′	5′ AACTGACTAGCAACGGCCTG 3′	62-45°C
Heminested FL-PCR	5′ FAM TGTCAGTCTCTTCCCTCCCC 3′	5′ AACTGACTAGCAACGGCCTG 3′	63°C
FAA (FANCA gene)	5′ AGCCAACGGGTGTGCG 3′	5′ GCTCTGGCGGGAAGGGA 3′	62-45°C
Met1Lys (T – A) Exon1 Nested PCR	5′ CAATAGGAAGGCAGCGCG 3′	5′ GAGTTCGGGACCCACGAG 3′	47°C
FAA (FANCA gene)	5′ GATGGTGGGTTTCTCCGC 3′	5′ GGTCCTGATGGCTTCGCA 3′	62-45°C
Exon 2 – 14 bp deletion Heminested PCR	5′ GGGACCCCGTGTGTGAAT 3′	5′ GGTCCTGATGGCTTCGCA 3′	
FAA (FANCA gene) Thr1131Ala (A—G) Exon 34 Fnu4H I cuts mutant allele	5′ CTGCCTTGAACTCTTTTGC 3′	5′ GGAGGTCAGCGGTTTGTG 3′	62-45°C
TspRI cuts normal allele Heminested PCR	5′ CTGCCTTGAACTCTTTTGC 3′	5′ TCAGCGGTTTGTGAGGAC 3′	55°C
FAA-D16S520	5′ CGGTATTAGCAATCAGGG 3′	5′ CTTACCAACCTCCACAGC 3′	62-45°C
Heminested FL-PCR	5′ FAM TTTCCAGAGAAAAAGAACAC 3′	5′ CTTACCAACCTCCACAGC 3′	55°C
FAA-D16S3021	5′ TATTCCCCACAGTGCTAAG 3′	5′ AGCTTTTATTTCCCAGGG 3′	62-45°C
Heminested FL-PCR	5′ HEX TGCCACATCTGGTCACTTA 3′	5′ AGCTTTTATTTCCCAGGG 3′	55°C
FAA-D16S3026	5′ CTCCCTGAGCAACAAACA 3′	5′ CGCCTGATTTAGGCTTT 3′	62-45°C
Heminested FL-PCR	5′ FAM CGTGGCTTCACTAAAACA 3′	5′ CGCCTGATTTAGGCTTT 3′	55°C
FAA-D16S3407	5′ TCAGCCACACAGATAAACC 3′	5′ TTGTTTCCAATGTGTCTCC 3′	62-45°C
Heminested FL-PCR	5′ HEX TGAGTTTGTTTTTTTCTGCT	5′ TTGTTTCCAATGTGTCTCC 3′	55°C
FAA-D16S408	5′ TCTTATCCCAGGAACCC 3′	5′ CATCTGAGCAACCACCA 3′	62-45°C
Heminested FL-PCR	5′HEX CATGGTGATTGGGTCAAG 3′	5′ CATCTGAGCAACCACCA 3′	55°C

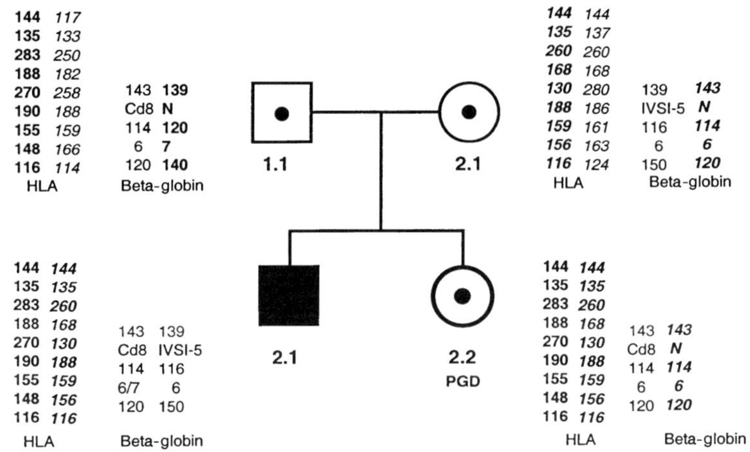

Embryo #	1	2	3	4	5	6	8	9
Beta-globin markers:								
BSTR	139 143	143 143	139 143	143 139	139 139	143 143	143 143	143 143
Mutation site	N N	Cd8 N	N N	Cd8 IVSI-5	N IVSI-5	Cd8 N	Cd8 N	Cd8 N
BTG	120 114	114 114	120 114	114 116	120 116	114 114	114 114	114 114
HPFH	7 6	6 6	7 6	6 6	7 6	6 6	6 6	6 6
D11S1997	140 120	120 120	140 120	120 150	140 150	120 120	120 120	120 120
HLA markers:								
D6S276	117 144	144 144	117 144	144 144	144 144	144 144	117 144	117 144
D6S258	133 135	135 135	133 137	135 137	135 137	135 135	133 135	133 135
D6S248	250 260	283 260	250 260	283 260	283 260	283 260	250 260	250 260
Mog A	182 168	188 168	182 168	188 168	188 168	188 168	182 168	182 168
RF	258 130	270 130	258 280	270 280	270 280	270 130	258 130	258 130
D6S1629	188 188	190 188	188 186	190 186	190 186	190 188	188 188	188 188
Ring 3CA	159 159	155 159	159 161	155 161	155 161	155 159	159 159	159 159
D6S1560	166 156	148 156	166 163	148 163	148 163	148 156	166 156	166 156
D6S291	114 116	116 116	114 124	116 124	116 124	116 116	114 116	114 116
	NORMAL Nonmatch	CARRIER Full match	NORMAL Nonmatch	AFFECTED Nonmatch	CARRIER Nonmatch	CARRIER Full match	CARRIER Nonmatch	CARRIER Nonmatch
		ET				ET		

Figure 4.4. Preimpantation HLA matching combined with PGD for thalassemia. (*Top*) Family pedigree with HLA haplotype analysis based on parental (1.1 and 1.2) and affected child's (2.1) genomic DNA testing. HLA markers order is presented on the upper left for father and upper right for mother. Paternal and maternal matching HLA haplotypes are shown in **bold**. (*Middle*) Maternal mutation and linked polymorphic markers were first assessed by sequential multiplex polar body analysis. Two oocytes (# 4 and 5) had affected allele, while the remaining six (#1, 2, 3, 6, 8, and 9) were normal (data not shown). Based on blastomere results, one embryo was affected (#4), two were homozygous normal (#1 and 3), and five were carriers of paternal (#2, 6, 8 and 9) or maternal (#5) mutations. As seen from HLA typing below, embryos #2 and 6 are also fully HLA-matched to the sick sibling (2.1). (*Bottom*) HLA typing by short tandem repeats along with mutation analysis was performed on blastomeres from eight embryos, two of which (#2 and 6) were predicted to be HLA matched to that of the affected sibling (2.1), although carrying the paternal mutation (also see above). Prenatal testing confirmed these results, and a healthy baby girl with HLA type matching that of the sick sibling was born. Cord blood stem cells were collected during the delivery and transplanted to the sibling (2.1) resulting in a successful hemapoietic reconstitution.

mainly for selection of the paternal mutation-free embryos (embryos #1, 3, and 5) and HLA matching, simultaneously with linked marker analysis (see Figure 4.4). In addition, HLA typing was also done in PB1 and PB2 to identify the oocytes with maternal HLA match, which is useful for interpretation of HLA matching results in bastomeres. Only embryos #2 and 6 appeared to be HLA matched to the affected sibling, and were transferred, resulting in birth of a healthy child, who was confirmed to be thalassemia-free as well as HLA matched to the affected sibling with thalassemia (see Figure 4.4). The diagnosis of unaffected HLA-matched status was also confirmed in other two children, including the stillbirth, resulting from the other clinical pregnancies mentioned (data not shown).

Of 73 embryos tested in 11 cycles for FANCA and FANCC, 68 embryos were with conclusive data, of which 20 were predicted to contain two copies of the mutant gene, 48 were either homozygous normal or heterozygous, and nine of these (13.2%) were also HLA matched to the affected siblings in the family. These embryos, plus one with only maternal HLA match, were transferred in 8 of 11 cycles, resulting in the birth of one full HLA-matched child and one maternal-matched child free of FA. The result of the successful cord blood transplantation in one of the cases was reported earlier, and the other one is still in preparation [2].

Of the remaining five PGD cycles, performed for couples at risk for WAS, X-ALD, HIGM and HED-ID, the last two resulted in births of unaffected HLA-matched children (Figures 4.5 and 4.6). In one of them performed for HIGM, of 15 oocytes tested by PB1 and PB2, 5 of 11 oocytes with conclusive results appeared to be free of maternal mutation, but only one was maternal HLA matched (embryo #2 in Figure 4.5). In addition, three of five oocytes with maternal mutation were HLA matched (see embryos #11, 13, and 15 in Figure 4.5). However, embryos 13 and 15 were affected and non-paternal matched, while only maternal mutant chromosome was detected in embryo #11. Only one embryo (embryo #2), predicted to be mutation-free and maternal matched by PB analysis, appeared to be a normal female with also a paternal match. The transfer of this single embryo resulted in a singleton pregnancy, confirmed to be unaffected and HLA matched by amniocentesis, yielding birth of a healthy, HLA-matched baby girl.

In a single cycle performed for HED-ID, 16 embryos were with conclusive results, of which six were free of maternal mutation, based on PB1 and PB2 testing, but none of these were maternal HLA matched. As seen from Figure 4.6, of 16 resulting embryos, for which blastomere biopsy results were available both for mutation analysis and HLA typing, three were affected males (embryos #17, 20, and 21; only the last being HLA matched), four females carriers, two of which were nonmatched (embryos #3 and 4), one HLA recombinant (embryo #13), and one HLA matched (embryo #12). The remaining seven embryos were unaffected, including two male nonmatched embryos (embryos #16 and 24), the former containing extra maternal X-chromosome, and five normal female embryos, of which only one (embryo #26) was HLA matched. This embryo, together with the embryo #12, which was a normal female carrier, were transferred, resulting in a singleton pregnancy and birth of an unaffected child who was confirmed to be HLA matched to the affected sibling. The other normal embryos, which were not HLA matched to the affected sibling, were frozen for future use by the couple.

The presented data demonstrate the feasibility of practical application of PGD with HLA typing, despite the expected low probability of selecting HLA matched unaffected embryos. As predicted in the first case report on PGD with HLA typing for FA, the approach appeared extremely attractive for those couples who have an affected child with lethal congenital disease, for whom HLA-matched stem cell transplantation is the only hope [20, 22]. Overall, eight unaffected HLA-matched pregnancies have been established, representing the world's first practical experience in the field, complementing the most recent experience on preimplantation HLA typing without testing for causative gene [20]. Because the embryo selection in the latter series was based solely on HLA matching, 23% of the embryos were transferred (see Section 4.3), in agreement with the predicted probability, in contrast to the present data, involving the search for the HLA-matched embryos among only those that were also unaffected. Although the sample size is too small for comparisons, only 13.7% of the HLA-matched embryos were also unaffected, which is lower than the expected 18.7%, and of course much lower than that in a sole preimplantation HLA typing [20].

A.

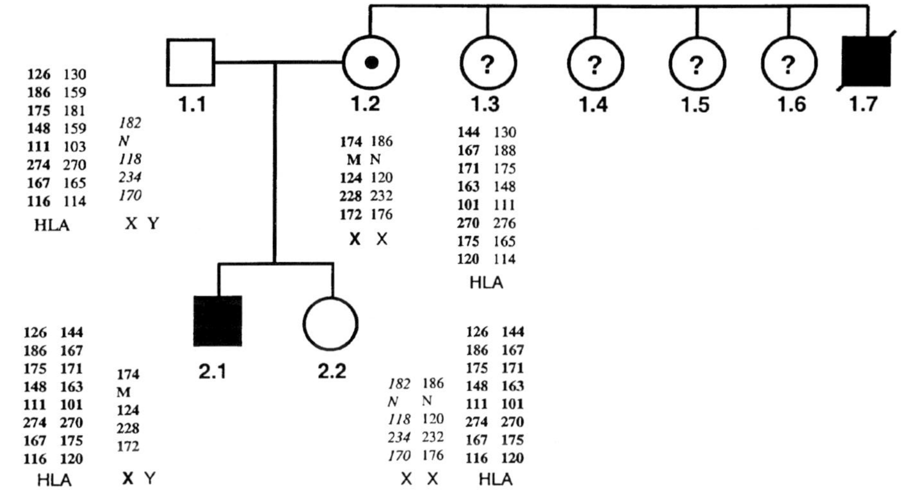

B.

Embryo #	1	2	3	4	5	6	7	9	10	11	13	15
X-chromosome markers:												
DXS1187	186	182 186	182 186	182 174	182 174	186	182 186	186	182 186	174	174	174
CD40	N	N N	N N	N M	N M	N	N N	N	N N	M	M	M
(CA)n	120	118 120	118 120	118 124	118 124	120	118 120	120	118 120	124	124	124
DXS8094	232	234 232	234 232	234 228	234 228	232	234 232	232	234 232	228	228	228
DXS1062	176 Y	170 176	170 176	170 172	170 172	Y 176	170 176	Y 176	170 176	172	Y 172	Y 172
HLA markers:												
D6S276	130 130	126 144	126 130	130 130	126 130	126 130	130 144	126 130	126 130	126 144	130 144	130 144
Mog a	159 188	186 167	186 188	159 188	186 188	186 188	159 167	186 188	186 188	186 167	159 167	159 167
D6s265	181 175	175 171	175 175	181 175	175 175	175 175	181 171	175 175	175 175	175 171	181 171	181 171
D6S510	159 148	148 163	148 148	159 148	148 148	148 148	159 163	148 148	148 148	148 163	159 163	159 163
TNF a	103 111	111 101	111 111	103 111	111 111	111 111	103 101	111 111	111 111	111 101	103 101	103 101
D6S273	270 276	274 270	274 276	270 276	274 276	274 276	270 270	274 276	274 276	274 270	270 270	270 270
LH1	165 165	167 175	167 165	165 165	167 165	167 165	165 175	167 165	167 165	167 175	165 175	165 175
D6S291	114 114	116 120	116 114	114 114	116 114	116 114	114 120	116 114	116 114	116 120	114 120	114 120
Predicted genotype	Normal male Nonmatch	Normal female Match ET	Normal female Nonmatch	Carrier Nonmatch	Carrier Nonmatch	Normal male Nonmatch	Normal female Nonmatch	Normal male Nonmatch	Normal female Nonmatch	*X O Match*	Affected Male Nonmatch	Affected Male Nonmatch

Figure 4.5. Preimplantation HLA typing combined with PGD for X-linked HIGM syndrome. (*A*) Family pedigree in three generations. Marker order is located next to maternal haplotypes. Paternal (2.1), maternal (2.2), and the affected sibling (3.1) CD 40 gene haplotype assignment is based on genomic DNA testing. Paternal and Maternal matching HLA haplotypes are shown in **bold**. (*B, upper*) PCR analysis of blastomeres removed from 12 embryos showed that all but three embryos (#11, 13, and 15) were predicted to be unaffected. (*B, lower*) HLA typing was performed simultaneously with mutation analysis of all blastomeres. Embryo #2 was predicted to be normal female and to have the same HLA profile as the affected sibling (3.1). The transfer of this embryo resulted in pregnancy and birth of a healthy unaffected HLA-matched baby girl (3.2). Cord blood stem cells were collected at birth for stem cell transplantation.

A.

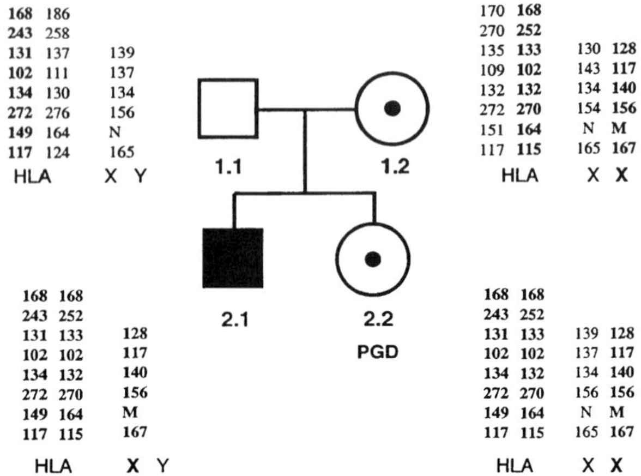

B.

Embryo #	2	3	4	6	8	12	13	16	17	18	20	21	24	25	26	27
X-markers:																
DXS 9929	139 130	139 **128**	139 **128**	139 130	139 130	139 **128**	139 **128**	139 130	**128**	130	**128**	**128**	130	139 130	139 130	139 130
DXS8103	137 143	137 **117**	137 **117**	137 143	137 143	137 **117**	137 **117**	137 143	**117**	143	**117**	**117**	143	137 143	137 143	137 143
DXS1684	134 134	134 **140**	134 **140**	134 134	134 134	134 **140**	134 **140**	134 134	**140**	134	**140**	**140**	134	134 134	134 134	134 134
DXS8087	156 154	156 **156**	156 **156**	156 154	156 154	156 **156**	156 **156**	156 154	**156**	154	**156**	**156**	154	156 154	156 154	156 154
NEMO Gene	N N	N **M**	N **M**	N N	N N	N **M**	N **M**	N N	**M**	N	**M**	**M**	N	N N	N N	N N
DXYS154	165 165	165 **167**	165 **167**	165 165	165 165	165 **167**	165 **167**	**Y** 165 165	**Y** **167**	**Y** 165	**Y** **167**	**Y** **167**	**Y** 165	165 165	165 165	165 165
HLA markers:																
MOG a	186 170	**168** 170	186 170	**168** 170	186 170	**168** **168**	**168** **168**	**168** 170	**168** 170	186 170	186 **168**	**168** **168**	186 170	186 170	**168** **168**	**168** 170
RF	258 270	**243** 270	258 270	**243** 270	258 270	**243** **252**	**243** **252**	**243** 270	**243** 270	258 270	258 **252**	**243** **252**	258 270	258 270	**243** **252**	**243** 270
D6S1568	137 135	**131** 135	137 135	**131** 135	137 135	**131** **133**	**131** **133**	**131** 135	**131** 135	137 135	137 **133**	**131** **133**	137 135	137 135	**131** **133**	**131** 135
TNF	111 109	**102** 109	111 109	**102** 109	111 109	**102** **102**	**102** **102**	**102** 109	**102** 109	111 109	111 **102**	**102** **102**	111 109	111 109	**102** **102**	**102** 109
9N2	130 132	**134** 132	130 132	**134** 132	130 132	**134** **132**	**134** **132**	**134** 132	**134** 132	130 132	130 **132**	**134** **132**	130 132	130 132	**134** **132**	**134** 132
D6S273	276 272	**272** 272	276 272	**272** 272	276 272	**272** **270**	**272** **270**	**272** 272	**272** 272	276 272	276 **270**	**272** **270**	276 272	276 272	**272** **270**	**272** 272
LH 1	164 151	**149** 151	164 151	**149** 151	164 151	**149** **164**	**149** 151	**149** 151	**149** 151	164 151	164 **164**	**149** **164**	164 151	164 151	**149** **164**	**149** 151
D6S291	124 117	**117** 117	124 117	**117** 117	124 117	**117** **115**	**117** 117	**117** 117	**117** 117	124 117	124 **115**	**117** **115**	124 117	124 117	**117** **115**	**117** 117
Predicted genotype	Normal Nonmatch	Carrier Nonmatch	Carrier Nonmatch	Normal Nonmatch	Normal Nonmatch	**Carrier Match**	Carrier Recombinant	**XXY** Nonmatch	Affected Nonmatch	Normal Nonmatch	Affected Nonmatch	**Affected Match**	Normal Nonmatch	Normal Nonmatch	**Normal Match**	Normal Nonmatch
						ET									ET	

Figure 4.6. PGD for the mutation in NEMO gene with preimplantation HLA typing. (*A*) Family pedigree showing maternal and paternal matching HLA haplotypes in **bold**. Marker order for testing NEMO gene is located next to maternal haplotypes. Paternal (1.1), maternal (1.2), and the affected sibling (2.1) NEMO gene haplotype assignment is based on genomic DNA testing. (*B, upper*) Results of blastomere DNA analysis from 16 embryos showing that three embryos were affected (embryos #17, 20, and 21), one with extra X chromosome, suggesting the XXY genotype, with the remaining being either carriers or unaffected. (*B, lower*) HLA typing was performed simultaneously with mutation analysis of all blastomeres, showing that two of the unaffected embryos (embryos #12 and 26) were also HLA matched to the affected sibling. The transfer of these embryos resulted in pregnancy and birth of a healthy unaffected HLA matched baby girl (2.2). Cord blood stem cells were collected at birth and transplanted to affected sibling (2.1) resulting in a successful hematopoietic reconstitution.

This relatively low number of unaffected HLA-matched embryos detected in the present series may be influenced by advanced reproductive age of the women involved (approximately 35 years), known to affect significantly the number of available embryos for testing and also the success rate of assisted reproduction. The mean number of embryos available per cycle was under 10 from these women, which clearly limits the chances for finding a sufficient number of unaffected embryos that are also HLA matched to a sibling. Assuming that one in four embryos is expected to be HLA matched and three of four unaffected, the overall probability of HLA-matched embryos to be also unaffected could not be expected to be higher than 1 in 5.3 embryos. So with the availability of only 8.5 embryos on the average with conclusive results in our material, only 1.6 HLA-matched unaffected embryos may have been detected. Assuming also that not all embryos develop to the status acceptable for transfer, approximately one embryo per cycle could have been available for transfer, which is of course below the optimal number of embryos to be replaced to ensure a clinical pregnancy and birth outcome. Overall, 53 unaffected HLA-identical embryos were transferred in 34 of 46 cycles performed (1.6 on the average), which resulted in 23.5% pregnancy rate, representing the range expected in women of advanced reproductive age.

The other limitation associated with advanced reproductive age is a higher prevalence of aneuploidies, which should be tested at least for those chromosomes in which the causative genes are located, and for chromosome 6, to which HLA genes are assigned, as this may affect the accuracy of PGD and HLA typing. Testing for chromosome 6 and for those chromosomes in which the causative genes for the above conditions are localized, revealed 39 (9.8%) embryos with aneuploidies, including a total of 14 trisomies, 24 monosomies, and 1 uniparental disomy, which may have affected the accuracy of diagnosis if not detected. This suggests that aneuploidies should be routinely detected by adding primers for microsatellite markers for chromosome 6 and other chromosomes in which causative gene is localized to the multiplex PCR system. This allows detecting the copy number of chromosomes as well as uniparental disomies of chromosome 6, which may result in misdiagnosis of the HLA matching.

Finally, one of the important phenomena affecting the accuracy of preimplantation HLA typing is a possible recombination in the HLA region, which may lead to misdiagnosis if not detected. As demonstrated in this study, the microsatellite markers linked to the HLA alleles may be tested simultaneously in the multiplex PCR system, allowing detection of recombination in the HLA region, observed in 4.3% of cases in our data (data not shown). Because preimplantation HLA typing may never be able to identify the HLA match for recombinant siblings, haplotype analysis prior to initiation of the actual cycle is required so that the couples may be informed about their possible options. For example, in one of our cases performed for thalassemia, the fact that the child was recombinant became obvious only during PB1 analysis, without which maternal hyplotypes cannot be established. While paternal haplotypes may be identified through sperm typing, the testing for maternal haplotypes requires the maternal somatic cell haploidization, which may be performed by somatic cell nuclei transfer and fusion with matured oocytes, as described in Chapter 2 [23].

Despite the need for further improvement of the technique as mentioned, the presented results shows that couples with affected children requiring HLA-compatible stem cell transplantation have a realistic option to undergo IVF and PGD with combined preimplantation HLA typing, so as to have an unaffected HLA-matched child as potential donor of compatible stem cells for sibling. Together with the previously reported application of PGD for FANCC, and the most recent report on preimplantation HLA typing not involving testing for causative gene, at least 10 HLA-matched children have already been born, confirmed to be HLA matched for affected siblings. Although not all these sibling have yet been treated, preliminary results provide encouraging results. At the present time, umbilical cord blood stem cells collected from these children have already been transplanted to the siblings with FANCC [2] thalassemia, HED-ID, and Diamond-Blackfan anaemia [20], resulting in a successful hematopoietic reconstitution in all of these conditions, which suggests that it is an acceptable approach for the couples at risk, despite being highly controversial. On the other hand, it also allows overcoming an important controversy of therapeutic cloning, as it avoids the use of surplus human embryos for obtaining embryonic stem cells (see Chapter 8). Also, no embryo is discarded based on the results of preimplantation HLA typing, as all unaffected embryos are frozen for future use by the couple.

So, couples at risk of having children with congenital bone marrow disorders have to be informed about the presently available option not only for avoiding the birth of an affected child, but also for selecting a suitable stem cell donor for their affected siblings, which may presently be the only hope for treating of siblings with congenital bone marrow failures.

However, patients should be fully aware of limits of the expected successful outcome of the above testing, which was shown to result in preselection and transfer of the HLA-matched unaffected embryos in only 13.7% of the embryos tested, which is even a bit lower than may have been predicted. Despite such a relatively moderate success rate, the number of PGD requests in combination with HLA typing has been increasing overall, with the recent emergence of a considerable proportion of cases involving preimplantation HLA typing without PGD. More than a dozen cases have already been performed for HLA typing without PGD, i.e., with the only objective being preselection of an HLA-matched progeny for transplantation treatment of siblings with bone marrow disorders, which are presented in the section below.

4.3 Preimplantation HLA Matching Without PGD

The first experience for preimplantation HLA matching without testing for the causative gene included 13 IVF cycles initiated for nine couples wishing to have another child who may also be a potential cord blood donor for the affected siblings with leukemia, or Diamond-Blackfan anaemia (DBA) [20]. These conditions require bone marrow transplantation and have also been successfully treated by cord blood transplantation [24]. Although mutation testing may also be required in combination with HLA typing in DBA, which can be caused by mutations in the gene encoding ribosomal protein S19 on chromosome 19 (19q13.2) or another gene mapped to chromosome 8 (8p23.3-p22), the majority of DBA are sporadic with no mutation detected, such as in both cases performed in this study [14].

A standard IVF protocol was combined with micromanipulation procedure to remove single blastomeres from the day 3 cleaving embryos, as described above. HLA genes were tested simultaneously with the short tandem repeats in the HLA region [12, 21] using a multiplex heminested PCR system [19, 25]. A haplotype analysis for father, mother, and the affected child was performed for each family prior to preimplantation HLA typing, using a set of polymorphic STR markers located throughout the HLA region, as shown in Figure 4.7. This allowed detecting and avoiding misdiagnosis due to preferential amplification and ADO, exceeding 10% in PCR of single blastomeres [19, 26], potential recombination within the HLA region, and a possible aneuploidy or uniparental disomy of chromosome 6, which may also affect the diagnostic accuracy of HLA typing of the embryo.

Two to three embryos predicted to be HLA compatible with the affected sibling were transferred back to the uterus, while the other HLA-matched embryos were frozen, to be available for future transfer. With consent of the parents, all HLA-incompatible embryos and those with inconclusive results were exposed to PCR analysis of the whole embryo to confirm the blastomere diagnosis. Follow-up analysis was also performed in the established pregnancies using chorionic villus sampling (CVS) or amniocentesis.

At the present time, a total of 237 embryos were tested in 16 clinical cycles from 12 couples (14.8 embryos per cycle on the average), of which 51 were predicted to be HLA matched (21.5%) to the affected siblings (Table 4.3). The HLA-matched embryos were available for transfer in all the cycles. Overall, 33 HLA-matched embryos were transferred in 16 clinical cycles resulting in eight singleton clinical pregnancies and eight HLA-matched children born. These results suggest that testing of approximately 10 embryos per cycle allows preselecting a sufficient number of the HLA-matched embryos for transfer to achieve a clinical pregnancy and birth of an HLA-matched progeny.

The usefulness of detection of recombination within the HLA region is demonstrated in Table 4.4, describing the results of HLA typing of one of the cycles resulting in the birth of an HLA-matched child to a sibling with acute lymphoid leukemia (ALL). Of 10 embryos tested simultaneously for 11 alleles within the HLA region in this family, crossover between D6S426 and Ring alleles was observed in embryos #4, and 9, and between D6S276 and D6S510 in embryo #7 and 9. Of the remaining seven embryos, three were fully matched (embryos #2, 6, and 8), while the other four were HLA-incompatible to the affected sibling, as seen from the haplotypes of the mother, father, and affected child, presented in Table 4.4.

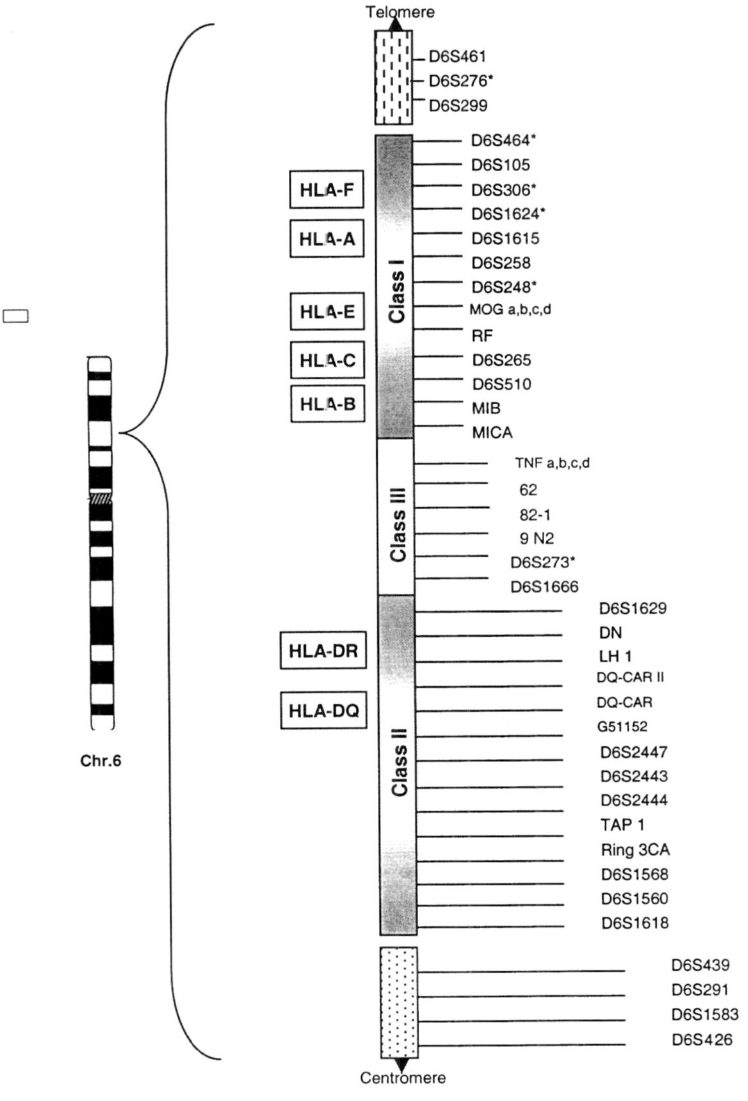

Figure 4.7. Polymorphic markers in HLA region applied for preimplantation HLA typing [20, 21, 28, 34].

As seen from Figure 4.8, presenting the results of preimplantation HLA matching for DBA, one embryo (embryo #8) with maternal and another (embryo #16) with both maternal and paternal crossover (in Ring and S6S426 alleles) were detected in testing of 16 embryos (only eight embryos shown). In addition, there was another embryo with trisomy 6 (embryo #5) with extra maternal

Table 4.3. Results and outcomes of preimplantation HLA typing

Patients/Cycles	Total Embryos Tested	HLA Match	No. Embryos Transferred	No. Transfers	Pregnancy/Birth
12 / 16	237	51	33	16	8 / 8*

*Ongoing pregnancies.

Table 4.4. Preimplantation HLA typing resulting in birth of an HLA-matched baby for affected sibling with acute lymphoid leukemia

HLA Genes and STRs	Embryo No. 1	2	3	4	5	6	7	8	9	10	Father	Mother	Affected baby
D6S276*	*130/118*	**144/139**	*130/118*	*130/118*	*130/* **139**	**144/139**	*130/118*	**144/139**	*130/***139**	*130/* **139**	**144/** *130*	*118/***139**	**144/139**
HLA A	*3/2*	**1/32**	*3/2*	*3/2*	*3/***32**	**1/32**	*FA*	**1/32**	*3/***32**	*3/***32**	**1/***3*	*2/***32**	**1/32**
D6S510	*163/167*	**148/163**	*163/167*	*163/167*	*163/***163**	**148/163**	**148** */167*	**148/163**	*163/***163**	*163/***163**	**148/***163*	*167/***163**	**148/163**
HLA B	*AD 27*	**27/18**	*ADO/27*	*FA*	**FA**	**27/18**	*ADO/27*	**27/18**	*ADO/***18**	*ADO/***18**	**27/** *57*	*27/***18**	**27/18**
MIB	*118/114*	**114/124**	*118/114*	*118/114*	*118/***124**	**114/124**	*118/114*	**114/124**	*118/***124**	*118/***124**	**114/***118*	*114/***124**	**114/124**
TNF a	*94/102*	**110/102**	*94/102*	*94/102*	*94/***102**	**110/102**	*94/102*	**110/102**	*94/***102**	*94/***102**	**110/** *94*	*102/***102**	**110/102**
D6S273*	*275/273*	**273/271**	*275/273*	*275/273*	*275/***271**	**273/271**	*275/273*	**273/271**	*275/***271**	*275/***271**	**273/** *275*	*273/***271**	**273/271**
HLA-DRB1	*ADO/1*	**11/10**	*ADO/1*	*ADO/1*	*ADO/***10**	**11/10**	*ADO/1*	**11/10**	*ADO/***10**	*FA*	**11/** *7*	*1/***10**	**11/10**
G51152	*181/208*	**181/191**	*181/208*	*181/208*	*181/***191**	**181/191**	*181/208*	**181/191**	*181/***191**	*181/***191**	**181/** *181*	*208/***191**	**181/191**
Ring 3CA	*162/162*	**160/155**	*162/162*	*162/162*	*162/***155**	**160/155**	*162/162*	**160/155**	*162/***155**	*162/***155**	**160/** *162*	*162/***155**	**160/155**
D6S426	*144/130*	**140/144**	*144/130*	**140** */130*	*144/***144**	**140/144**	*144/130*	**140/144**	*144/130*	*144/***144**	**140/** *144*	*130/***144**	**140/144**
PREDICTED GENOTYPE	*Nonmatch*	**Match**	*Nonmatch*	*Nonmatch Recombinant*	*Nonmatch*	**Match**	*Nonmatch Recombinant*[c]	**Match**	*Nonmatch Recombinant*	*Nonmatch*	**NA**	**NA**	**NA**

ADO, allele drop out; FA, failed amplification.

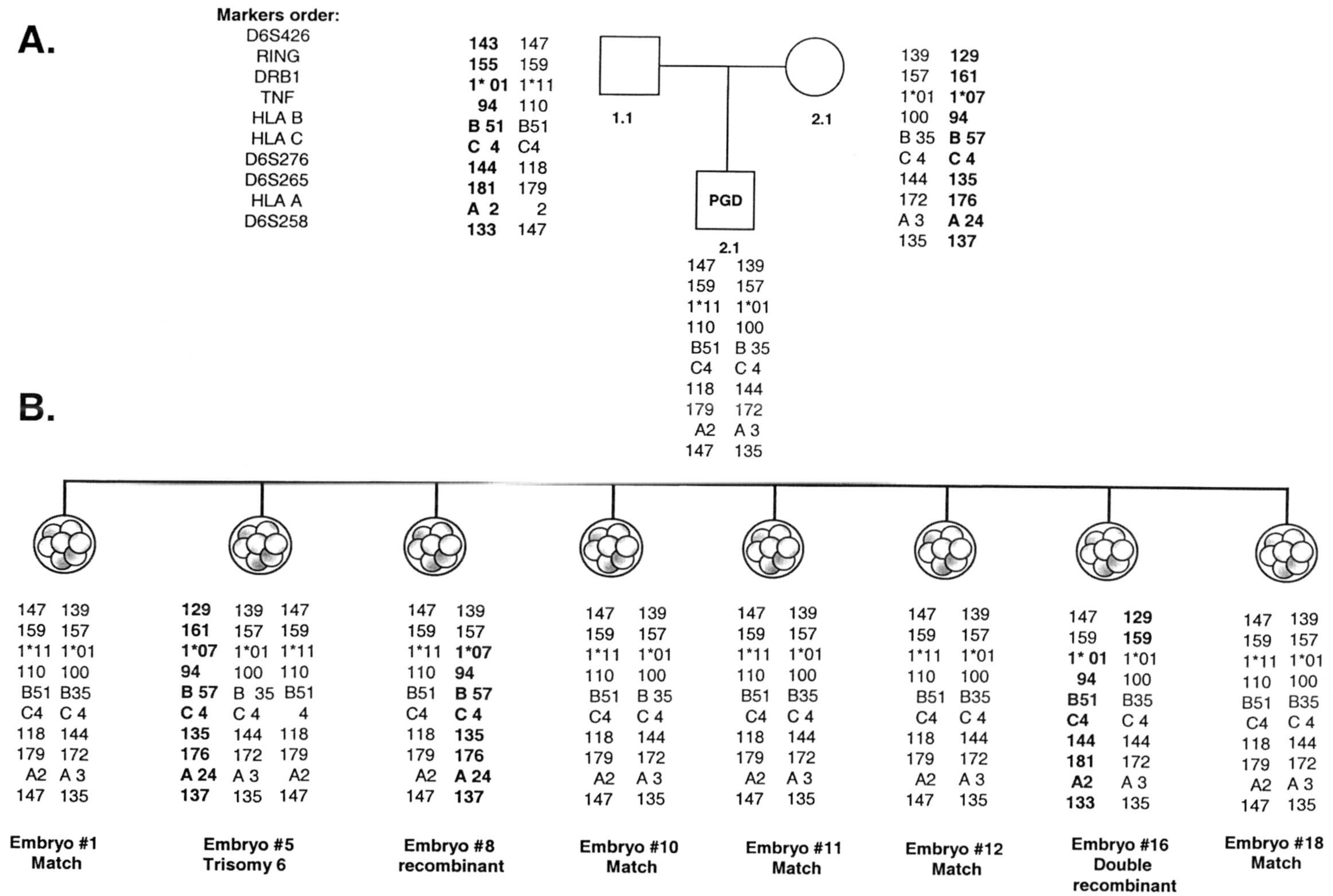

Figure 4.8. Preimplantation HLA typing for Diamond-Blackfan anemia, resulting in birth of HLA-matched child. (*A*) Family pedigree with marker order and haplotypes of mother, father, and the affected child. HLA Matching haplotypes are not bold. (*B*) Results of HLA typing of biopsied blastomeres from eight embryos (other eight embryos that were also tested are not shown). Embryos #1, 10, 11, 12, and 18 are HLA-matched to the affected sibling (see *A*). Embryo #8 is maternal nonmatched due to maternal recombination, as well as embryo #16, which is both paternal and maternal nonmatched, due to double recombination in the paternal and maternal Ring and D6S426 alleles. Embryo #5 is also nonmatched due to extra maternal chromosome, suggesting trisomy 6.

chromosome 6, making this and the other two above also unacceptable for transfer. However, five embryos appeared to be HLA matched, of which two were transferred back to the patient, resulting in the birth of an HLA-matched baby (see Figure 4.8).

The relevance of aneuploidy testing for chromosome 6 is seen from the results of HLA typing in the other cycle, resulting in birth of a baby who was HLA matched to the sibling with ALL (Figure 4.9). Two of ten embryos tested in this case (of which only eight embryos are shown in Figure 4.9) appeared to have only maternally derived chromosome 6, one with only one maternal chromosome (embryo #1), and the other with two maternal chromosomes, representing uniparental maternal disomy of chromosome 6 (embryo #2). In addition, crossover between D6S291 and class II HLA alleles was evident in the embryo #7, making this embryo unacceptable for transfer. Of the remaining seven embryos, only two were HLA matched to the affected sibling and were transferred (embryos #4 and 9), resulting in the birth of an HLA-matched baby.

Of only seven embryos available for testing in another case of DBA, three appeared to be HLA matched to the affected sibling (embryos #2, 3 and 9) and were transferred, resulting in a singleton pregnancy and birth of an HLA-matched baby (Table 4.5). In the other cycle performed for ALL, resulting in the birth of an HLA-matched child, six HLA-matched embryos were preselected from 11 tested, with no evidence for recombination or aneuploidy.

Presented data show the feasibility of preimplantation HLA matching for families with affected children with bone marrow disorders who may wish to have another child as a potential HLA-matched donor for stem cell transplantation treatment of the affected sibling. As seen from our data, HLA-matched embryos were preselected and transferred in all cycles, resulting in clinical pregnancies and birth of HLA-matched children in almost every second transfer cycle. Despite the high rate of preferential amplification and ADO in PCR analysis of single blastomeres, a potential recombination within the HLA region described in our material, and a high rate of mosaicism for aneuploidies at the cleavage stage (see Chapter 5), the introduced approach appeared to be highly accurate in preselecting HLA-matched embryos for transfer. This approach involves multiplex PCR analysis involving simultaneous testing for HLA alleles together with STR markers within HLA and flanking regions, allowing avoidance of misdiagnosis due to ADO, aneuploidy, or recombination of HLA alleles, which cannot be detected so efficiently by other currently used methods for HLA typing. Because of the detected recombination in the HLA region, with a possible few hot spots, this should be detected in order to avoid the risk for misdiagnosis. On the other hand, accummulated data on aneuploidy of chromosome 6 further continue the value of testing for copy number of chromosome 6.

As mentioned, preimplantation HLA matching has initially been used together with PGD for FA to preselect and transfer unaffected embryos, allowing unaffected pregnancy and also producing a potential donor progeny for bone marrow transplantation for the affected sibling with FA. Although controversial, this appeared to be a reasonable option, because only a limited number of the embryos resulting from a hormonal hyperstimulation in IVF are actually selected for transfer (see Chapter 8). Therefore, instead of a selection of embryos for transfer based on morphological criteria, those unaffected embryos representing an HLA match are preselected.

The results also demonstrate the prospect for the application of this approach to other conditions requiring an HLA-compatible donor for stem cell transplantation. This provides a realistic option for those couples who would like to have another child anyway, as they may potentially provide an HLA-matched progeny for an affected sibling. In addition to leukemias and sporadic forms of DBA, the method may be applied for any other condition, such as for couples having affected children with different cancers awaiting an HLA-compatible donor with no success for years. These new indications make preimplantation testing a genuine alternative to conventional prenatal diagnosis, providing patients the important prospect not only to avoid an inherited risk without facing termination of pregnancy, but also to establish a pregnancy with particular genetic parameters to benefit the affected member of the family.

4.4 Worldwide Experience on Preimplantation HLA Typing

Preimplantation HLA typing opens an important possibility of PGD application for stem cell therapy. Because of the limited availability of

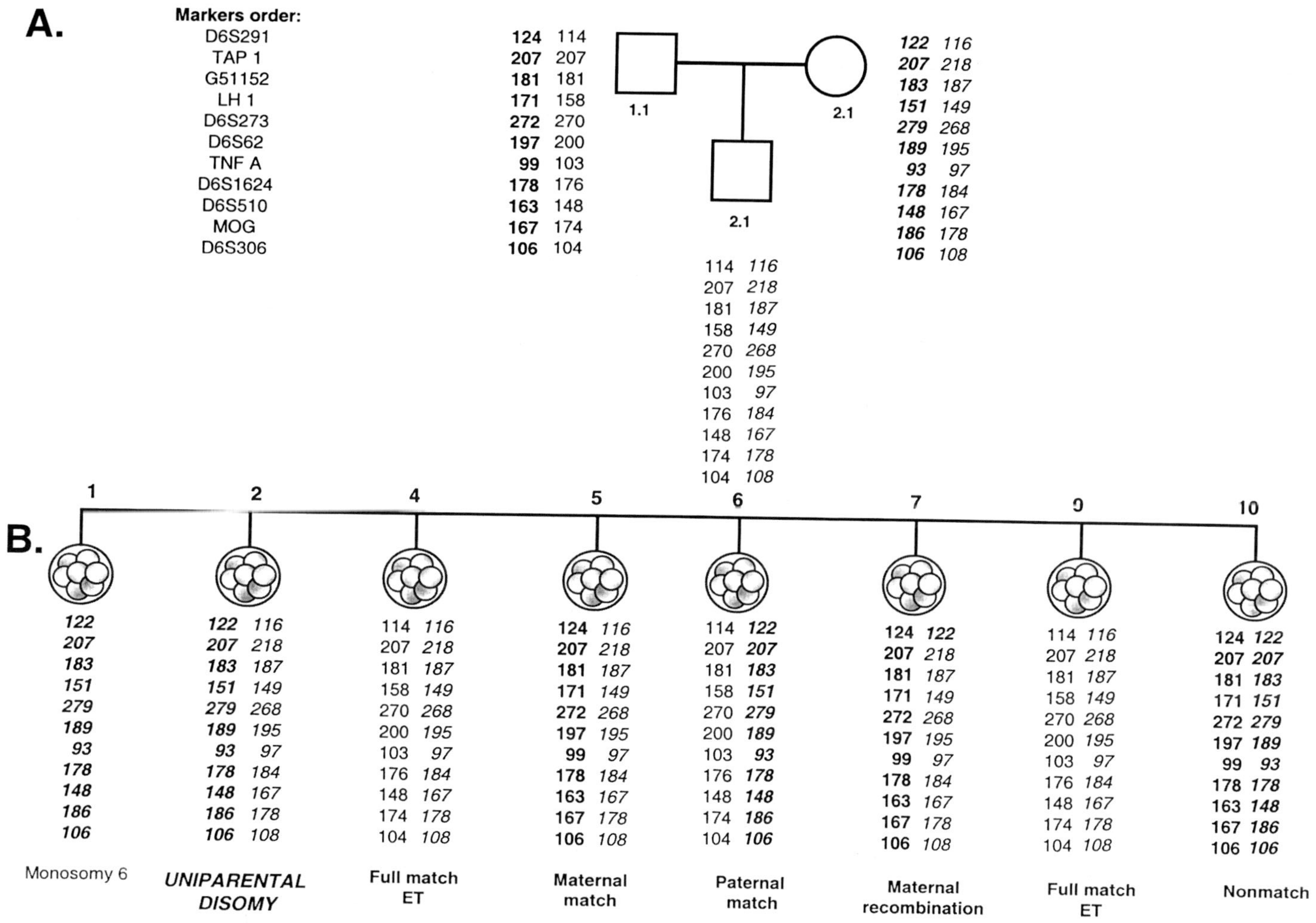

Figure 4.9. Preimplantation HLA typing for ALL, resulting in birth of an HLA baby. (*A*) Family pedigree with marker order and haplotypes of mother, father, and the affected child. Matching maternal and paternal haplotypes are not bold (*B*) Results of HLA typing of biopsied blastomeres from eight embryos. Embryos #4 and 9 are HLA matched to the affected sibling (see *A*). Embryo #1 is maternal nonmatched with no paternal chromosome present (monosomy 6). Embryo #2 is also nonmatched due to only maternal chromosomes present (uniparental disomy). Embryo #7 is both paternal and maternal nonmatched, the latter being due to maternal recombination between 96S291 and class II HLA alleles. Other two embryos are nonmatched, embryos #6 being maternal nonmatch and embryo #10 both maternal and paternal nonmatched.

Table 4.5. Preimplantation HLA typing resulting in birth of an HLA-matched baby for affected sibling with Diamond-Blackfan anemia

HLA Genes and STRs	Embryo No. 2	3	7	8	9	10	11	Father	Mother	Affected Child
D6S461*	225/227	225/227	229/227	229/227	[b] 229/227	229/229	229/229	225/229	229/227	225/229[b]
D6S276*	143/143	143/143	141/143	141/ADO	143/143	141/117	141/117	143/141	117/143	143/143
D6S258	144/132	144/132	135/132	135/132	144/132	135/135	135/135	144/135	135/132	144/132
D6S248*	253/278	253/278	276/278	276/278	253/278	276/278	276/278	253/276	278/278	253/278
MOG a	159/169	159/169	167/169	167/169	159/169	167/169	167/169	159/167	169/169	159/169
RF	275/269	275/269	257/269	257/269	275/269	257/263	257/263	275/257	263/269	275/269
9N-2	129/133	129/133	131/133	131/127/133	129/133	131/127	131/127	129/131	127/133	129/133
D6S273	276/270	276/270	274/270	274/270	276/270	274/270	274/270	276/274	270/270	276/270
LH 1	163/168	163/168	163/168	163/168	163/168	163/179	163/179	163/163	179/168	163/168
D6S2447	147/159	147/159	152/159	152/159	147/159	151/151	151/151	147/151	151/159	147/159
TAP 1	205/220	205/220	205/220	205/220	205/220	205/207	205/207	205/205	207/220	205/220
Ring 3CA	159/155	159/155	155/155	155/159[a]	159/155	155/159	155/159	159/155	159/155	159/155
D6S439	125/125	125/125	123/125	123/125	125/125	123/125	123/125	125/123	125/125	125/125
D6S291	114/123	114/123	116/123	116/114	114/123	116/114	116/114	114/116	114/123	114/123
D6S426	129/129	129/129	144/129	FA	129/142[c]	144/142	144/142	129/144	142/129	129/129
PREDICTED GENOTYPE	Match	Match	*Nonmatch*	*Nonmatch Recombinant*[c]	Match	*Nonmatch*	*Nonmatch*	NA	NA	NA

[a] Maternal recombination in Ring 3CA allele.

[b] Paternal recombination between alleles for D6S461 and D6S276 markers in embryo 9, as well as maternal recombination between the same alleles in the affected child. However, this is outside the HLA region and may not be taken in consideration.

[c] Maternal recombination between D6S291 and D6S426, which is also outside the HLA region, so embryo 9 was considered to be HLA matched to the affected child.

HLA-matched donors even among family members, this approach appeared to be attractive for couples with children requiring HLA-matched bone marrow transplantation [1, 20, 22, 27, 28].

It is well known that to achieve an acceptable engraftment and survival in stem cell therapy requires the finding of an HLA-identical stem cell transplant. However, there remain a large number of patients for which no HLA-matched family member exists, so the search is extended to haplotype-matched unrelated donors. This allowed successful application of stem cell transplantation to some individuals without a matched related donor, despite resulting in severe complications in more than half of the patients [29].

The experience of bone marrow transplantation for haemoglobinopathies presently comprises thousands of patients, showing significant progress of cure in the world's largest centre [30, 31]. The success rate is reported as 87% for Class 1 patients (with regular iron chelation therapy, who have neither hepatomegaly nor liver fibrosis), 85% for Class 2 patients (with regular/irregular chelation, borderline hepatomegaly, and fibrosis), and 80% for Class 3 patients (with irregular chelation, hepatomegaly, and fibrosis) under age 17, suggesting that this may have wider implications for congenital bone marrow failures, depending primarily on the availability if HLA-matched donors.

Due to the small number of children per family, only one third of patients are able to find an HLA-identical sibling, which may further be improved by 3% using an extended family search for a matched related donor with one or two identical ancestral haplotypes [32]. In the remaining patients, the only resort is the identification of a matched unrelated donor, which might be maximized by establishing national registries. These registries allow overcoming to some extent the genetic heterogeneity in the populations, which may affect the frequency of unique haplotypes, so improving the donor selection process.

Stem cell transplantation obtained from umbilical cord blood provides the potential for further expanding the donor pool to patients without a suitable family match, and has presently been performed for 2500 children and over 1000 adults [29], allowing for a greater degree of HLA disparity in choosing donor cord blood units, so increasing the likelihood that a suitable unit can be identified for any particular patient. However, the advantages of this alternative source of stem cells can be more fully realized in the setting of the availability of matched related donor cord blood units, such as from siblings, providing the advantages of earlier transplantation, lower risk of complications, and lower treatment-related mortality.

HLA-identical sibling donors of cord blood stem cells are the only solution for patients affected with FA, allowing successful treatment in up to 85% of cases. So, over two decades, these families were offered the option of prenatal diagnosis combined with HLA typing on cells derived from chorionic villus sampling or amniocentesis [33]. Of more than 80 pregnancies conceived during this time, one resulted in the first successful use of cord blood transplant in 1988 [3], opening the era of an alternative to bone marrow transplantation. However, because the probability of having an unaffected child who may be also an HLA match for an affected sibling is only 1 in 5, these families often went through multiple cycles of pregnancy before conceiving an unaffected HLA match. So PGD provides a much more attractive approach, because sufficient number of embryos may be tested at one time, increasing the chances to identify an appropriate match. In addition, PGD allows identification of the match before pregnancy, obviating the risk for termination of pregnancy for the HLA type alone, which cannot be acceptable on ethical grounds.

As described above in sections 4.1 and 4.2, the presently available PGD protocols provide an option not only for avoiding an affected pregnancy, but also for preselection of HLA-compatible donors for affected siblings. The approach involves testing single biopsied blastomeres by multiplex nested PCR analysis to test simultaneously for mutations, linked markers, some chromosomal errors, and HLA genes, allowing preselection of unaffected embryos, which may be also an HLA match to the affected siblings requiring stem cell transplantation [28]. As seen from Figure 4.7, multiple STR markers throughout the HLA region are applied [21, 28, 34], allowing 100% accuracy of HLA typing, through picking up possible recombination in the HLA region as well as the copy number of chromosome 6, both affecting the accuracy of preimplantation HLA typing [20, 28].

Preimplantation HLA typing has presently been applied in at least five different PGD centres from four countries, including Australia [35], Belgium [22], Turkey [36], and the United States [28], which have performed HLA genotyping in 151 cycles overall (Table 4.6). This includes 110 cycles in combination with PGD for

Table 4.6. Preimplantation HLA typing: Worldwide experience [20, 22, 28, 33, 35, 36]

Disease	Cycles	Pregnancies
Fanconi anemia	39	6
Thalassemia	62	8
X-linked HGIM	5	2
Wiscott-Aldrich syndrome	2	1
X-linked HED-ID	1	1
X-linked ADL	1	0
Diamond-Blackfan anaemia	4	2
Aplastic anaemia	2	0
Leukemia	35	10
TOTAL	151	30*

*16 children born, 3 pregnancies, lost and others ongoing.

thalassemia (62 cycles), FA (39 cycles), HGIM (five cycles), HED-ID (one cycle), X-ALD (one cycle), and WAS (two cycles). Preimplantation HLA typing as a sole purpose was performed in 41 cycles, including for leukemias (35 cycles) and aplastic (two cycles) and sporadic Diamond-Blackfan anaemia (four cycles).

Sixty three of these cycles were described in sections 4.1, 4.2 and 4.3 [2, 20, 28], involving HLA typing in 667 embryos, including 73 in combination with FA, 311 in combination with thalassemia, 46 in combination with WAS, IgM, HED, and X-ALD, and 237 for leukemia and DBA without testing for causative gene. This resulted in preselection of HLA-matched embryos in 13.2% of unaffected embryos for FA, 14.2% of thalassemia-free embryos, 11.6% for embryos free of WAS, IgM, HED, and X-ALD, and 21.5% in preimplantation HLA typing for leukemia and DBA. A total of 86 HLA-matched embryos were transferred in 50 cycles (1.72 HLA-matched embryos per cycle on the average), yielding 16 (32%) clinical pregnancies and birth of 14 HLA-matched healthy children as potential HLA-compatible donors for their siblings requiring bone marrow transplantation.

Presented data show that couples undergoing preimplantation HLA typing may be expected to require a repeated cycle to be able to preselect and transfer HLA-matched embryos. Even with the probability of selecting only one HLA-matched embryo from five tested, an acceptable pregnancy rate was observed, despite transferring only one or two embryos on the average, suggesting the usefulness of preimplantation HLA matching as part of PGD. The data provide a realistic option for couples desiring to establish a pregnancy potentially providing an HLA-matched progeny for treatment of the affected family members. However, preimplantation HLA typing raises important ethical, legal, and social issues [37, 38], which are discussed in detail in Chapter 8.

References

1. Verlinsky Y, Rechitsky S, Schoolcraft W, Strom C, Kuliev. Designer babies—are they reality yet? Case report: Simultaneous preimplantation genetic diagnosis for Fanconi anemia and HLA typing for cord blood transplantation. Reprod BioMed Online 2000;1:31.
2. Verlinsky Y, Rechitsky S, Schoolcraft W, Strom C, Kuliev A. Preimplantation diagnosis for Fanconi anemia combined with HLA matching. JAMA 2001;285:3130–3133.
3. Gluckman E, Broxmeyer HA, Auerbach Ad, et al. Hematopoietic reconstitution in a patient with Fanconi's anemia by means of umbilical-cord blood from HLA-identical sibling. N Engl J Med 1989;321:1174–1178.
4. Strathdee CA, Duncan AMV, Buchwald M. Evidence for at least four Fanconi anaemia genes including FACC on chromosome 9. Nat Genet 1992;1:196–198.
5. Strathdee CA, Gavish H, Shannon WR, Buchwald M. Cloning of cDNAs for Fanconi's anaemia by functional complementation. Nature 1992;356:763–767.
6. Whitney MA, Saito H, Jakobs PM, Gibson RA, Moses RE, Grompe M. A common mutation in the FACC gene causes Fanconi anaemia in Ashkenazi Jews. Nat Genet 1993;4:202–205.
7. Gluckman E, Devergie A, Schaison G, et al. Bone marrow transplantation in Fanconi anemia. Br J of Haematol 1980;45:557–564.
8. Wagner J, Davies SM, Auerbach AD. Hemotopoietic stem cell transplantation in the treatment of Fanconi anemia. In: Forman SJ, Blum KG, Thomas ED (eds.). Hematopoietic Cell Transplantation, ed. 2. Blackwell Science, Malden, MA, 1999, pp. 1204–1219.
9. Gibson RA, Morgan NV, Goldstein LH, Pearson IC, Kesterton IP. Novel mutations and polymorphisms in the Fanconi anemia group C gene. Hum Mutat 1996;8:140–148.
10. Mitsunaga S, Tokunaga K, Kashiwase K, Akaza T, Tadokoro K, Juji T. A nested PCR-RFLP method for high-resolution typing of HLA-A alleles. Eur J Immunogenet 1998;25:15–27.
11. Blasczyk R, Huhn U, Wehling J, Huhn D, Salama A. Complete subtyping of the HLA-A locus by sequence-specific amplification followed by direct sequencing or single-strand confirmation polymorphism analysis. Tissue Antigens 1995;46:86–95.
12. Foissac A, Salhi M, Cambon-Thomsen A. Microsatellite in the HLA region: 1999 update. Tissue Antigens 2000;55:477–509.
13. Yoshida M, Kimura A, Numano F, Sasazuki T. Polymerase-chain-reaction-based analysis of polymorphism in the HLA-B gene. Hum Immunol 1992;34:257–266.
14. Online Mendelian Inheritance in Man (OMIM). 2001. John Hopkins University http://www.ncbi.nlm.nih.gov/Omim

15. Online Human Gene Mutation Database (HGMD). 2004. CELERA http://archive.uwcm.ac.uk/uwcm/mg/search/1119297.html
16. Modell B, Kuliev A. History of community genetics: The contribution of hemoglobin disorders. Commun Genet 1998;1:3–11.
17. Lucarelli G, Andreani M, Angelucci E. The cure of thalassemia by bone marrow transplantation. Blood 2002;16:81–85.
18. Kuliev A, Rechitsky S, Verlinsky O, et al. Birth of healthy children after reimplantation diagnosis of thalassemias. J. Assist Reprod Genet 1999;16:207–211.
19. Verlinsky Y, Kuliev A. Atlas of Preimplantation Genetic Diagnosis. Parthenon, N.Y., 2000.
20. Verlinsky Y, Rechitsky S, Sharapova T, Morris R, Tharanissi M, Kuliev A. Preimplantation HLA typing. JAMA 2004;291:2079–2085.
21. Foissac A, Salhi M, Cambon-Thomsen A. Microsatellite in the HLA region: 1999 update. Tissue Antigens 2000;55:477–509.
22. Van deVelde H, Georgiou I, De Rycke M et al. Novel universal approach for preimplantation genetic diagnosis of β-thalassemia in combination with HLA matching of embryos. Hum Reprod 2004;19:700–708.
23. Tesarik J, Mendoza C. Somatic cell haploidization: An update. Reprod BioMed Online 2003;6:60–65.
24. WHO News and Activities. Cord blood banking. Bull WHO 1998;76:313–314.
25. Rechitsky S, Strom C, Verlinsky O, et al. Accuracy of preimplantation diagnosis of single-gene disorders by polar body analysis of oocytes. J Assist Reprod Genet 1999;16:192–198.
26. Rechitsky S, Strom C, Verlinsky O, et al. Allele drop out in polar bodies and blastemeres. J Assist Reprod Genet 1998;15:253–257.
27. Kuliev A, Verlinsky Y. Thirteen years' experience of preimplantation diagnosis: Report of the Fifth International Symposium on Preimplantation Genetics. Reprod BioMed Online 2004;8:229–235.
28. Rechitsky S, Kuliev A, Tur-Kaspa I, Morris R, Verlinsky Y. Preimplantation genetic diagnosis with HLA matching. Reprod BioMed Online 2004;9:210–221.
29. Baker S, Wagner J. Advantages of umbilical cord blood transplantation. Abstracts of International Seminar on Preimplantation HLA Typing and Stem Cell Transplantation, Limassol, Cyprus, 2004, pp. 7–8 (www.pgdis.org).
30. Lucarelli G, Andreani M, Angelucci E. The cure of thalassemia by bone marrow transplantation. Blood 2002;16:81–85.
31. Gaziev J, Lucarelli G. Stem cell transplant for thalassemia. Reprod BioMed. Online 2005;10:111–115.
32. Costeas PA. Bone marrow donor registry and availability of HLA matched donors. Abstracts of International Seminar on Preimplantation HLA Typing and Stem Cell Transplantation, Limassol, Cyprus, 2004, p. 9 (www.pgdis.org).
33. Auerbach AD. Preimplantation genetic diagnosis combined with HLA typing: The Fanconi anemia experience as a model. Abstracts of International Seminar on Preimplantation HLA Typing and Stem Cell Transplantation, Limassol, Cyprus, 2004, p. 13 (www.pgdis.org).
34. Rechitsky S, Verlinsky O, Kuliev A, Verlinsky Y. Preimplantation non-disease testing. Reprod BioMed Online 2002 4(suppl. 2):23–24.
35. Marshall J, Beaton A, deBoer K, Leigh D. McArthur S. Preimplantation HLA matching: Sydneys' strategy. Abstracts of International Seminar on Preimplantation HLA Typing and Stem Cell Transplantation, Limassol, Cyprus, 2004 p. 17 (www.pgdis.org).
36. Kahraman S, Karlilaya G, Sertiel S, Karadayi H, Findicli N, Oncu N. Clinical aspects of preimplantation genetic diagnosis of single gene disorders combined with HLA matching. Reprod BioMed. Online 2004;9:529–532.
37. Fost NC. Conception for donation. JAMA 2004;291:2125–2126.
38. Edwards RG. Ethics of PGD: Thoughts on the consequences of typing HLA in embryos. Reprod BioMed Online 2004;9:222–224.

5

Preimplantation Genetic Diagnosis for Chromosomal Disorders

It is well known that chromosomal abnormalities originate predominantly from female meiosis. As demonstrated by DNA polymorphism studies performed in families with aneuploid spontaneous abortions or liveborn babies with trisomy syndromes, these abnormalities derive mainly from meiosis I [1–3]. It was suggested that the age-related increase of common trisomies is probably determined by the age-related reduction of meiotic recombination, resulting in premature separation of bivalents and chromosomal nondisjunction. Meiosis II errors were also postulated to derive from meiosis I, as a result of the increased meiotic recombination rate, which may lead to a separation failure of bivalents [4].

With the advent of PGD for aneuploidies it has become possible to directly test the outcome of the first and second meiotic divisions, using PB1 and PB2, as described in Chapter 2. PB1 is extruded following maturation of oocytes, representing a by-product of meiosis I, while PB2 is a by-product of meiosis II and is extruded following the exposure of oocytes to sperm or ICSI. As will be described below, the frequency and types of chromosomal errors detected by this approach are different from what was described in the traditional studies of meiotic chromosomes in metaphase II (MII) oocytes, according to which chromosomal anomalies in oocytes originate mainly from the errors of whole bivalents as a result of chromosomal nondisjunctions [5]. In contrast, direct testing of meiotic outcome using PB1 and PB2 analysis showed not only a higher prevalence of meiotic errors, but also significant contribution of chromatid, rather than chromosomal errors. The discrepancy may be due to a poor quality of meiotic chromosome preparations in earlier studies, and also to the lack of testing of the corresponding chromosome set extruded in PB1, without which the resulting oocyte karyotype could not be reliably evaluated, particularly in the cases of missing chromosomes or chromatids. This was demonstrated in the study of simultaneous testing of MII oocytes with their corresponding PB1, which showed that the normal chromosome pattern is represented by paired fluorescent signals for each chromosome, while the lack or addition of one or both signals in either oocyte or PB1 reflects an exactly opposite pattern in the corresponding MII oocytes or PB1, suggesting a high accuracy of PB1 testing for prediction of the oocyte genotype [6–10].

Based on the above data, PB1 testing was applied clinically for the preselection of aneuploidy-free oocytes, which have demonstrated the practical relevance of PB1 testing for IVF patients of advanced reproductive age [11–16]. The data also demonstrated that the genotype of the resulting zygotes could not be accurately predicted without information about the outcome of the second meiotic division, which may be inferred from PB2 testing. The present experience includes FISH analysis of approximately 15,000 oocytes, presented below, demonstrating the accuracy of evaluation of oocyte karyotype by testing PB1 and PB2, and also providing an attractive approach for the study of the origin of human aneuploidies.

This makes possible the preselection of oocytes and embryos with the highest developmental potential, currently based on morphological criteria, which is of obvious interest for improving efficiency of assisted reproduction technology (ART). However, morphological parameters have no value for excluding aneuploid embryos from transfer, which requires chromosomal studies of oocytes and embryos, still limited to a few IVF centres [17–21].

In addition to aneuploidy testing, different approaches have been tested for possible prediction of the developmental potential of oocytes, such as pronuclear morphology scoring [22], microtubule and microfilament organization assessment [23], and PB1 grading [24–26]. As will be described below, approximately 42% of oocytes were shown to have PB1 aneuploidies, which also allows predicting approximately a half of the second meiotic division errors detected by PB2 testing, this making it possible to exclude almost two thirds of aneuploid oocytes from ICSI.

However, because chromosomal studies are still not readily available and require specialized expertise, the possibility of PB1 morphological grading, as a potential means for preselection of viable oocytes and embryos for transfer may seem to be an attractive approach, since it could be easily adopted in any IVF practice. Preliminary reports on the utility of PB1 morphology have been controversial, suggesting a positive correlation between well-shaped, round PB1 within a cohort and fertilization rates, embryo quality, and implantation rates [24, 25], or no correlation with fertilization but a positive correlation between implantation rates and the presence of fragmented PB1 within a cohort [26]. Furthermore, these were retrospective studies in which chromosomal status of the oocytes and embryos was not investigated.

So a prospective study of PB1 morphology in relation to fertilization, chromosomal status, development potential of the resulting embryos, and outcome of embryo transfer was carried out to investigate if such testing could be used for predicting developmental potential and chromosomal status of oocytes and embryos without cytogenetic analysis.

Oocytes were obtained from 90 patients in 91 IVF cycles randomly for different indications, using a standard IVF protocol. PB1 morphology was assessed in 831 mature oocytes prior to ICSI (day 0) and at fertilization assessment (day 1), and placed into three major categories: grade 1, representing a round or oval shape which may sometimes be flattened; grade 2, representing a nonfragmented PB1 of irregular shape; and grade 3, representing a partially or totally fragmented PB1 (Figure 5.1). PB1 observation was performed using an inverted microscope (Diaphot, Nikon, Garden City, NY) with Hoffman modulation contrast optics and a magnification of 200×. Prior to ICSI,

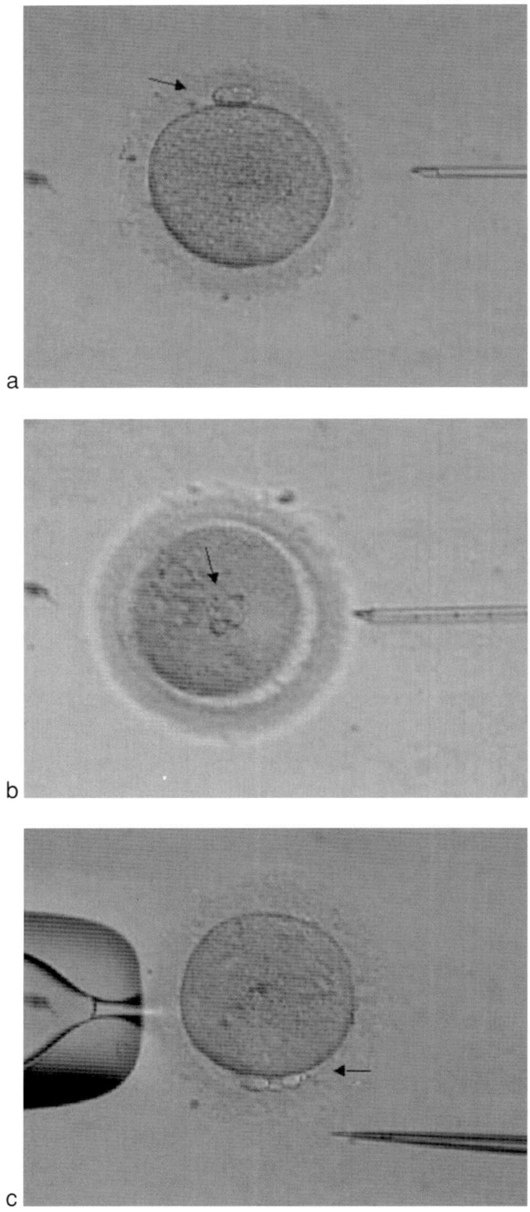

Figure 5.1. First polar body grading based on morphology. (*A*) Grade 1—intact, round or oval (side view); (*B*) Grade 2—nonfragmented irregularly shaped (top view; magnification does not present actual working magnification of ×200); (*C*) Grade 3—fragmented.

Table 5.1. Morphological characteristics of the first polar body of human oocytes before ICSI and after 16 hours in culture

PB1 Grade	Characteristics of PB1	No. of oocytes (%) Total*	with changed grade after 16 hrs
	Patients with number of oocytes ≤ 10		
1	Regular shape	77	36 (46.8)[a]
2	Irregular shape	67	53 (79.1)[b]
3	Fragmented	68	0
Total		212	89 (42.0)
	Patients with number of oocytes > 10		
1	Regular shape	200	95 (47.5)[a]
2	Irregular shape	208	147 (70.7)[b]
3	Fragmented	211	0
Total		619	242 (39.1)

*PB grade on day 1 was not known for all fertilized oocytes.
[a,b] $P < 0.001$.

oocytes were rotated so that both the side view and top view of PB1 were observed. Oocytes were then followed up after ICSI from fertilization throughout preimplantation development and embryo transfer. Data on PB1 grading for each oocyte was correlated to fertilization rate, day-3 embryo quality as to cell number and degree of fragmentation present, the development capability to the blastocyst stage, and embryo transfer outcome.

Due to the differences in patient response to hormonal stimulation, patients were divided into two groups, consisting of 42 poor responders (50 cycles, group I) in which the number of retrieved oocytes did not exceed 10, and 48 good responders (50 cycles, group II) with greater than 10 available oocytes.

Aneuploidy testing was requested by 49 patients in 50 cycles in which 395 oocytes and embryos were tested on day 1 (PB analysis), day 3 (blastomere analysis), or both, using micromanipulation techniques and FISH analysis as previously described with the application of a five-color probe specific for chromosomes 13, 16, 18, 21, and 22 (Vysis, Downers Grove, IL). Chi-square analysis was performed on all data.

The overall distribution of oocytes according to PB1 grades on day 0 and day 1 in group I and group II is presented in Table 5.1. As seen from this table, the number of oocytes with PB1 of different grades on day 0 was similar and distributed equally in both patient groups. PB1 grading on day 1 revealed the changes of PB1 morphology in both patient groups for each grade category except grade 3. Overall, the grade changes were observed in 331 (36.2%) of 831 PB1, and were similar in both groups I and II. Significant differences for such changes were observed only for the oocytes with grade-2 PB1, 79.1% and 66.3% of which, in groups I and II respectively, became grade 3 on day 1 ($P < 0.001$).

Analysis of fertilization rates suggested lower rates for oocytes with grade-1 PB1 (69.5% in group I and 73.6% in group II), compared with grade-2 PB1 (85.5%) in group 1 ($P < 0.05$), and with grade-3 PB1 (83.6%) in group II ($P < 0.001$). Table 5.2 presents the cleavage rate of embryos, deriving from oocytes with different PB1 grades, evaluated by the number of blastomeres present on day 3. The rates were similar in each PB1 category for each patient group, evidenced by the number of embryos with 5–10 cells: 68.3, 68.5, and 73.9% in grades 1, 2, and 3, respectively, in group I, and 67.7, 70.9, and 71.9% in group II. The morphological quality of the embryos with respect to the degree of fragmentation observed on day 3 in relation to the oocytes with different PB1 grading showed no difference in the three PB1 grades or patient groups.

Table 5.2. Embryo cell # on day 3 in relation to PB1 grades

PB1 Grade	Total no. of zygotes[A]	Group I* zygotes	2–4-cell	5–10-cell	Group II** zygotes	2–4-cell	5–10-cell
1	276	82	26 (31.7)	56[a] (68.3)	194[B]	62 (32.0)	130[a] (67.0)
2	277	71[B]	22 (31.0)	48[b] (67.6)	206	60 (29.1)	146[b] (70.9)
3	280	69	18 (26.1)	51[c] (73.9)	211[B]	59 (28.0)	151[c] (71.5)
Total	833	222	66 (29.7)	155 (69.8)	611	181 (29.6)	427 (69.9)

*Patients with number of oocytes ≤10.
**Patients with number of oocytes >10.
[A]Some zygotes were cryopreserved [15].
[B]Some zygotes did not develop [4].
[a,b,c]No significant difference.

Table 5.3. Implantation rates in relation to PB1 grades

PB1 Grade	# Embryos transferred	# Implanted
1	30	13[a] (43.3)
2	25	9[b] (36)
3	31	11[c] (35.5)
1+2	63	12[d] (19)
2+3	44	12[e] (27.3)
1+3	66	19[f] (28.8)
1+2+3	58	5 (8.6)
Total	317	81 (25.6)

[a–f]No significant difference.

The proportion of grade 1–1.5 embryos (little or no fragmentation present), resulting from oocytes with PB1 grades 1, 2, and 3 were 68.3, 67.1, and 79.7% in group I, and 64.6, 65.1, and 61.9% in group II, respectively.

Similar results were observed in the embryos' potential to reach the blastocyst stage on days 5 and 6. Of 109 group-I embryos, 35(32%) reached the blastocyst stage of development with 24 (22%) at a gradable expanded stage, with no significant difference in any of the PB1 grade groups. The same was true for 555 group-II embryos, of which 249(44.9%) reached the blastocyst stage with 148 (26.7%) at a gradable expanded stage, again with no significant difference in any of the PB1 grade groups.

Table 5.3 presents implantation rates of 317 embryos, 124 from group I, 122 from group II, and 71 from the patients, in whom aneuploidy testing was performed. Embryos were grouped according to the oocytes' corresponding PB1 grade in which they were derived. No difference was seen between PB1 grades 1, 2, and 3 as well as embryo transfers consisting of a combination of embryos in which there were two different PB1 grades given. The overall implantation rates for group I, group II, and the group of PGD patients were 20.2, 32.8, and 22.5%, respectively.

Results of aneuploidy testing in 395 embryos, resulting from oocytes with different PB1 grades obtained from 49 patients (50 cycles) of the mean age of 37 ± 3.6 years, are summarized in Table 5.4. The data show no difference in aneuploidy rates in any PB1 grade category, which were 75.9, 65.3, and 66.4% for PB1, grades 1, 2, and 3, respectively. The data also failed to reveal any difference in the frequency of error for each chromosome tested in relation to the PB1 grade (Figures 5.2 and 5.3). The higher prevalence of aneuploidy for chromosomes 21 and 22 for all PB1 grades is in agreement with our findings for women of advanced reproductive age (see below).

The data show that PB1 morphology may not be a useful predictor of developmental potential or chromosomal normalcy of the resulting embryos, either in good or poor responders. Because PB1 morphology was graded in sequence before and after fertilization, it was possible to detect the grading changes we observed in greater than one third of the oocytes in both patient groups. These changes were significant for the grade-2, intact, irregular-shaped PB1, in which the majority on day 0 became grade 3 by day 1. Although this may be representative of postextrusion changes, they seem to have no practical relevance, since all the clinical parameters studied showed no correlation with any of PB1 grades. The relevance of the observed higher fertilization rate in oocytes with the irregular PB1 shape, compared with those

Table 5.4. Oocyte and embryo chromosome abnormalities in relation to PB1 grades (mean age of patients 37 ± 3.6 years)

PB1 Grade	# of embryos	Normal	Abnormal	Chromosome				
				13	16	18	21	22
1	137	33[a] (24.1)	104 (75.9)	29 (20.7)	29 (20.7)	24 (17.1)	31 (22.1)	26 (18.6)
2	118	41[b] (34.7)	77 (65.3)	17 (14.4)	20 (16.9)	13 (11.0)	22 (18.6)	26 (22.0)
3	140	48[c] (33.6)	93 (66.4)	22 (15.6)	17 (12.1)	23 (16.3)	30 (21.3)	24 (17.0)
Total	395	121 (30.6)	276 (69.4)	68 (17.0)	66 (16.5)	60 (15.0)	83 (20.8)	76 (19.0)

[a,b,c]No significant difference.

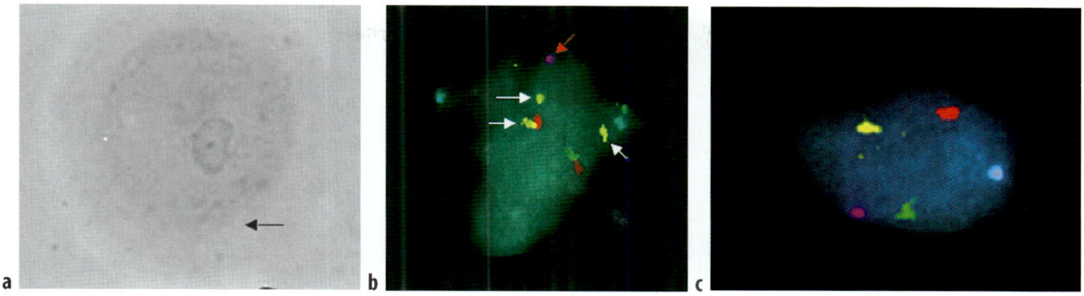

Figure 5.2. Monosomy 18 and trisomy 22 in grade 2 first polar body. (*A*) First polar body with a slightly irregular shape, grade 2 morphology on day 0. (*B*) Corresponding FISH image of chromosomes 13,16,18,21, and 22 (day 1) showing a missing violet-blue signal (chromatid) for chromosome 18 and an extra gold signal (chromatid) for chromosome 22. (*C*) Corresponding FISH image of the second polar body (day 1) showing a normal number of signals for each of the five chromosomes tested.

with the regular-shaped PB1 in poor responders, as well as to those with the fragmented PB1 in good responders is not clear, since these findings conflict with other data. Previous retrospective studies suggested either no correlation [26], or even the possibility for a positive relationship of PB1 morphology with higher fertilization rate [24, 25]. These conflicting data may be attributed to the timing of the ICSI procedure in relation to the cytoplasmic maturation of the oocytes.

On the other hand, in contrast to previous reports, we found no relationship of PB1 morphology to the embryo quality and cleavage rate, which makes PB1 grading questionable for reliable preselection of the cleavage-stage embryos for transfer. Furthermore, in contrast to the earlier findings of a positive relationship of PB1 morphology to blastocyst formation potential [24], our data showed no relationship either in good or poor responders. No relationship was observed either between PB1 grading and the outcome of embryo transfer, which was similar in different patient groups, based on PB1 morphology. This is not in agreement with previous data, some of which suggested higher implantation and pregnancy rates for embryos derived from oocytes with intact, round versus fragmented PB1 [24, 25], nor does it agree with another report which describes an association between higher implantation and pregnancy rates in cohorts of oocytes in which a greater percentage of fragmented PB1 are present [26].

It is of interest that the aneuploicy testing of embryos resulting from oocytes with different PB1 grading failed to reveal any relationship between PB1 morphology and karyotype, suggesting that PB1 morphology is not useful for screening of chromosomal aneuplodies in preimplantation development. Thus, the data provide no evidence for any relationship of PB1 morphology with chromosomal normalcy, embryo quality and

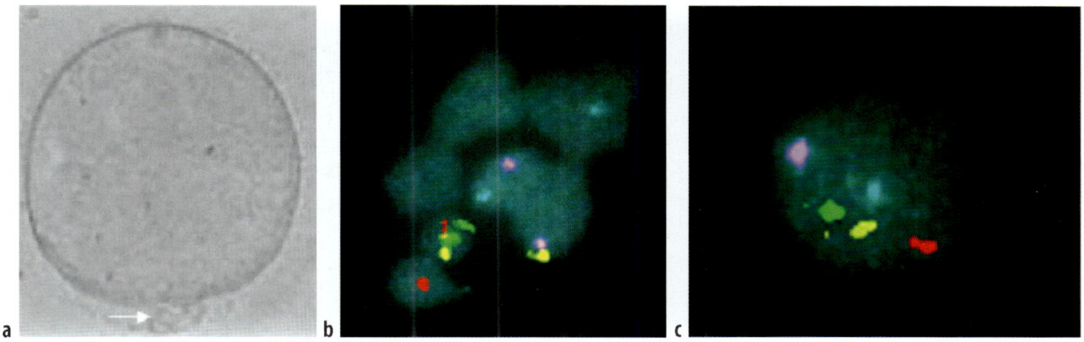

Figure 5.3. Normal karyotype of fragmented (grade 3) first polar body. (*A*) Fragmented first polar body. (*B*) FISH image of chromosomes 13, 16, 18, 21, and 22 (day 1) in the first polar body showing normal pattern. (*C*) Normal FISH image for the second polar body.

Table 5.5. Overall nuclear abnormalities detected by PB1 and PBII FISH analysis using probes for chromosomes 13, 16, 18, 21, and 22

Couples	Cycles	Total oocytes studied	Oocytes with FISH results	Normal oocytes	Abnormal oocytes
852	1297	8382	6733	3224 (47.9%)	3509 (52.1%)

developmental potential, and outcome of embryo transfer, suggesting that PB1 grading is of no prognostic value for the developmental potential of embryos to be used in preselection of embryos for transfer.

5.1 Aneuploidies

As mentioned, the direct testing of meiosis errors has become possible with the introduction of PGD for chromosomal disorders, based on the use of PB1 and PB2 sampling, which are removed simultaneously following fertilization and fixed and analysed by FISH on the same slide, as described in Chapter 2. As mentioned, because PB1 and PB2 are extruded from oocytes as a normal process of maturation and fertilization, their removal is not expected to have any biological effect on the embryo development, which is currently obvious from the outcomes of hundreds of pregnancies resulting from PGD [27]. The biopsied and fixed PB1 and PB2 were studied using fluorescent probes specific for chromosomes 13, 16, 18, 21, and 22 (Abbott, Downers Grove, IL). The results of such studies are presently available for 17,397 oocytes, overall, obtained from 2458 IVF cycles performed in 1678 patients of average age of 38.5 years.

These data are in agreement with our previous analysis of 8382 oocytes obtained in 1297 IVF cycles. In this study FISH results were available in 6733 (80.3%) oocytes [28]. The results for both PB1 and PB2 in this analysis were obtained in 4929 (73.2%) oocytes, with the remaining oocytes being either with PB1 (909, 13.5%) or PB2 (895, 13.3%) results. A total of 3509 (52.1%) oocytes were with aneuploidies (Table 5.5), of which 965 (27.5%) had errors in both PB1 and PB2, 1467 (41.8%) in only PB1, and 1077 (30.7%) in only PB2. Of 5831 PB1 with results, 2432 (41.7%) were with errors, comparable to 2042 (35.2%) errors detected in 5808 PB2 (Table 5.6). On one hand, the observed rates should be an overestimate, as the average maternal age of the IVF patients from whom the oocytes were obtained was 38.5 years.

Table 5.6. Frequency and types of meiosis I and meiosis II errors

	Meiosis I (PB1)		Meiosis II (PB2)	
FISH Data	No.	%	No.	%
Normal	3399	58.3	3766	64.8
Abnormal				
Disomy	387	15.9	842	41.2
Nullisomy	1313	54	748	36.6
Complex	732	30.1	452	32.1
Total Abnormal	2432	41.7	2042	35.2
TOTAL	5831	100	5808	100

On the other hand, this may be an underestimate, because 3630 of these oocytes were tested using a tricolor probe specific for chromosomes 13, 18, and 22, with only 3103 analysed using a five-color probe specific for chromosomes 13, 16, 18, 21, and 22.

The types of errors observed in PB1 and PB2 are presented in Figures 5.4 and 5.5. At least three times higher frequency was observed for missing signals (nullisomy) compared with extra signals (disomy) in PB1 (approximately 3:1 ratio), in contrast to a comparable distribution of missing and extra signals in PB2. PB1 data also showed a 63.5 % chromatid errors in the overall PBI abnormalities (48.1% missing and 15.4% extra chromatids), compared with a 6.4% chromosome errors (5.9% missing and 0.5% extra chromosomes) (Table 5.7). Therefore, as in chromatid errors, missing chromosomes were more frequent than extra chromosomes, suggesting a possible maintenance of the extra chromatid or chromosome

Table 5.7. Chromosome and cromatid errors in meiosis I

Types of errors	Number	%
Extra chromatid	375	15.4
Missing chromatid	1169	48.1
Extra chromosome	12	0.5
Missing chromosome	144	5.9
Complex	732	30.1
TOTAL	2432	100

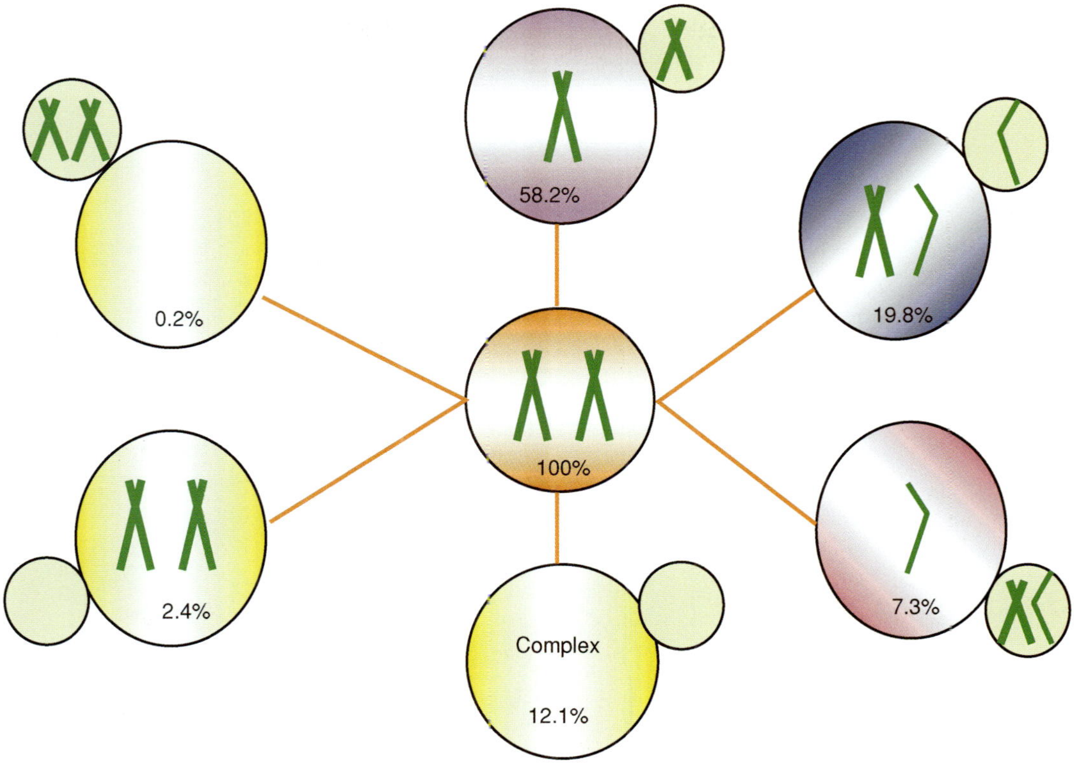

Figure 5.4. Meiosis I errors observed in first polar body FISH analysis. (*Centre*) Primary oocyte containing diploid set of chromosomes with the doubled amount of chromatin (4n) prior to maturation. (*Top*) Normal segregation of homologues in the first meiotic division, resulting in the extrusion of the first polar body (PB1) (smaller circle) containing one of the homologues. Accordingly, the resulting secondary (metaphase II) oocyte contains the remaining homologue with two chromatids. (*Left*) Chromosome nondisjunction leading to the extrusion of PB1 containing both homologues, which results in nullisomy of this chromosome in MII oocyte (*top left*). Similarly, the extruded PB1 may not contain any material, so both homologues will be left in MII oocyte, resulting in disomic oocyte (bottom left). (*Right*) Chromatid malsegregation, leading to either a single chromatid extrusion with PB1 (*top right*), which results in the extra chromatid material in MII oocyte, or to extra chomatid extrusion with PB1 resulting in the lack of one chromatid in MII oocyte (*bottom right*). (*Bottom*) Cromatic or chromosome malsegregation involving different chromosomes, resulting in different types of errors of different chromatids or chromosomes in MII oocyte.

material in MII oocytes, which is in agreement with a higher frequency of trisomies over monosomies in postimplantation embryos, detected by testing of spontaneous abortions. Only 15.9% of PB1 abnormalities were disomies, compared with 54% nullisomies, with the remaining being of complex origin. Although the observed excess of missing signals in PB1 may be also attributable to technical errors, such as hybridization failure, it is also possible that a certain meiosis I mechanism exists, preventing an extra chromosome material extrusion into PB1 if meiotic errors occur during the oocyte maturation process.

A significant proportion of abnormalities in PB1 and PB2 (30.1% and 32.1%, respectively) were of complex origin, represented by different types of errors, or errors of different chromosomes. Of 1582 complex abnormalities, 1242 (78.5%) involved simultaneously two or more chromosomes, and 340 (21.5%) the same chromosome(s) in both PB1 and PB2 (Table 5.8).

Of 965 oocytes with both PB1 and PB2 being abnormal, different chromosomes were involved in 487 (56.1%). In the oocytes with the same chromosome(s) involved, chromosome 21 appeared to be most frequent (167, 19.2%), followed by chromosome 22 (59, 6.8.1%), chromosome 13 or 18 (44 and 43, 5.1 and 5%, respectively), and chromosome 16 (26, 3%); two or more same chromosomes were observed in 42 (4.8%)

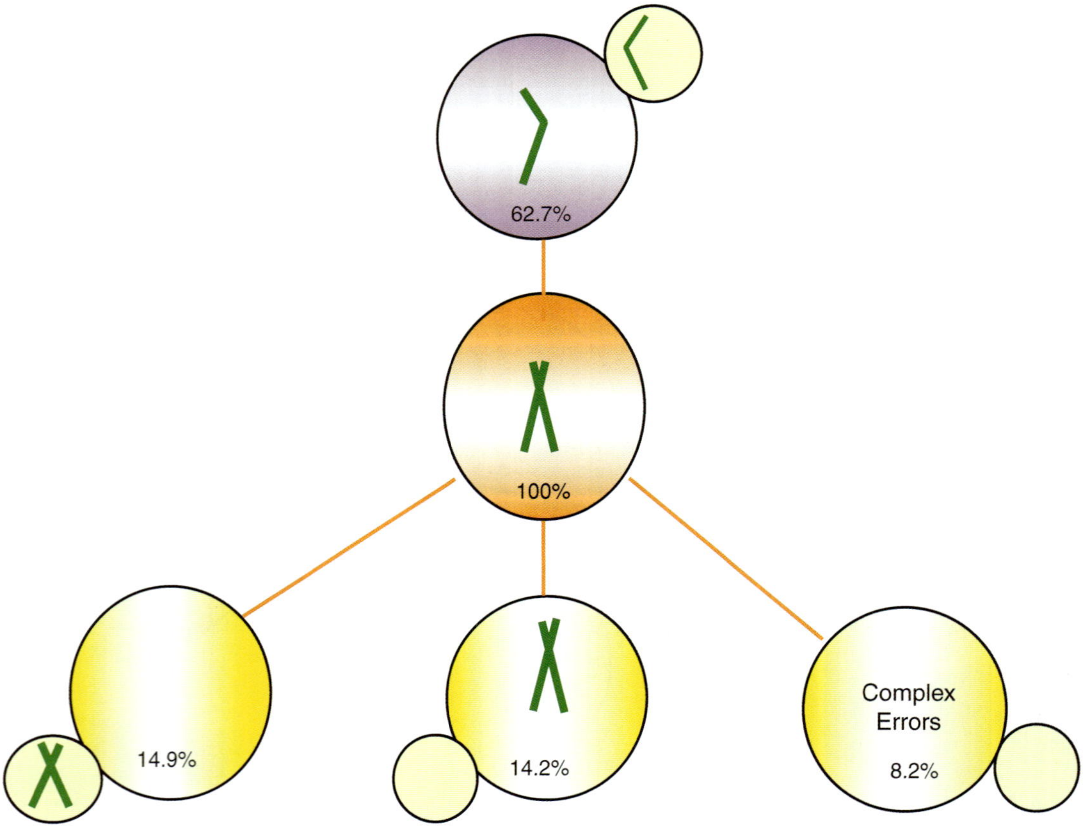

Figure 5.5. Meiosis II errors in second polar body FISH analysis. (*Centre*) Secondary (metaphase II) oocyte containing haploid set of chromosomes (2n) prior to fertilization. (*Top*) Normal segregation of chromatids in the second meiotic division (62.7%), resulting in the extrusion of the second polar body (PB2) (smaller circle) containing one of the chromatids. Accordingly, the resulting maternal contribution to zygote contains the remaining sestrid chromatid. (*Bottom left*) Cromatid nondisjunction leading to the extrusion of PB containing both chromatids, which results in nullisomy of this chromosome in maternal pronucleus (14.9%). (*Bottom centre*). Similarly, the extruded PB2 may not contain any material, so both chromatids will be left in oocyte, resulting in disomic oocyte (14.2%). (*Bottom right*) Complex chromatid malsegregation, leading to different errors of different chromatids.

cases. Although 282 (32.5%) of these zygotes in the latter group appeared to be balanced, the preselection and transfer of the embryos resulting from such oocytes is yet to be justified, as incidental abnormalities of other chromosomes cannot be excluded.

Chromosome-specific patterns of meiotic errors is shown in Table 5.9. The most frequent chromosomes involved in meiotic error were chromosomes 21 and 22 (10.9% and 11.8%, respectively), deriving from both meiosis I and meiosis II. Involvement in meiotic error of chromosomes 13, 16, and 18 was less frequent (6.0, 6.4, and 6.8%, respectively), and their error patterns were not similar. Chromosomes 16 and 22 errors originated predominantly in meiosis II (52.4% and

Table 5.8. Complex aneuploidies

Total abnormal oocytes	Total complex errors	Involving same chromosome in both PB	Involving > 1 chromosome		
			2 Chromosomes	> 2 Chromosomes	Total
3509	1582 (45.1%)	340 (21.5%)	929	313	1242 (78.5%)

Table 5.9. Origin of nondisjunction in aneuploidies by PB FISH analysis of chromosomes 13, 16, 18, 21, and 22

Chromosome	Total oocytes studied	Total abnormal	MI errors	MII errors	MI & MII errors
13	5907	354 (6.0%)	165 46.6%	128 36.2%	61 17.2%
16	4583	294 (6.4%)	77 26.2%	154 52.4%	63 21.4%
18	6648	455 (6.8%)	274 60.2%	135 29.7%	46 10.1%
21	6648	725 (10.9%)	314 43.3%	269 37.1%	142 19.6%
22	4583	539 (11.8%)	166 30.8%	274 50.8%	99 18.4%

50.8% meiosis II vs. 26.2% and 30.8% meiosis I respectively), in contrast to chromosomes 13 and 18 errors, deriving more frequently from meiosis I (46.6 and 60.2% vs. 36.2 and 29.7% in meiosis II, respectively).

The data show that testing only for five chromosomes has revealed as high as 52% aneuploidy rate. Although some overestimate attributable to limitations of FISH technique cannot be excluded, the majority of PB1 and PB2 abnormalities were true errors as were confirmed by the follow-up studies of the embryos resulting from oocytes with meiosis I and meiosis II errors [12]. In contrast to the well-established concept of female meiosis I origin of chromosomal abnormalities, our results show that the observed errors originate from both meiosis I and II as per the expected patterns of segregation illustrated in Figures 5.4 and 5.5. The results are of clinical significance, suggesting that the genotype of the resulting zygote cannot be predicted without testing of the outcomes of both meiotic divisions, inferred from PB1 and PB2. For example, testing of meiosis I errors alone should reduce aneuploidy rate in the resulting embryos at least by two thirds. Despite the fact that approximately one third of these oocytes still will be aneuploid following the second meiotic division, PB1 testing still could sufficiently improve the implantation and pregnancy rates in poor-prognosis IVF patients or ICSI patients by applying ICSI selectively to the oocytes with aneuploidy-free PB1. On the other hand, only half of the abnormalities deriving from second meiotic division may be detected by PB1 analysis as complex errors; therefore, to avoid the transfer of all the embryos resulting from aneuploid oocytes, testing of both PB1 and PB2 is still required. As PB1 and PB2 are extruded in a normal process of oocyte maturation and fertilization, having no biological significance in preimplantation and postimplantation development, their removal and testing may become a useful tool in assisted reproduction practices to identify the oocytes without nuclear abnormalities, which should help in the preselection of oocytes with the highest potential for establishing a viable pregnancy, improving significantly the IVF efficiency.

As can be seen from the types of the errors shown in the above table and figures, the majority of abnormalities in meiosis I are represented by chromatids errors, in contrast to the expected chromosomal nondisjunction suggested by most of the previous studies mentioned. However, we still observed chromosomal errors in 6.4% of oocytes, which does not support the other extreme claim that all abnormalities in MII oocytes are of chromatid origin [29]. So, probably both chromatid and chromosomal errors are involved in producing MII abnormalities, with the frequency of chromatid errors being much higher than chromosomal ones (chromatid/chromosome error ratio 9:1). There is no doubt that both of these meiosis I errors lead to aneuploidy in the resulting embryos, as demonstrated by the follow-up study of the embryos resulting from these oocytes, the transfer of which were avoided. However, differences in the effect of chromatid and chromosomal errors on preimplantation and postimplantation development cannot be excluded.

As mentioned, as much as 41.8% of abnormal oocytes deriving from meiosis I errors may be detected by testing of PB1 (see Figure 5.6). In addition, this information allows predicting a significant proportion of oocytes with the second meiotic division errors: in our data, of 2041 oocytes with meiosis II errors, 965 (47.3%) had also PB1 aneuploidies. However, the remaining 1076, which represent approximately one third of the overall

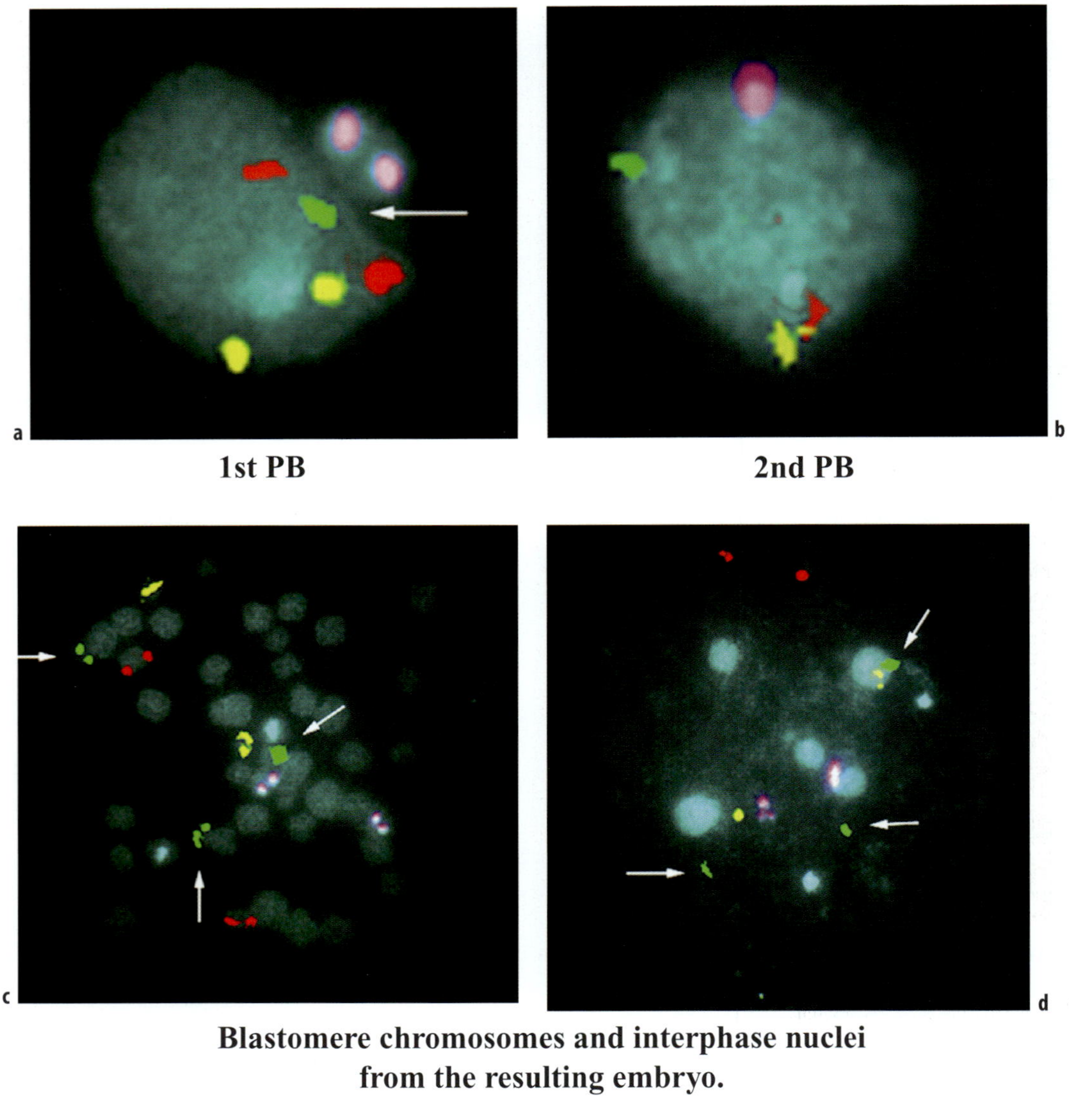

Figure 5.6. Meiosis I error resulting in trisomy 21. PB1 and PB2 were simultaneously removed on day 1 at the pronuclear stage of development following fertilization assessment. (*A*) FISH image of PB1 after a 3-hr hybridization with MultiVysion PB panel probe for chromosomes 13 (red), 16 (aqua), 18 (violet blue), 21 (green), and 22 (gold), showing one chromosome 21 signal (*white arrow*) instead of two, indicating an extra chromatid 21 was retained in the oocyte. (*B*) FISH image of PB2 showing a normal number of signals for each of the chromosomes tested. Information obtained by the PB testing indicates trisomy 21 is present in the embryo due to the chromatid error in meiosis I. (*C*) FISH image of metaphase chromosomes obtained after embryo biopsy and fixation on day 3, confirming the presence of three chromosomes 21 (*white arrows*). Some double dot (paired) signals are seen representing each chromatid of these metaphase stage chromosomes. (*D*) FISH image of an interphase nucleus isolated from a second cell from the corresponding embryo, also showing three signals for chromosome 21 (*white arrows*).

number of abnormal oocytes, became abnormal only following the second meiotic division, which could not be predicted by PB1 results, and would be missed if testing were limited to PB1 (Figure 5.7). This may suggest that in order to identify all oocytes with chromosomal abnormalities, the outcome of both the first and second meiotic divisions should be studied, using PB1 and PB2.

In contrast to meiosis I errors, there was no difference in the frequency of missing or extra chromatid errors following the second meiotic

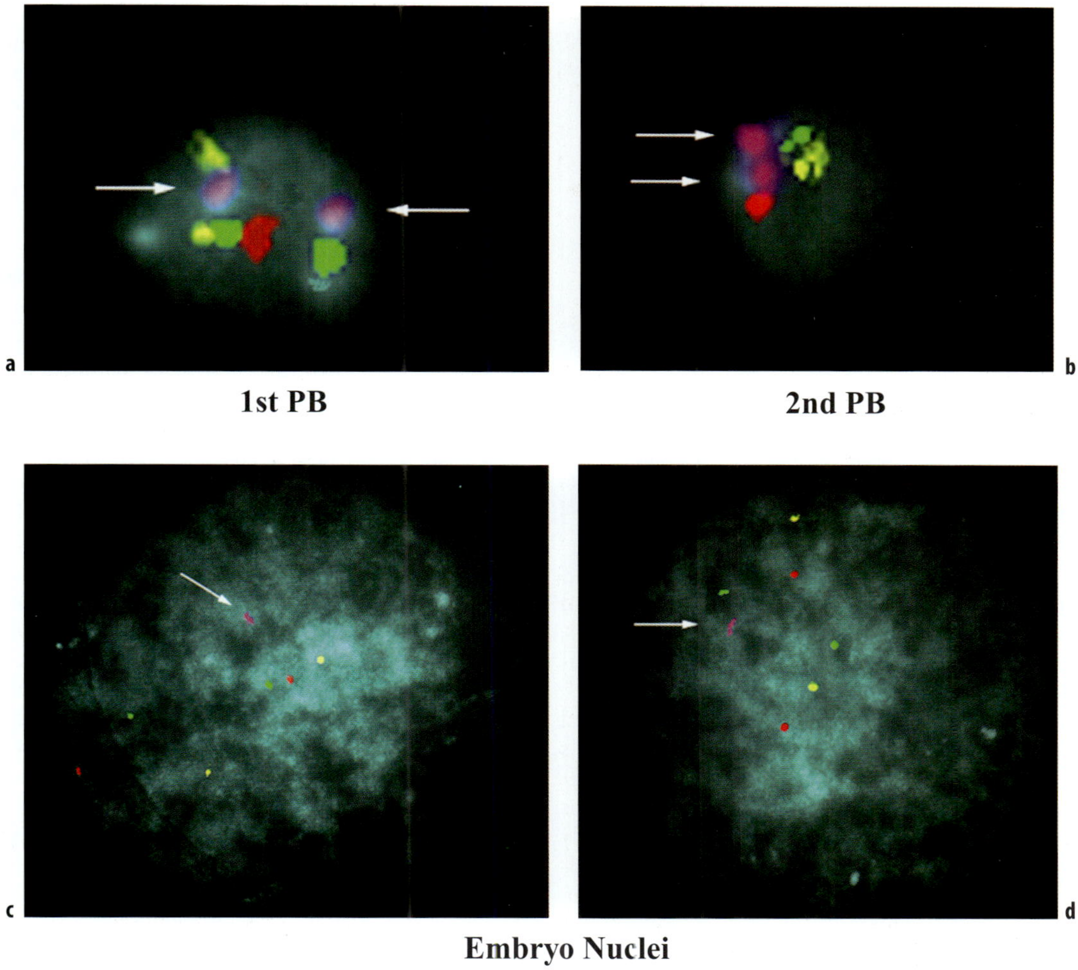

Figure 5.7. Meiosis II error resulting in monosomy 18. (*A*) FISH image of PB1 after a 3 hour hybridization with MultiVysion PB panel probe for chromosomes 13 (red), 16 (aqua), 18 (violet blue), 21 (green), and 22 (gold), showing two chromosomes 18 signal (*white arrows*), indicating a normal pattern of signals. (*B*) FISH image of PB2 showing a normal number of signals for each of the chromosomes tested, except for chromosome 18, which contain two instead of one signal (*white arrows*). Information obtained by the PB testing indicates that monosomy 18 is present in the embryo due to the chromatid error in meiosis II. (*C* and *D*) FISH images of metaphase chromosomes obtained after embryo biopsy and fixation on day 3, confirming the presence of sigle chromosome 18 (*white arrows*).

division (Figure 5.5). Overall, 2042 (35.2%) of 5807 oocytes tested had meiosis II errors, of which 41.2% were with extra chromatid, 36.6% with missing chromatid, and 32.1% with complex errors.

The data also show that the direct testing of meosis I and meisis II errors allows avoiding the transfer of at least 50% of embryos resulting from aneuploid oocytes, which should clearly contribute to the pregnancy outcome of the IVF patients participating in this study, as will be demonstrated Chapter 6.

As mentioned, collecting more data, presently involving testing of 17,397 oocytes, further confirms that at least half of the oocytes obtained from IVF patients of advanced reproductive age are aneuploid, originating either from meiosis I, meiosis II, or both, as detected by PB1 and PB2 testing. Although, as mentioned, some overestimate attributable to limitations of FISH technique cannot be excluded, the majority of these abnormalities in oocytes seems to represent the true errors, obvious from the increase of aneuploidy rate with maternal age (Figure 5.8) and also from the

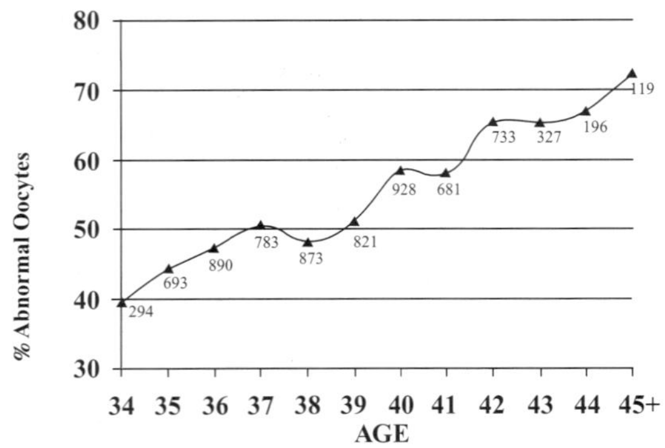

Figure 5.8. Line graph demonstrating the linear relationship between increasing maternal age and the percentage of aneuploid oocytes. Aneuploid rates range from 39% at age 34 to as high as 72% for women age 45 and older based on studies of up to five chromosomes (i.e., chromosomes 13, 16, 18, 21, and 22) (n = number of oocytes analysed).

follow-up studies of the embryos resulting from the oocytes with meiosis I, and meiosis II errors. In contrast to the traditional concept that aneuploidies mainly originate from female meiosis I, the direct testing of the outcome of meiosis I and meiosis II, presently performed by FISH analysis of 7103 PB1, and 7125 PB2, indicates comparable error rates in meosis I (41.8%) and meiosis II (37.3%). In addition, although the aneuploidy rate is expected to be higher with the testing for additional chromosomes, available data indicate an increase of the complex abnormalities rather than the overall aneuploidy rate [15]. The fact that the meiotic error of one chromosome may effect the segregation of other chromosomes was demonstrated also in XO female mice [30], and this was also observed in our data on the follow-up of meiosis I errors through meiosis II and cleavage of the resulting embryos (see Sections 5.1.1 and 5.1.2).

The other difference of these data from the results of the traditional meiotic studies is related to the observation that the majority of aneuploidies are represented by chromatid errors (see Figures 5.4 to 5.7), as previously suggested in mouse model [30], and one of the traditional meiotic studies, which, however, failed to observe any instance of chromosomal nondisjunction [29]. As seen from Figure 5.4, however, 41.8% of oocytes are abnormal following meiosis I, which includes both chromosome (2.6%) and chromatid errors (27.1%), with the majority of chromatid errors (19.8%) resulting in an extra chromatid in the MII oocyte (see also Figure 5.6). Even if some of these errors are attributable to technical factors, such as a possible failure of hybridization resulting in a missing signal in PB1, a more than twofold difference observed between missing and extra chromatids in the resulting MII oocytes may indicate the maintenance (in oocyte) rather than extruding the extra chromosome (chromatid) material (with PB1) when the meiosis I errors happen. This was also the case with chromosomal errors, which resulted in an extra chromosome in the MII oocytes 10 times higher (2.4%) than missing a chromosome (0.2%) (see Figure 5.4).

The fact that both the chromosomal and chromatid meiosis I errors lead to aneuploidy in the resulting embryos has been also confirmed by the follow-up study of the embryos resulting from these meiosis I errors, as the transfer of these embryos were avoided and used for follow-up study. Although the differences in the effect of the chromatid and chromosomal errors on preimplantation and postimplantation development cannot be excluded, this, however, has not yet been evaluated, because the embryos resulting from such oocytes were neither transferred nor further cultured, due to their use for confirmation of PB1 diagnosis, according to the patients' consent. In contrast to meiosis I errors, there were no differences in the missing or extra chromatid error rates following meiosis II (14.9% and 14.2%, respectively) (see Figure 5.5), suggesting different mechanisms of meiosis I and meiosis II errors.

5.1.1 Complex Errors and Aneuloidy Rescue in Female Meiosis

The above data show that more than one third of oocytes are abnormal following meiosis I, based on testing of PB1 alone (see Figure 5.4). It is of interest that more than one third of these oocytes with meiosis I errors have sequential errors in meiosis II as well, as inferred from PB2 testing. On the other hand, meiosis II errors are observed in one third of oocytes detected as normal following meiosis I, with the overall meiosis II error rate comparable to the meiosis I error rate (see Figure 5.5). So, approximately one half of meiosis II errors originate from oocytes with meiosis I errors, and the other half derive from the euploid MII oocytes, confirming that the preselection of aneuploidy-free oocytes should be based on the testing of the outcome of both the first and second meiotic divisions. As will be shown

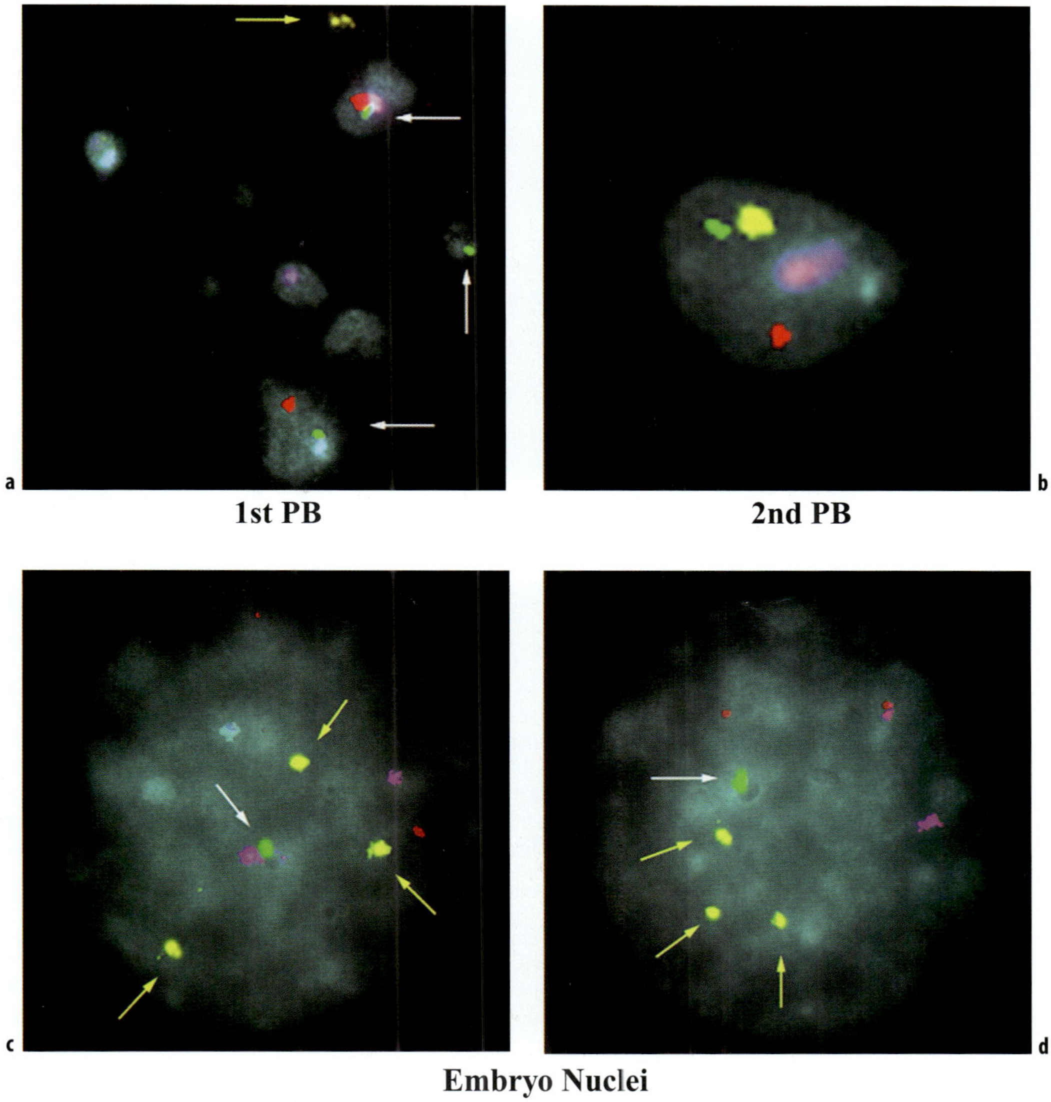

Figure 5.9. Complex errors in meiosis I resulting in monosomy 21 and trisomy 22. (*A* and *B*) FISH image of PB1 and PB2 after hybridization with the MultiVysion PB panel probe, revealing a double error in PB1, (*A*) represented by three (instead of two) chromosome 21 signals (*white arrows*) and only one split chromosome 22 signal (*yellow arrow*). (*B*) A normal number of signals (one each) are present in PB2. (*C* and *D*) FISH images of nuclei obtained from the resulting embryo, showing three chromosome 22 signals and one chromosome 21 signal, confirming the predicted double aneuploidy.

below (Section 5.1.2), there seems to be almost no chance of euploid embryos resulting from oocytes with sequential meosis I and meiosis II errors.

As seen from Figures 5.4 and 5.5, there was a comparable proportion of complex errors in meiosis I (12.1%) and meiosis II (8.2%), which included the oocytes with different chromosomes involved in the errors and different type of errors involved (Figure 5.9). This may be due to the age-related effects on the recombination frequency [1, 4, 31] or spindle formation errors also reported to increase with age [32, 33]. The data on the molecular mechanisms of cohesion of sister chromatids in meiosis may also be of relevance for understanding the nature of the age-related increase of meiotic errors resulting in complex abnormalities

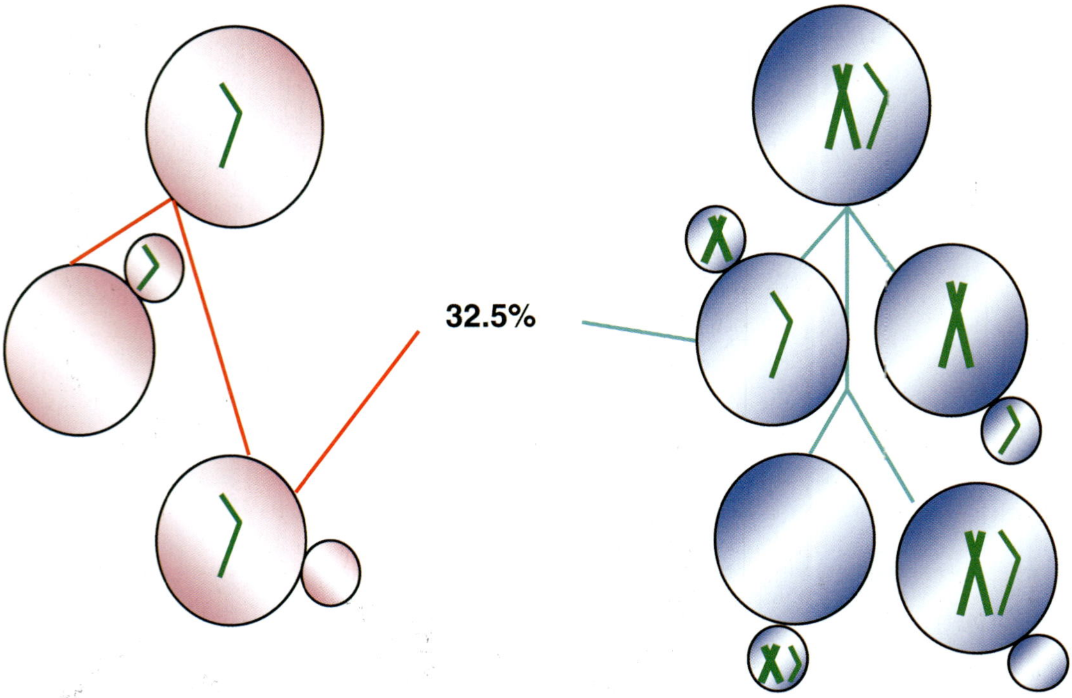

Figure 5.10. Schematic representation of segregation patterns of errors occurring in meiosis II from secondary oocytes with meiosis I errors resulting in balanced maternal contribution to zygote. (*Left*) Meiosis II "rescue" as a result of nonextrusion of chromatid to the second polar body from the secondary (metaphase II) oocytes with missing chromatid originating from a chromatid error in meiosis I. (*Right*) shows meiosis II "rescue" as a result of extrusion of two chromatids instead of one from a MII oocyte with extra chromatid resulting from meiosis I error. Overall, 32.5 % is an observed proportion balancing oocytes, originating from all types of sequential errors resulting in a normal (balanced) maternal contribution to zygote.

[34, 35]. Whatever the cause of the observed aneuploidy rates, approximately one third of chromosomal abnormalities in the resulting zygotes, overall, were represented by complex errors, suggesting that testing for only five chromosomes would probably detect the majority of chromosomal abnormalities resulting from meiosis I and II. This is not only because the abnormalities of these five chromosomes are most common, but also because the application of additional chromosome-specific probes will probably detect the proportional increase of the complex error rate, rather than a significant overall increase of chromosomally abnormal oocytes.

As seen from Figure 5.10, the follow-up of the outcome of meiosis I errors through meiosis II showed that almost one third (32.5%) of meiosis I errors appeared to result in apparently euploid zygotes, following a sequential error in meiosis II. Although the mechanism of the observed formation of such balanced zygotes is not yet understood, it may be similar to the well-known phenomenon of "trisomy rescue" in postzygotic embryo development, which may result in a uniparental disomy and imprinting disorders. As will be described in Section 5.1.2, the observed aneuploidy rescue mechanism in female meiosis cannot ensure the chromosomal normalcy of the resulting embryos to be useful for the embryos' transfer.

Analysis of the chromosomes-specific patterns of meiosis errors showed that, as expected, chromosome 21 and chromosome 22 errors were most prevalent, 10.9% and 11.8%, respectively, with the rates of 6, 6.4, and 6.8% observed for chromosomes 13, 16, and 18 errors, respectively (see Table 5.9). As demonstrated in Table 5.9, the chromosome-specific patterns of errors in meiosis I and meiosis II were not identical. In contrast to the previous reports on the predominant origin of chromosome 16 and 21 errors in meiosis I [1–3] and chromosome 18 errors in meiosis II [36], the direct data showed no significant difference in

Table 5.10. Chromosomes 13, 16, 18, 21, and 22 testing in day 3 embryos originating from oocytes with meiosis I and II errors resulting in balanced chromosome set

Balanced oocytes	#	Resulting embryos	#
1 Chromosome	70		
		Abnormal for the same chromosome	8
		Abnormal for 1 different chromosome	9
		Complex abnormality	35
		Normal for 5 chromosomes	18
2 Chromosomes	10	Complex abnormality	10
1 Balanced and 1 unbalanced	20	Complex abnormality	20
TOTAL	100	TOTAL NORMAL (%)	18

the origin of the chromosome 21 errors, and the opposite tendencies for chromosome 16 and 18 errors (see Table 5.9). The chromosome 13 error pattern was similar to that of chromosome 21, originating with a comparable frequency from meiosis I and meiosis II, and the chromosome 22-specific pattern was similar to the chromosome 16 pattern, originating predominantly in meiosis II (50.8% meiosis II errors vs. 30.8% meiosis I errors), in contrast to chromosome 18 errors, deriving more frequently from meiosis I (60.2% from meiosis I vs 29.7% from meiosis II). The comparison of these direct data to the data derived from DNA polymorphism studies, obtained from liveborn babies and spontaneous abortions with similar aneuploidies, will be of relevance for understanding of possible differences in viability of aneuploid embryos of different origin.

5.1.2 Mitotic Errors in Cleaving Embryos in Relation to Meiosis Errors

As shown above, approximately half of meiosis II errors are observed in the oocytes with prior errors in meiosis I. As a result of such sequential errors, almost one third of the resulting zygotes may have been considered normal (euplolid), provided that the preceding errors in meiosis I and meiosis II have no effect on the further preimplantation developments of the corresponding embryos. To investigate if these meiosis errors could affect the sequential mitotic divisions in the resulting zygotes and if these apparently euploid zygotes may develop into chromosomally normal embryo acceptable for embryo transfer in PGD cycles, follow-up testing of these embryos was carried out at the cleavage stage. As seen from Table 5.10 of 100 embryos tested overall, only 18%, deriving from the apparently balanced zygotes, were euploid for all five chromosomes analyzed, while the remaining majority were with chromosomal abnormalities [37, 38]

All of the chromosomally normal (euploid for five chromosomes tested) embryos appeared to result from zygotes with only one chromosomal error rescue, with none resulting from the zygotes balanced for two chromosomes. The fact that only a few resulting embryos (11%) were abnormal for the same chromosome, for which sequential meiosis I and meiosis II led to the balanced set, may suggest that the observed sequential errors in female meiosis may be attributable to the meiotic apparatus abnormality overall, rather than to a single chromosome segregation defect, which may further lead to a general defect of mitotic apparatus of the resulting embryos. This seems to be in agreement also with the observed types of aneuploidies detected in the resulting embryos, which in 79.3% of cases were represented by complex errors, including mosaicism (Figure 5.11), known to be highly prevalent at the cleavage stage [17, 18].

As the average reproductive age of the patients from whom the oocytes were obtained was approximately 38.5 years, the observed genomic instability in mitotic divisions of the apparently balanced zygotes following meiosis II rescue may be age-related. Although the mechanism by which the age factor may lead to these changes is not known, the underlying mechanisms of the aging process involve increasing errors in the mitotic machinery of dividing cells and chromosomal abnormalities. It was also suggested that deviations in the cytoplasmic organization, such as mitochondrial distribution, may reduce meiotic competence of oocytes and predispose the embryos to common cleavage abnormalities [23, 39–41]. The relationship between these cytoplasmic changes and the nuclear organization during maturation and fertilization of oocytes may

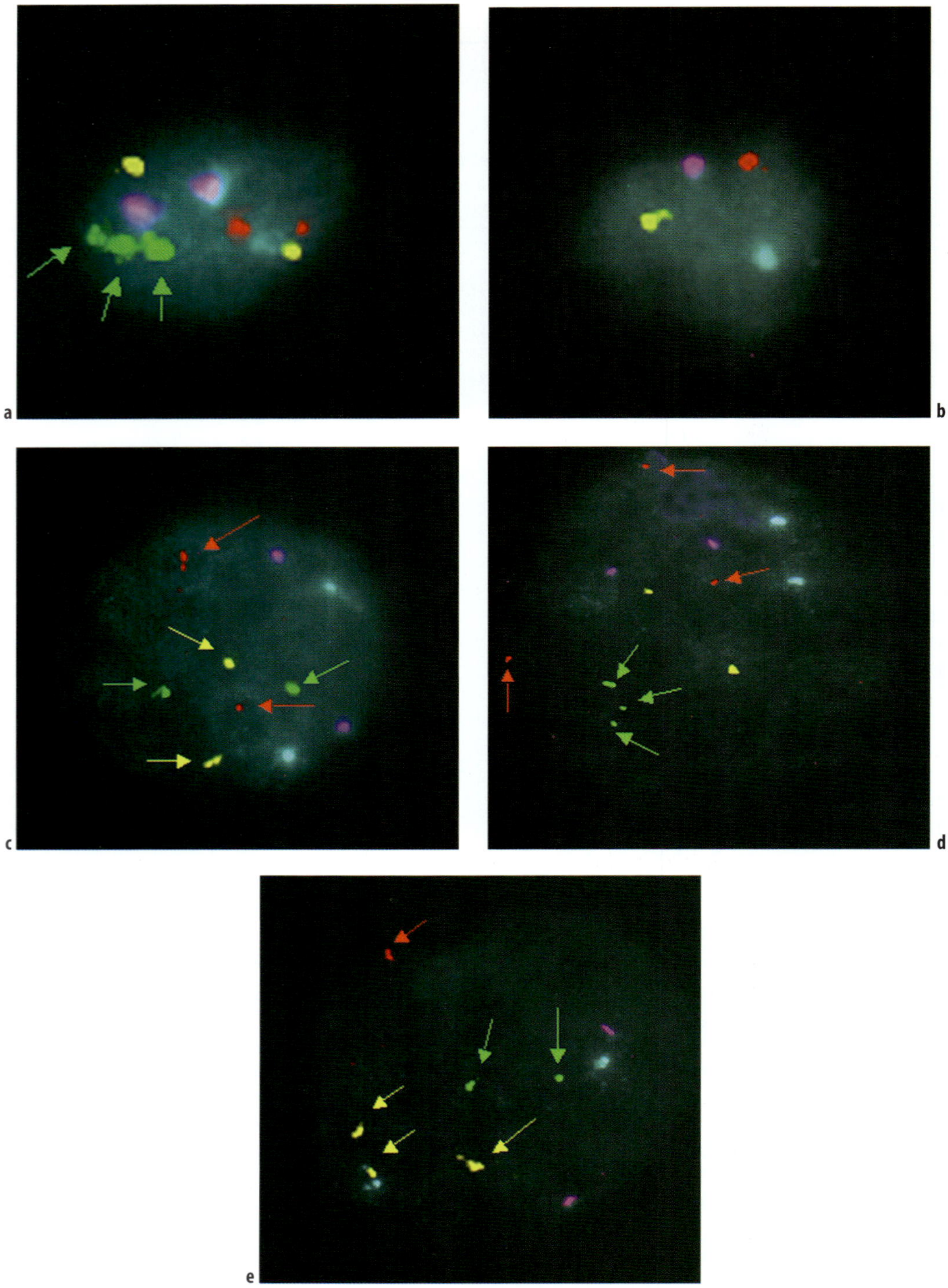

Figure 5.11. Sequential chromosome 21 errors in meiosis I and II resulting in an abnormal mosaic embryo. (*A* and *B*) FISH images of PB1 and PB2 after hybridization with the MultiVysion PB panel probe. (*A*) A normal number of signals for each chromosome (double dots) are detected in PB1 with the exception of three signals for chromosome 21 (*green arrows*). (*B*) Four instead of five signals are detected in PB2, with a missing green signal for chromosome 21, suggesting the normal number for all five chromosomes in the resulting oocyte. (*C* to *E*) Mosaicism in the embryo for chromosomes 13, 21, and 22. (*C*) FISH image of an interphase nucleus from the resulting embryo with normal number of signals, including chromosome 21 (green arrows). (*D*) FISH image of a second nucleus from the same embryo, in which three signals for chromosome 13 (*red arrows*) and chromosome 21 (*green arrows*) are present with a normal number of signals for chromosomes 16, 18, and 22. (*E*) FISH image of third interphase nucleus from the same embryo, in which three signals for chromosome 22 (*yellow arrows*) and only one signal for chromosome 13 (*red arrow*) are present, together with two signals for chromosomes 16, 18 and 21. The observed mosaicism may be associated with the sequential errors of chromosome 21 in meiosis I and meiosis II.

determine an abnormal development and mitotic errors at the cleavage stage, as suggested in prospective analysis of pronuclear zygote morphology in relation to chromosomal abnormalities detected in PGD for poor-prognosis IVF patients [22, 42].

According to the data on PGD for aneuploidies performed at the cleavage stage, at least 60% of embryos tested were with chromosomal abnormalities [17, 18, 43]. Although the reported types of aneuploidies may differ in different studies, there seems to be no doubt that approximately half of these abnormalities are represented by mosaicism. As there was not any information about the initial chromosomal set of the zygotes from which the mosaic embryos originated in any of these studies, the nature of mosaicism in preimplantation embryos is not known, despite its high prevalence and the potential clinical relevance. There were, however, some indirect observations, suggesting that the observed mosaicism at the cleavage stage may be of different nature, with some of mosaic types increasing with maternal age [44], and therefore, probably stemming from the female meiosis errors, and the others possibly attributable to an immaturity of centrosome structures in sperm, expected to be active from the first mitotic divisions of zygote, suggested for the cases of TESE patients [45].

It may be also suggested that a significant proportion of mosaic embryos originate from the oocytes that are aneuploid from the onset, through a process of "trisomy rescue." A possible high rate of further mitotic errors in cleaving embryos deriving from the oocytes with the complex aneuploidies may also explain the phenomenon of chaotic embryos, which make up almost a half of the embryos with mosaicism. A comparable prevalence of aneuploidies in oocytes and embryos with the differences of the types of chromosomal anomalies mainly attributable to a high frequency of mosaicism at the cleavage-stage embryos, may also support a prezygotic origin of the majority of the embryo chromosome abnormalities, including mosacism.

The comparison of the chromosome-specific aneuploidy rates in oocytes and embryos may be also of relevance to understanding the relationship between oocyte and embryo abnormalities (Table 5.11). As can be seen from these data, there is almost a two times higher rate for each chromosome error, except chromosome 16, in oocytes compared with that in embryos, which may indicate to a possible correction of some of the aneuploidies through the mechanism of "trisomy rescue," probably resulting in a certain proportion of mosaic embryos following the first three cleavage divisions. In fact, exact data on the mosaicism rate in preimplantation development are not known, because only a limited number of the preimplantation embryos were fully studied, with the majority available from PGD for aneuploidies performed through a single biopsied blastomere, which may not be representative of the whole embryo. Although the possibility of the postzygotic mitotic errors in the cleavage-stage embryos euploid from the onset cannot be excluded, the proportion of the aneuploidy and mosaicism stemming from these errors is not known, as well as the impact of these postzygotic errors on the preimplantation and postimplantation embryo development.

Table 5.11. Chromosome-specific aneuploidy rates in oocytes and cleavage-stage embryos* (chromosomes 13, 16, 18, 21, and 22)

Chromosome	Total oocytes studied	Total abnormal	Embryos studied	Total abnormal
13	5907	354 (6.0%)	882	21 (2.4%)
16	4583	294 (6.4%)	520	27 (5.2)
18	6648	455 (6.8%)	999	17 (1.7%)
21	6648	725 (10.9%)	882	38 (4.3%)
22	4583	539 (11.8%)	302	17 (5.6%)

*Data for embryos from cleavage stage taken from Munné [18].

There was, however, disagreement between specific types of aneuploidies detected by PB and blastomere testing. As mentioned, PB testing predicted 3:1 disomy/nullisomy ratio in oocytes following meiosis I, which is in agreement with predominance of trisomies over monosomies in spontaneous abortions. With the sole exception of monosomy 21, autosomal monosomies are not compatible with postimplantation development and have never been detected in recognized pregnancies or at birth. On the other hand, significantly higher prevalence of autosomal monosomies over trisomies has recently been reported in cleaving embryos [43], suggesting their possible postzygotic origin through mitotic nodisjunction or anaphase lag in the first cleavages. To investigate if this is an overestimate of monosomies due to a hybridization failure in FISH analysis, we compared this to the data obtained by PCR-based testing of blastomeres from women of advanced reproductive age, the results of which are presented below.

5.1.3 PCR-Based Aneuploidy Testing in Cleaving Embryos

To improve the accuracy of preimplantation genetic diagnosis, it is becoming a current practice to test for single-gene disorders together with chromosomal abnormalities, such as in cases of advanced reproductive age, so the same single cell is tested for both genetic and chromosomal disorders. One of possible approaches for PCR-based testing for chromosomal aneuploidies is DNA fingerprinting, which is based on the patterns of alleles that uniquely identify an individual, relying on a multiplex fluorescent PCR of low template DNA (see Chapter 2, Figure 2.12). Using STRs with a high heterozygosity, we performed single-cell DNA fingerprinting for detection of aneuploidies for chromosomes 2, 3, 4, 5, 6, 7, 8, 9, 11, 12, 13, 15, 16, 17, 18, 19, 21, and 22 in PGD for single-gene mutations mapped in these chromosomes, preimplantation HLA matching, or testing for aneuploidies. Overall, 2074 blastomeres were tested, with the number of blastomeres tested for each chromosome ranging from a few for chromosomes 3 and 9 to over a hundred for chromosomes 6, 7, 11, 13, 16, 18, 19, 21, and 22, so the individual aneuploidy prevalence was evaluated only for these chromosomes, and was 11, 5, 7, 10.1, 6.7, 10.2, 7, 16.9, and 7.6%, respectively. Overall, 42% aneuploidy prevalence was evaluated based in the series of 276 blastomeres (Figure 5.12), in which copy numbers of chromosomes 13, 16, 18, 21, and

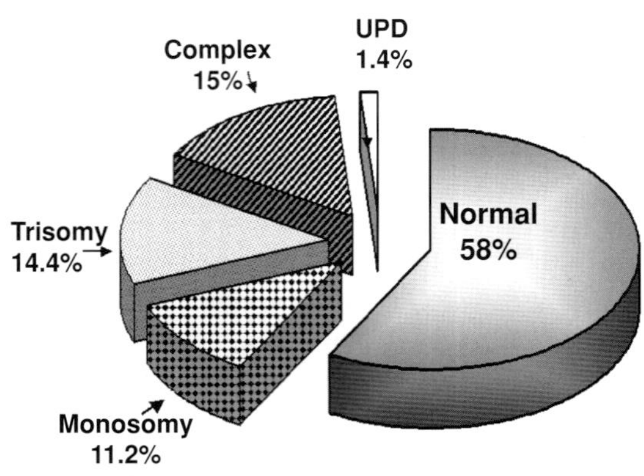

Figure 5.12. PCR-based aneuploidy rate for chromosomes 13, 16, 18, 21, and 22 in preimplantation embryos. Pie chart showing relative distribution and different types of aneuploid embryos. There is comparable prevalence of trisomies and monosomies, with more than one third of aneuploidies represented by complex errors, and 1.4% represented by uniparental disomies (UPD).

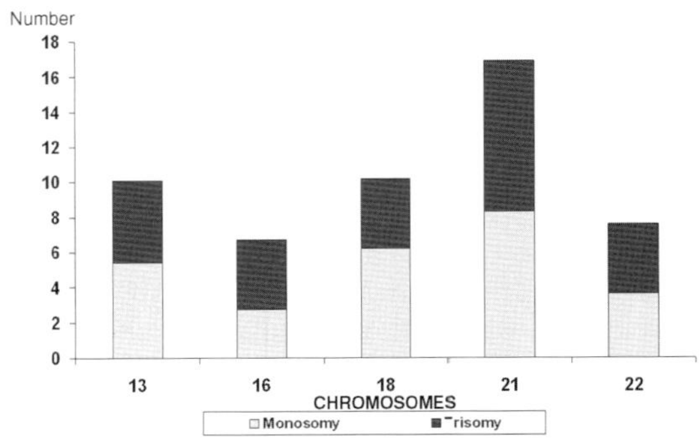

Figure 5.13. Relative distribution of trisomies and monosomies for different chromosomes (chromosomes 13, 16, 18, 21, and 22) detected by PCR-based aeuploidy testing of 276 preimplantation embryos. Bar graph demonstrating the observed distribution of each chromosome aneuploidy with relative proportion of trisomies and monosomies showing no significant difference, although there is a tendency of predominance of trisomy over monosomy for all chromosomes except chromosome 18.

22 were detected in the same blastomeres. No significant monosomy/trisomy ratio differences were observed, opposite to findings of FISH analysis at the cleavage stage, despite monosomy/trisomy ratio variations for individual chromosomes (Figures 5.12 to 5.13). As can be seen from these data, there is no significant excess of monosomies over trisomies for any of the chromosomes, which is also evident from Table 5.12 showing the comparison of prevalence of trisomy and monosomy for each chromosome obtained by different methods. Although the numbers of blastomeres tested are not yet comparable, preliminary data fail to confirm the higher prevalence of monosomies in preimplantation embryos. The data also showed errors of two or more chromosomes observed in one third of aneuploid blastomeres, in agreement with previous FISH data in oocytes and blastomeres. Further data collection will be required to exclude a possible monosomy overestimate due to allele drop out in single-blastomere PCR, as well as trisomy underestimate due a detection failure of extra maternal or paternal chromosomes because of indistinguishable homologues sharing the same polymorphic markers. The data show that, in addition to PGD for single-gene disorders, simultaneous DNA fingerprinting for copy numbers of chromosomes allows avoiding the transfer of chromosomally abnormal embryos in couples of advanced reproductive age.

Table 5.12. Comparison of FISH and PCR-based aneuploidy prevalences

	#Embryos tested		Monosomic		Trisomic		Total		Total (%)	
Chromosome	FISH*	*PCR***	FISH	*PCR*	FISH	*PCR*	FISH	*PCR*	FISH	*PCR*
XY	1741		8		13		21		1.2	
1	559		8		6		14		2.5	
4	327	*42*	4	*1*	3		7	*1*	2.1	*2.3*
6	194	*625*	2	*53*	1	*16*	3	*69*	1.5	*1.1*
7	244	*278*	4	*13*	3	*1*	7	*14*	2.9	*5*
13	1801	*276*	35	*15*	18	*13*	53	*28*	2.9	*1*
14	280		1		2		3		1.1	
15	1066	*59*	31	*6*	19	*4*	50	*10*	4.7	*16*
16	1665	*396*	49	*11*	37	*13*	86	*24*	5.2	*6.5*
17	609	*65*	9	*2*	7	*0*	16	*2*	2.6	*3*
18	2058	*276*	23	*17*	24	*11*	47	*28*	2.3	*10*
21	2011	*276*	56	*23*	38	*24*	94	*47*	4.7	*17*
22	1274	*312*	50	*10*	34	*14*	84	*24*	6.6	*7*

*Munné et al. [43].
***Rechitsky et al., unpublished data.*

5.1.4 Practical Relevance of Autosomal Monosomy Detection

Discordance of the trisomy/monosomy ratio detected in oocytes and embryos also may be explained by the fact that a certain proportion of monosomies may not be true monosomies, but represent mosaic embryos, which will actually form euploid embryos in the process of preimplantation or postimplantation development. To investigate the fate of monosomies detected prezygotically and postzygotically, these embryos were followed up to the blastocyst stage and the resulting blastocysts were reanalyzed for chromosomal status by FISH analysis using commercial probes specific for five chromosomes, including chromosomes 13,16, 18, 21, and 22. A total of 3140 nuclei were analyzed from oocytes and embryos, obtained from women of average age of 38.6 (±3.6) years. A total of 134 monosomic embryos were followed up overall, 51 (38%) of which were detected by PB testing (monosomy 13, 3; monosomy 16, 10; monosomy 18, 5; monosomy 21, 14; monosomy 22, 17; and complex monosomies, 2), and 83 (62%) by blastomere analysis (monosomy 13, 10; monosomy 16, 11; monosomy 18, 10; monosomy 21, 15; monosomy 22, 21, and complex monosomies, 16). The proportion of monosomy confirmation in the resulting blastocysts for prezygotic monosomics was 88.1%, which is significantly higher compared with postzygotic monosomies, confirmed only in 59.6% of cases, the remaining showing the normal karyotype or mosaicism (Table 5.13). It is also of note that only 47 of 83 postzygotic monosomics developed to blastocyst stage (56.6%), compared with 41 of 51 prezygotic ones (80.4%) suggesting that, although the proportion of monosomies that achieve balstocyst depends on the monosomy origin, autosomal monosomies may be compatible with preimplantation development, irrespective of the origin, probably being lost during implantation (Figure 5.14).

Based on the above data, it may be suggested that the most accurate preselection of embryos for transfer in PGD for aneuploidies may be performed by a sequential testing of meiosis I, meiosis II, and mitotic errors, through sequential PB1, PB2, and blastomere sampling. This may allow the avoidance of the transfer of embryos with prezygotic chromosomal errors, which seems to be the major source of chromosomal abnormalities in the embryo, and also the detection of possible mitotic errors in embryos resulting from the euploid zygotes, the proportion of which cannot be evaluated at the present time. The accumulated data on such sequential sampling will help to evaluate possible differences in viability of the embryos with chromosomal abnormalities of meiotic and mitotic origin.

Table 5.13. Follow-up of monosomies to blastocyst stage

Origin of monosomies	Total #	Blastocyst	Confirmation	
			#	%
Prezygotic	51	41	37	88.1
Postzygotic	83	47	28	59.6

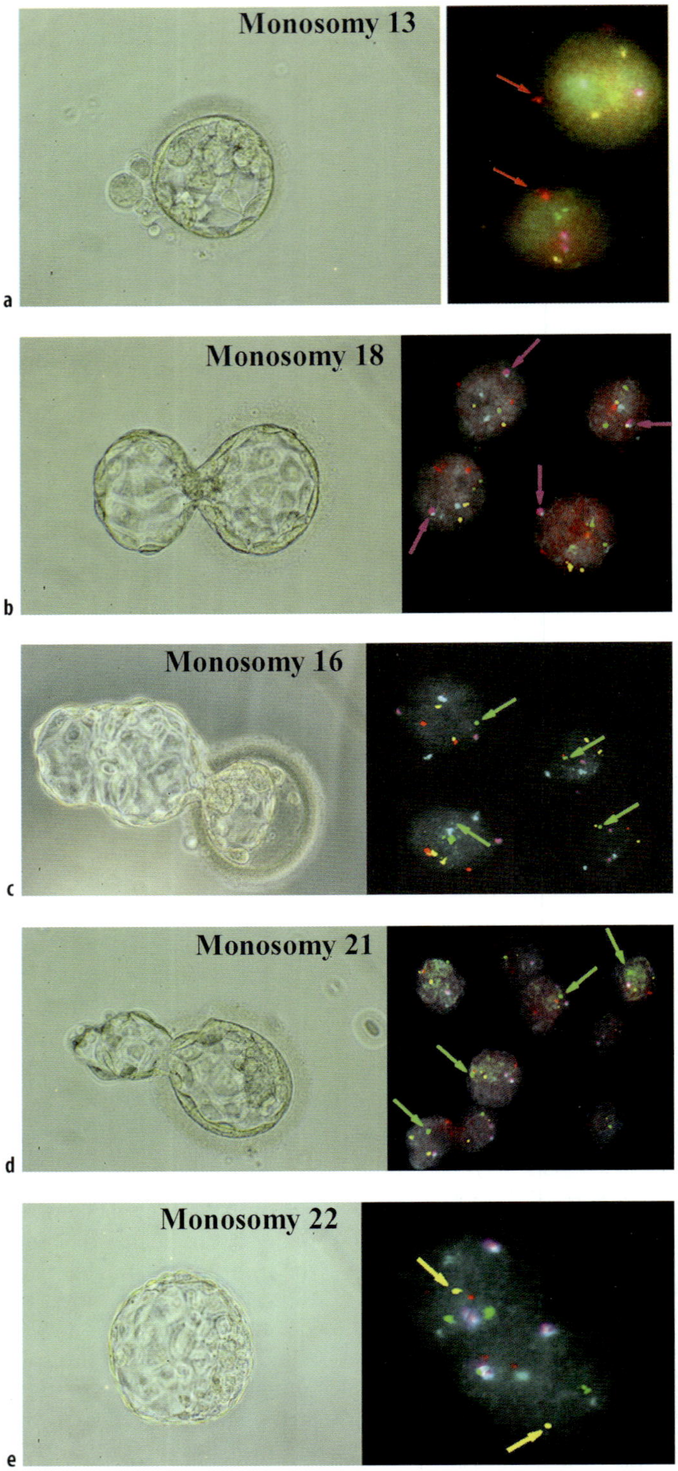

Figure 5.14. Development of embryos with monosomy for different chromosomes to blastocyst stage. (*A*)-Image of poor quality cavitating morula, resulting from monosomy 13 embryo originating from meiosis II error (a single signal for chromosome 13 in blastomere is shown by *red arrow*) (*B*) Image of hatching blastocyst derived from monosomy 18 embryo, originating from meiosis I error (a single signal for chromsome 18 in blastocyst cells is shown by *violet-blue arrow*). (*C*) Image of a hatching blastocyst through the artificial opening creating during biopsy procedure with a compact inner cell mass derived from monosomy 16 embryo originating from meiosis II error (a single chromsome 16 signal in nuclei isolated from this embryo is shown by *green arrow*). (*D*) Day 5 image of the developing embryo with trisomy 21, originating from meiosis I error (a single signal for chromosome 21 in blastocyst cell is shown by *green arrow*). (*E*) Image of fully hatched blastocyst deriving from embryo with monosomy 22, detected by blastomere biopsy (a single signal for chromosome 22 in nuclei isolated from this embryo is shown by *yellow arrow*).

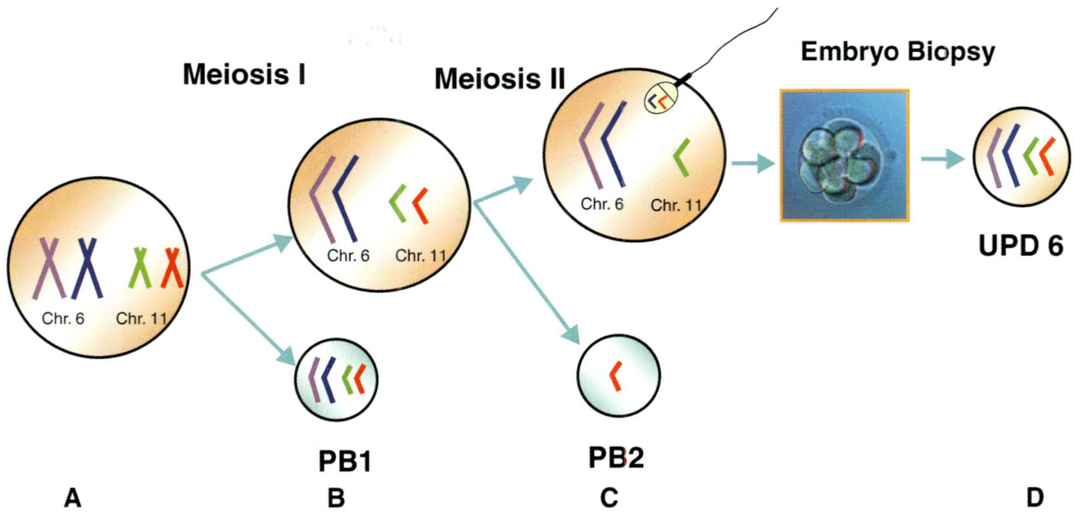

Figure 5.15. Uniparental disomy (UPD) for chromosome 6 detected through HLA matching of oocytes. (*A*) A primary oocyte with normal number of chromosomes 6 and 11 before maturation. (*B*) Testing of the first polar body (smaller circle) for thalassemia mutation and HLA shows the normal copies number of both chromosomes, suggesting the normal result for metaphase II oocyte (larger circle). (*C*) Sequential testing of the second polar body (smaller circle) revealed no chromosome 6 material while the normal (single copy) of chromosome 11 is present, suggesting extra chromosome material left in oocyte (larger circle, also showing penetrated sperm material with a of paternal copy of chromosome 6 and 11 to zygote). (*D*) Blastomere testing from day 3 embryo, revealing normal copy number of chromosomes 6 and 11, however both chromosome 6 copies are from maternal origin, with no paternal contribution, suggesting uniparental disomy 6 of maternal origin, originating from trisomy 6 zygote, as a result of trisomy rescue with loss of paternal chromosome 6.

5.1.5 Uniparental Disomies

The introduction of meiosis error testing as a possible integral component of IVF may be also useful in avoiding some of the imprinting disorders, which have recently been reported to be associated with IVF procedures [46–50]. More than two dozen cases of imprinting disorders, including Beckwith-Wiedemann and Algelman syndromes, have been described in association with ART, which may be caused by epigenetic errors, uniparental disomies, or other unknown factors. While epigenetic factors have been extensively reviewed [49, 51], the role and contribution of uniparental disomies have not been sufficiently understood. According to recent reviews, uniparental disomies of a few chromosomes were summarized to potentially lead to imprinting disorders, including chromosomes 6, 7, 11, 15, 19, and 20 [51]. There are not sufficient data on the prevalence of uniparental disomies in preimplantation development either, as they cannot be distinguished from normal set of chromosomes by presently used FISH technique. Uniparental disomy may be suspected in euploid embryos predicted to be trisomic by PB1 and PB2 testing, because one third of the embryos originating from trisomic zygotes will be with uniparental disomy as a result of "trisomy rescue." Also, uniparental dysomy may be accurately detected using DNA fingerprinting, as demonstrated in Figure 5.15 (see also Figure 4.9 in Chapter 4), showing uniparental disomy 6 detected in preimplantation HLA typing. Preliminary data on DNA fingerprinting for only five chromosomes, including chromosomes 13, 16, 18, 21, and 22 suggest 1.4% prevalence of uniparental disomies (see Figure 5.12). However, based on the prevalence of different types of aneuploidies, presented above, one of two oocytes (50%) obtained from women of advanced reproductive age may be expected to be aneuploid, one of three trisomic (approximately 35%, overall), and one of three trisomy rescued (approximately 12%), resulting in at least one of three to be uniparental disomy, suggesting as high as 4% expected uniparental disomies of maternal origin. On the other hand, despite low prevalence of paternally derived autosomal aneuploidies, the available data indicate of occurrence of paternal uniparental disomies as well (Figure 5.16).

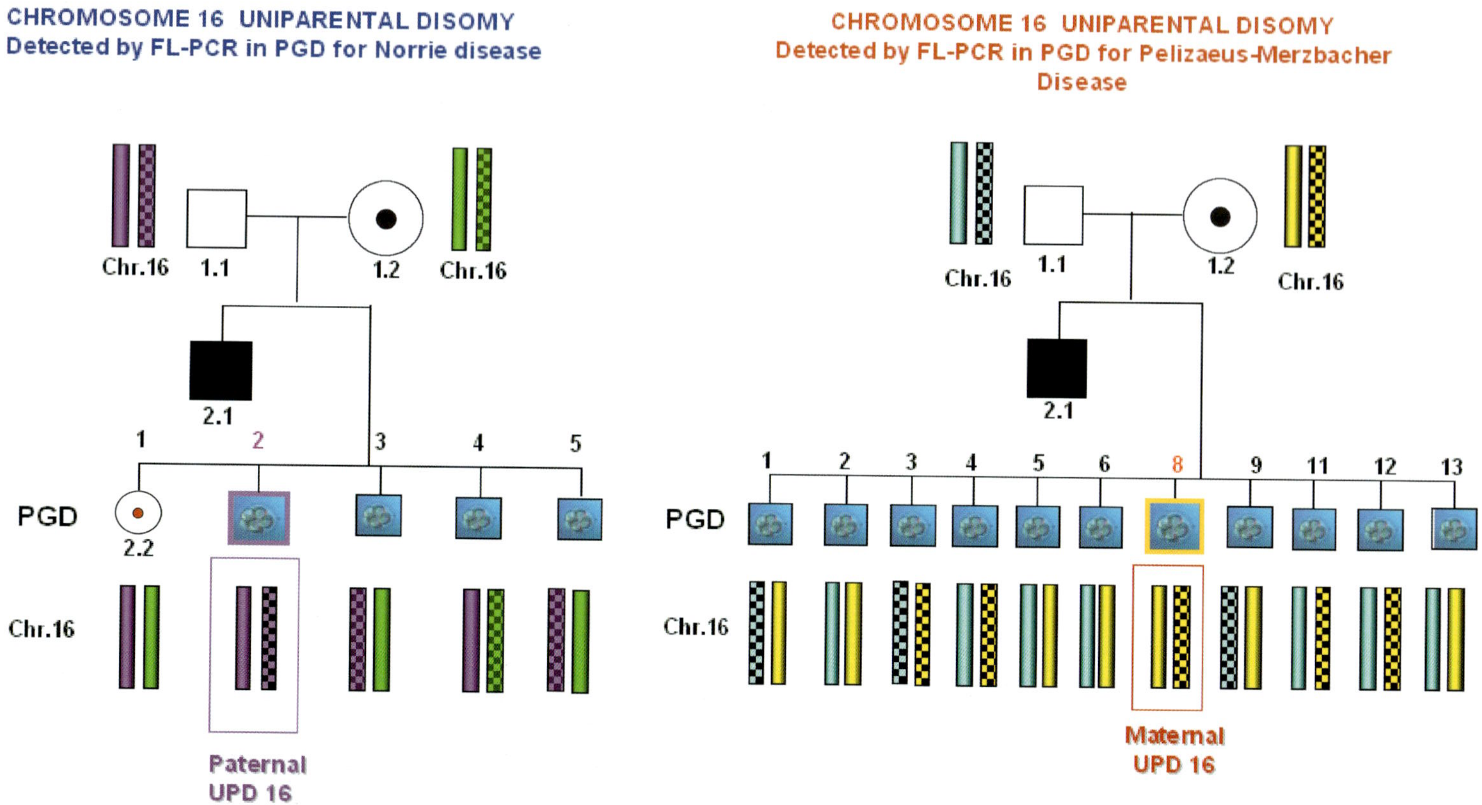

Figure 5.16. Chromosome 16 uniparental disomy of paternal and maternal origin. (*Left*) Chromosome 16 uniparental disomy detected by FL-PCR in PGD for Norrie disease, in which PCR based aneuploidy testing was performed because of advanced reproductive age. No maternal chromosome 16 was present, being lost in the process of trisomy rescue of paternal origin. (*Right*) Chromosome 16 uniparental disomy detected by FL-PCR in PGD for Pelizaeus-Merzbacher disease, in which PCR based aneuploidy testing was performed because of advanced reproductive age. No paternal chromosome 16 was present, being lost in the process of trisomy rescue of maternal origin.

The fact that more than half of IVF patients are age 35 or older and that more than half of their oocytes may be with aneuploidies, avoiding the transfer of the embryos resulting from these oocytes through PGD for aneuploidies should be clinically useful, in addition to potentially improving implantation and pregnancy rates, by avoiding the transfer of embryos with uniparental disomies as possible contributors to the imprinting disorders. Although the biological significance of uniparental disomies in preimplantation development is not known, it is possible that the detection and avoidance of uniparental disomies may also contribute in improvement of implantation and pregnancy rates.

The above data indicate the practical relevance of PGD for poor-prognosis patients, as more than half of the tested oocytes or embryos were with aneuploidies, which may clearly affect the developmental competence and the embryo's potential to implant if not removed from transfer. In contrast to the data obtained in traditional meiotic studies, the direct testing of the meiotic outcomes in patients of advanced reproductive age shows that chromosomal abnormalities originate comparably from meiosis I and meiosis II, and are predominantly of chromatid origin. Although isolated errors in meiosis I and meiosis II were also observed (39% and 31.9%, respectively), 42.7% of oocytes with meiosis I errors, overall, also had sequential meiosis II errors, resulting in apparently balanced zygotes in 32.5% of cases, which may represent a phenomenon of aneuploidy rescue in female meiosis. However, the resulting embryos from such balanced zygotes were predominantly aneuploid, suggesting the inherent predisposition of these zygotes to postzygotic chromosomal errors, following sequential errors in meiosis I and meiosis II. The chromosome-specific patterns of errors in meiosis I and meiosis II were different for each chromosome tested, and these patterns were not in agreement with the previously reported data based on DNA polymorphism in liveborn trisomies or spontaneous abortions. Comparison of the types of chromosomal aneuploidies and the prevalence of each chromosome-specific error in oocytes and embryos detected by FISH and DNA fingerprinting suggests that the majority of chromosomal aneuploidies in embryos originate from female meiosis, predisposing to further sequential postzygotic errors, which may explain the high rate of mosaicism in preimplantation embryos. The introduction of both PB and blastomere testing and the technique of DNA fingerprinting also demonstrated the occurrence of uniparental disomies, the possible impact of which is still to be documented. This may also indicate requirements for both oocyte and embryo testing in PGD for aneuploidies to exclude the transfer of embryos with aneuploidies orignating from meiosis and mitotic errors.

5.2 Chromosomal Rearrangements

Although PGD for chromosomal translocations was introduced only in 1996, it is currently one of the most practical applications of PGD, which appeared to have a major impact on the clinical outcome of carriers of balanced chromosomal translocations. Initially PGD was done for maternally derived translocations and performed by using PB1 [52], based on the fact that PB1 represent metaphase chromosomes. However, because PB1 alone does not always provide complete information, it is currently combined with PB2 analysis or performed by testing single cells obtained from preimplantation embryos, using, in the majority of cases, the interphase FISH analysis, which allows chromosome enumeration on the interphase cell nuclei. However, the number of chromosomes studied by FISH is limited to the number of chromosome-specific probes available. Even with the presently available methods for rehybridization of interphase nuclei for the second and the third time, complete karyotyping is not realistic in the present time. There are also limitations for the detection of some translocations, due to unavailability of certain segment-specific probes, making it highly important to develop the methods for visualization of chromosomes in single cells, including PBs and individual blastomeres, which may clearly improve the accuracy of PGD for chromosomal disorders [53].

As mentioned, one of the approaches to PGD for translocations is PB1 testing, based on the fact that PB1 never forms an interphase nucleus and consists of metaphase chromosomes. It has been shown that PB1 chromosomes are recognizable when isolated 2–3 hr after in vitro culture, with degeneration beginning 6–7 hr after extrusion [54]. Therefore, whole chromosome painting or chromosome-segment-specific probes were applied for testing of maternally derived chromosomal translocations in PB1 [52]. Although the method resulted in a significant reduction of spontaneous abortions in patients carrying translocations, yielding unaffected pregnancies and births of healthy children, it has shown to be sensitive to malsegregation and/or recombination between chromatids, requiring a further follow-up analysis of PB2 in order to accurately predict the meiotic outcome following the second meiotic division [55, 56]. However, despite the progress in transforming PB2 into metaphase chromosomes via electrofusion of PB2 nucleus with foreign one-cell human embryo, the proportion of metaphase plates did not exceed 64%, even after enucleation of the recipient one-cell-stage mouse embryo, to be useful in clinical practice [57]. Therefore, alternative methods were developed to convert single blastomere nuclei into metaphase chromosomes, following the fusion of single blastomeres with murine or bovine zygotes [58, 59]. These appeared to be useful for PGD of both maternally and paternally derived translocations.

One hundred eighty three PGD cycles for translocations were performed for 130 patients, using a standard IVF protocol combined with PB1 and PB2 removal and/or embryo biopsy and FISH analysis (some of this experience was reported previously [56]. The type of translocation and the maternal or paternal origin is presented in Table 5.14, showing that the majority were reciprocal translocations, deriving predominantly maternally, and so were performed using PB1 and PB2 testing. An increasing number of cases are now performed also by blastomere analysis, described in Chapter 2.

In each case, the PGD strategy depends mainly on the chromosomes involved in the rearrangement and the size of the segments, as well as on the origin of translocations, as paternally derived translocations may be performed only on blastomeres. In maternally derived translocations, PB1 and PB2 testing may be a method of choice, in which centromeric and locus-specific probes are used together with whole chromosome painting to determine the number of chromatids. Then rehybridization with subtelomeric probes is performed to identify a possible chromatid excange, as shown in Figure 5.17.

In paternally derived translocation, the visualization of single blastomere chromosomes may be the method of choice, as it has advantages of differentiation of normal from unbalanced chromosomes and avoids diagnosis based on split and/or faint locus-specific or subtelomeric-specific signals. It also allows combining different probes to improve accuracy and follow-up chromatid exchange identified by PB analysis. To visualize blastomere chromosome, the human single blastomeres were fused with enucleated or intact mouse zygotes at pronuclear stage, known to be at the S phase of the cell cycle, and fixing the resulting heterokaryons at the metaphase of the first cleavage division [56, 60] as described in section 2.2.2. In brief one to two hours before electrofusion with the human blastomeres, mouse zygotes were thawed, freed of zonae pellucidae, and pipetted to separate PB2. An intact single blastomere was brought into contact with the mouse zygote by agglutination with phytohemagglutinin, followed by induction of cell fusion using a single direct-current pulse. Four hours after fusion, heterokaryons were monitored for signs of the disappearance of pronuclei and fixed at mitosis following hypotonic treatment. To avoid monitoring and maintain the heterokaryons in mitosis, they were cultured in the presence of microtubuli inhibitors. Heterokaryons with persisting pronuclei left in culture by the ninth hour after fusion were fixed following an hour of treatment with ocadaic acid (OA). This was followed by 10–15-min incubation in a hypotonic solution and then the resulting mitotic heterokaryons were fixed, air-dried, and analysed by FISH.

Prior to FISH, the slides were pretreated with a 1% formaldehyde solution (Vysis Inc., Downers Grove, IL), followed by digestion of residual cytoplasmic proteins with 0.5 mg/ml pepsin. As mentioned, the strategy of FISH analysis depended on chromosomes involved in the structural rearrangement and the size of the translocated segment, and usually involves a combination of the locus-specific, centromeric, and/or subtelomeric probes with whole chromosome painting, as well as required rehybridization. The list of translocations and the probes applied is presented in Table 5.14, presenting experience of 183 PGD cycles performed for 130 carriers of balanced translocations.

One hundred ten PGD cycles for translocations were performed utilizing the nuclear conversion method. Overall, the technique was applied to 906 blastomeres, resulting in successful nuclear metaphase conversion in as many as 804 (88.7%). This made possible the preselection of

Table 5.14. List of translocations, FISH probes used, and cycle outcomes

Patients	Cycles	Karyotype	FISH probes	Embryo transfers/ pregnancies/ deliveries (# of children born) (# of miscarriages)
12	15	45,XX,rob(13;14)(q10;q10)	WCP 13, WCP 14, LSI 13 (13q14), Tel 14q	13 / 4 / 2 (2) (2*)
1	1	45,XX,rob(13;22)(q10;q10)	WCP 13, WCP 22, PB panel (chromosomes 13,16,18,21,22)	1 / 0
2	2	45,XX,rob(14;21)(q10;q10)	WCP 14, WCP 21, LSI 21(21q22.13-q22.2)	2 / 1 / 1 (1)
1	5	45,XX,rob(14;22)(q10;q10)	WCP 14, WCP 22, Tel 14 q, LSI 22 (22q11.2)	3 / 1 / 1 (1)
14	16	45,XY,rob(13;14)(q10;q10)	WCP 13, WCP 14, LSI 13 (13q14), Tel 14q	14 / 5 / 3 (4) (1) 1 ongoing
2	3	45,XY,rob(13;15)(q10;q10)	WCP13, WCP 15, LSI 13 (13q14), Tel 15q	3 / 1 ongoing
2	3	45,XY,rob(14;21)(q10;q10)	WCP 14, WCP 21, LSI 21	3 / 0
1	1	45,XX,der(4),t(4;13)(q35;q12),-13	WCP 4, WCP 13, CEP 4, Tel 4q, Tel 13q	0 / 0
1	1	46,XX,t(1;4)(q23;q31.1)	WCP 1, WCP 4, (CEP 18 as control)	1 / 0
1	1	46,XX,t(1;7)(q23;p11)	WCP 1, WCP 7, CEP 1	1 / 1 / 1 (1)
1	3	46,XX,t(1;8)(q42;p11.2)	WCP 1, WCP 8, CEP 1, CEP 8, Tel 1q, Tel 8p	3 / 2 / 1 (1) (1)
1	2	46,XX,t(1;15)(q32;q26)	WCP 1, WCP 15, CEP 15, CEP 1, Tel 15 q, Tel 1q	2 / 1 / 1 (1)
1	1	46,XX,t(2;3)(p13;q29)	WCP 2, CEP 3, Tel 2p, Tel 3q	1 / 0
1	1	46,XX,t(2;5)(q33;p14)	WCP 2, WCP 5, Tel 2q, Tel 5p	1 / 0
1	2	46,XX,t(2;10)(q31;q11.2)	WCP 2, WCP 10, CEP 2, Tel 2q	1 / 0
1	1	46,XX,t(2;13)(q22;q33)	WCP 2, WCP 13, CEP 2, Tel 13 q, LSI 13 (13q14)	1 / 1 / 1 (1)
1	1	46,XX,t(2;16)(q37;p13.1)	WCP 2, WCP 16, CEP 16, Tel 2q, Tel 16p	0 / 0
1	1	46,XX,t(2;20)(q37;q11.2)	WCP 2, WCP 20, CEP 2, Tel 20q	0 / 0
1	1	46,XX,t(3;5)(p25;q35.2)	WCP 3, WCP 5, Tel 3p, Tel 5q	1 / 0
1	1	46,XX,t(3;6)(p26.2;q21)	WCP 3, WCP 6, CEP 6, Tel 3p, Tel 6q	1 / 1 / 1 (1)
1	1	46,XX,t(3;7)(p21;q22)	WCP 3, WCP 7, Tel 7q, Tel 3p	1 / 1 / 1 (1)
1	1	46,XX,t(3;8)(q27;q22)	WCP 3,WCP 8, Tel 8q, CEP 8, CEP 3	1 / 0
1	1	46,XX,t(3;10)(p25;q15.1)	WCP 3, WCP 10, Tel 3p, Tel 10q, CEP 3, CEP 10	0 / 0
1	2	46,XX,t(3;12)(q21;q24.33)	WCP 3, WCP 12, CEP 3, CEP 12, Tel 3q	2 / 0
1	1	46,XX,t(3;19)(p25;p13.3)	WCP 3, WCP 19	1 / 0
1	1	46,XX,t(3;22)(q13.2;q11.2)	WCP 3, WCP 22, CEP 3, Tel 3q, Tel 22q	1 / 1 / 1 (1)
1	1	46,XX,t(4;8)(p16;q24)	CEP 4, Tel 4p, Tel 8q	0 / 0
1	2	46,XX,t(4;13)(p14;q31)	WCP 4, WCP 13, Tel 13q	0 / 0
1	2	46,XX,t(4;13)(q23;q21)	WCP 4, WCP 13, LSI 13 (13q14), CEP 4, Tel 4q	1 / 0
1	2	46,XX,t(4;14)(q21.3;q24.3)	WCP 4, WCP 14, Tel 14 q, CEP 4	2 / 0
1	1	46,XX,t(5;7)(p14;p21)	WCP 5, WCP 7, Tel 7p	1 / 0
1	2	46,XX,t(5;10)(q11;q22.1)	WCP 5, WCP 10, CEP 10	2 / 0
1	1	46,XX,t(5;14)(q35;q32.2)	WCP 5, WCP 14, Tel 5q	0 / 0
1	2	46,XX,t(5;21)(q11.2;q22)	WCP 5, WCP 21, LSI 21 (21q22.13-q22.2), Tel 21q, Tel 5p, Tel 5q	1 / 0
2	3	46,XX,t(5;21)(q31;q22.1)	WCP 5, WCP 21, LSI 5 (5q33-q34), LSI 21 (21q22.13-q22.2)	1 / 0
1	2	46,XX,t(6;13)(q21;q22)	WCP 6, WCP 13, LSI 13, Tel 6q, Tel 13q	1 / 1 / 1 (1)
1	1	46,XX,t(6;14)(q15;q13.1)	WCP 6, WCP 14, CEP 6, Tel 6q, Tel 14q	0 / 0
1	1	46,XX,t(6;15)(p21.3;q26.1)	WCP 6, WCP 15, CEP 15, Tel 15q	1 / 1 / 1 (1)
1	1	46,XX,t(7;11)(p11.2;q24.2)	WCP 7, WCP 11, CEP 11, Tel 11q	1 / 1 / 1 (1)
1	1	46,XX,t(7;13)(q36;q12)	WCP 7, WCP 13, LSI 13 (13q14), CEP 7, Tel 7q	0 / 0
1	1	46,XX,t(7;18)(q32;q23)	WCP 7, WCP 18, CEP 18	1 / 0
1	1	46,XX,t(7;19)(q32;p13.2)	WCP 7, WCP 19, CEP 7, Tel 7q, Tel 19p	1 / 1(1)
1	3	46,XX,t(7;22)(p22.3;p11.2)	CEP 7, Tel 22q, Tel 7p	1 / 1 / 1 (1)
2	2	46,XX,t(8;10)(q22.1;q22.1)	WCP 8, WCP 10, CEP 10, Tel 8q, Tel 10q	1 / 0
1	3	46,XX,t(8;10)(q24.3;q24.1)	WCP 8, WCP 10, CEP 10, Tel 10q, Tel 8p, Tel 8q	1 / 0
1	1	46,XX,t(8;10)(p10;p10)	WCP 8, WCP 10, CEP 10, Tel 10p, Tel 8p	1 / 0
1	1	46,XX,t(8;12)(q21.2;p13.3)	WCP 8, WCP 12, Tel 12 p, CEP 12	0 / 0
1	2	46,XX,t(8;22)(q24.1;q11.2)	WCP 8, WCP 22, CEP 8	1 / 0
1	1	46,XX,t(9;11)(q13;q13.5)	WCP 9, WCP 11, CEP 11, Tel 11q	1 / 1(1*)

(Cont.)

Table 5.14. *Continued*

Patients	Cycles	Karyotype	FISH probes	Embryo transfers/ pregnancie/ celiveries (# of children born) (# of miscarriages*)
1	1	46,XX,t(9;11)(p24;q23.1)	WCP 9, WCP 11, CEP 11, Tel 9p, Tel 11q	1 / 1 / 1 (1)
1	1	46,XX,t(9;13)(q22.3;q12.3)	WCP 9, WCP 13, CEP 9, LSI 13 (13q14), Tel 9q	0 / 0
1	1	46,XX,t(9;13)(p21;q21)	WCP 9, WCP 13, CEP 9, Tel 9p, Tel 13q	0 / 0
1	2	46,XX,t(9;13)(q22;q14)	WCP 9, WCP 13, CEP 9, LSI 13 (13q14), Tel 9q	0 / 0
1	1	46,XX,t(10;11)(p15;q23)	WCP 10, WCP 11, CEP 10, Tel 10 p, Tel 11q	1 / 1 / 1 (3)
1	1	46,XX,t(10;15)(q23;q21)	WCP 10, WCP 15, CEP 15	0 / 0
1	1	46,XX,t(11;14)(p15;q24)	WCP 11, WCP 14, CEP 11, Tel 11p	0 / 0
2	5	46,XX,t(11;22)(q23;q11.2)	WCP 11, WCP 22, CEP 11, Tel 11q, LSI 22 (22q11.2)	4 / 1 ongoing
1	1	46,XX,t(12;15)(p13.3;q13.3)	WCP 12, WCP 15, Tel 12p, Tel 15q, CEP 15	1 / 0
1	4	46,XX,t(12;18)(p13.31;q21.32)	WCP 12, WCP 18, CEP 18, Tel 12p, Tel 18q	2 / 1(1*)
1	1	46,XX,t(14;18)(q24.3 ;p11.31)	WCP 14, WCP 18, CEP 18, Tel 14q, Tel 18p	0 / 0
1	2	46,XX,t(14;21)(q24.1;q22.1)	WCP 14, WCP 21, Tel 14q	1 / 1 / 1 (1)
1	1	46,XX,t(16;17)(p13.3;q11.2)	WCP 16, WCP 17, CEP 17, Tel 16p, Tel 17q	1 / 1 / 1 (2)
1	1	46,XX,inv(X)(p22.3;q13)	CEP X, Tel Xp/Yp, Tel Xq/Yq	1 / 1 / 1 (1)
1	1	47,XX,+fis(19)(p10),+fis(19)(q10)	Tel 19p, Tel 19q	1 / 0
1	1	46,XY,t(1;3)(p36.1;q27)	WCP 1, WCP 3, CEP 1, Tel 1p, Tel 3q	1 / 1 / 1 (1)
1	1	46,XY,t(1;3)(q42.3;q29)	WCP 1, WCP 3, CEP 3, Tel 1q, Tel 3q	1 / 0
1	1	46,XY,t(1;7)(q11;p13)	WCP 1, WCP 7, CEP 7, Tel 1q, Tel 7p	0 / 0
1	8	46,XY,t(1;8)(p13;q23)	WCP 1, WCP 8, CEP 8, CEP 1,Tel 1q	5 / 0
1	2	46,XY,t(1;21)(p13;q11.2)	WCP 1, WCP 21, CEP 1	2 / 0
1	1	46,XY,t(2;3)(q21;p13)	WCP 2, WCP 3, CEP 2,Tel 2q, Tel 3p	0 / 0
1	1	46,XY,t(2;11)(p13.3;p15.1)	WCP 2, WCP 11, CEP 2, CEP 11, Tel 2p	1 / 0
1	3	46,XY,t(2;18)(p13;q11)	WCP 2, WCP 18, CEP18	2 / 1 / 1 (1)
1	1	46,XY,t(2;19)(p25.1;p13.11)	WCP 2, WCP 19, Tel 2p, Tel 19p, CEP 4	1 / 1(1)
1	2	46,XY,t(3;6)(q27.1;p21.1)	WCP 3, WCP 6, CEP 6 Tel 3q, Tel 6p	2 / 1 / 1 (1)
1	1	46,XY,t(3;7)(p24;p21)	WCP 3, WCP 7, Tel 3p	1 / 0
1	2	46,XY,t(3;10)((q21;p14)	WCP 3, WCP 10, CEP 10	0 / 0
1	1	46,XY,t(3;10)(q21;q24.1)	WCP 3, WCP 10, CEP 10, Tel 3q, Tel 10q	1 / 0
1	1	46,XY,t(4;6)(q25;q21)	WCP 4, WCP 6, CEP 6, Tel 6q	0 / 0
1	1	46,XY,t(4;11)(q33.3;p13)	WCP 4, WCP 11, CEP 4, Tel 4q, Tel 11p	1 / 1 / 1 (1)
1	1	46,XY,t(4;17)(q31.1;p13)	WCP 4, WCP 17, CEP 4, Tel 4q, Tel 17p	1 / 0
1	1	46,XY,t(5;7)(q21.2;q36)	WCP 5, WCP 7, Tel 7q, CEP 7, Tel 5p	1 / 1 / 1 (1)
1	1	46,XY,(5;12)(p14;p13.33)	CEP 12, Tel 5p, Tel 12p	0 / 0
1	1	46,XY,t(6;7)(q23;q36)	WCP 6, WCP 7, CEP 7	1 / 1 / 1 (1)
1	2	46,XY,t(6;8)(q21;q32.1)	WCP 6, WCP 8	1 / 0
1	1	46,XY,t(7;10)(q36;p11.2)	WCP 7, WCP 10, CEP 10, Tel 7q, Tel 10p	1 / 0
1	1	46,XY,t(8;10)(q22.1;q22.1)	WCP 8, WCP 10, CEP 10, Tel 8q	1 / 1 / 1 (2)
1	2	46,XY,t(8;19)(p23.1;q13.3)	WCP 8, WCP 19, CEP 8, Tel 8p, Tel 19q	2 / 1 / 1 (1)
1	1	46,XY,t(9;11)(p22;q23)	WCP 9, WCP 11, CEP 11, Tel 9p, Tel 11q	1 / 0
1	1	46,XY,t(10;13)(p12;p11.2)	WCP 10, WCP 13, CEP 10	1 / 0
1	1	46,XY,t(10;22)(q24.1;p12)	WCP 10, WCP 22, LSI 22 (22q11.2), Tel 10q, CEP 10	1 / 1 / 1 (1)
1	1	46,XY,t(11;20)(p15;q13.12)	WCP 11, WCP 20, Tel 11p	1 / 1 / 1 (1)
1	1	46,XY,t(11;22)(q23.3;q11.2)	CEP 11, Tel 11q, LSI 22 (22q11.2)	0 / 0
1	2	46,XY,t(11;22)(q25;q12)	WCP 11, WCP 22, CEP 11, Tel 11q	1 / 0
2	3	46,XY,t(13;20)(q22;q11.2)	WCP 13, WCP 20	0 / 0
1	1	46,XY,t(15;16)(q13;q13)	WCP 15, WCP 16, CEP 15	1 / 0
1	1	46,XY,t(16;20)(q12.3;q13.1)	WCP 16, WCP 20, CEP 16, CEP 20, Tel 16q, Tel 20q	1 / 0
1	1	46,XY,t(18;20)(p11.2;q13.3)	WCP 18, WCP 20, CEP 18, Tel 18p, Tel 20q	1 / 1 / 1 (1)
1	2	46,XY,inv(5)(p13;q13)	Xcyte 5 (Metasystems), Tel 5p, Tel 5q	2 / 0
1	1	46,XY,t(9;6)(q22;q15q23.1)	G-Banding, WCP 6, WCP 9, CEP 6	0 / 0
130	183			126 / 45 / 34 (39) (8) 3 ongoing

WCP, whole chromosome paint; LSI, locus specific identifier; Tel, subtelomeric; CEP, centromeric enumeration probe; G-banding, Giemsa stain.

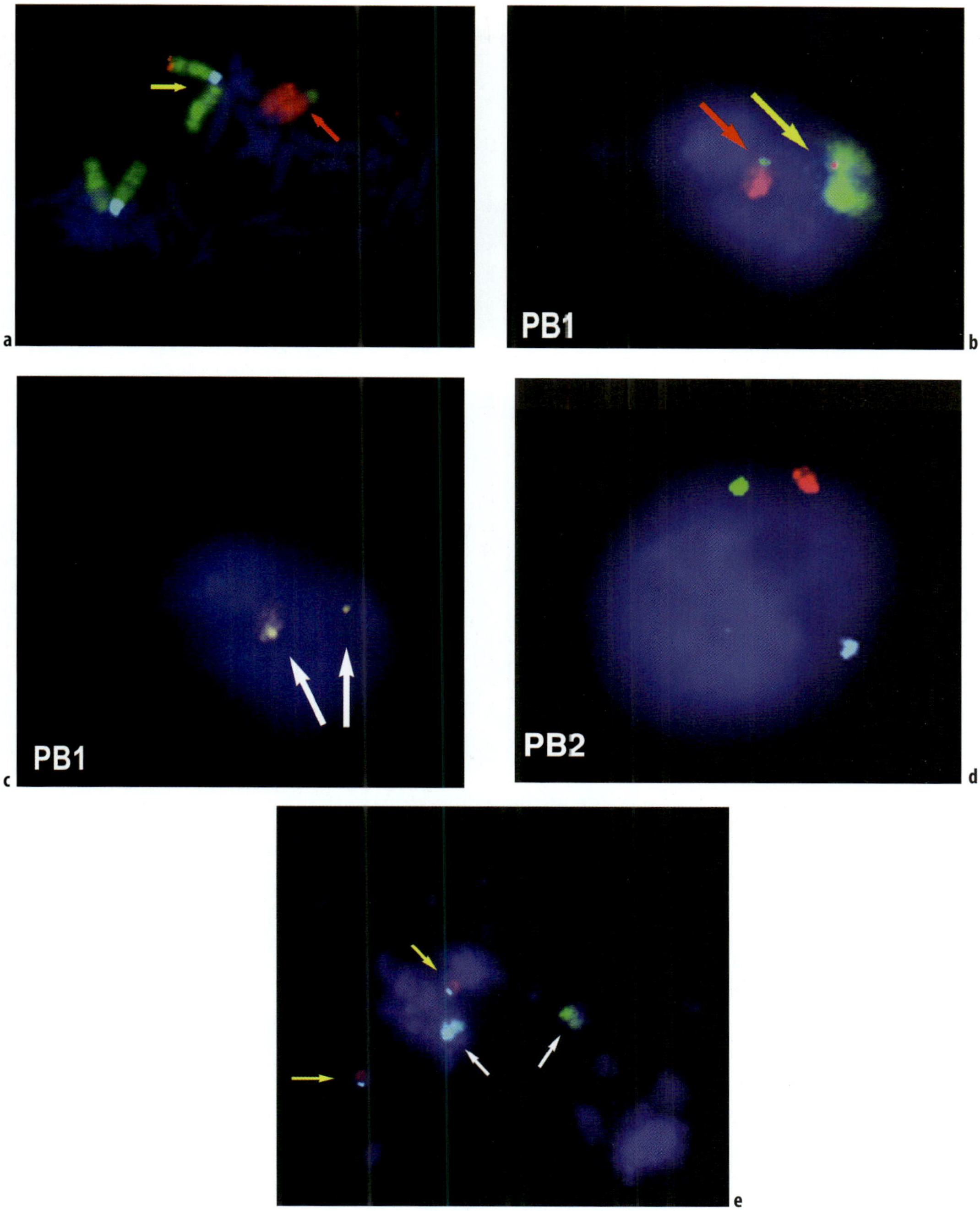

Figure 5.17. PGD for a maternally derived reciprocal translocation [46, XX, t(1; 15)(q32; q26)] by polar body analysis. (*A*) FISH analysis of metaphase chromosomes of a peripheral blood lymphocyte from the carrier. Chromosome 1 is identified with whole chromosome paint (WCP) in green in conjunction with a centromeric enumeration probe (CEP) 1 in aqua. Chromosome 15 is identified with WCP in orange, (visualization through a red single bandpass filter), in conjunction with a subtelomeric (Tel) 15q in orange since the translocated segment of chromosome 15 onto chromosome 1 is small. (*B*) FISH image of metaphase chromosomes from PB1 in which CE of both is observed. Chromosome 1 (chromatids) is identified by WCP in green; however, only one chromatid carries the signal for Tel 15q (derivative chromatid—yellow arrow) while the other chromatid does not (normal chromatid). Chromosome 15 (orange) is visualized in red with one chromatid seen as only red and one chromatid seen as red and green (der(15)—red arrow). (*C*) Rehybridization of PB1 metaphase chromosomes with Tel 15q (orange) only. Two signals are seen for Tel 15q indicating that both normal and derivative chromatids are present for each of the chromosomes of interest. (*D*) Hybridization of PB2 with CEP 1 (aqua), Tel 1q (orange) and CEP 15 (green) shows a balanced/normal number of signals, indicating that the oocyte carries either a normal or balanced chromosome complement. (*E*) Embryo follow-up analysis by embryo biopsy and blastomere nucleus conversion to metaphase chromosomes confirms the prediction by PB analysis showing a normal chromosome complement. Normal chromosomes 1 (white arrows) and chromosomes 15 (yellow arrows) are present.

Table 5.15. Preimplantation diagnosis for translocations by conversion of blastomere nuclei into metaphase or polar body analysis

Method	Blastomere conversion	PB1 & PB2	Interphase	TOTAL
Cycles	110	58	15	183
#Studied/Results	906/804(88,7%)	667/529 (79.3%)	93/86(92.5%)	1666/1419(85.2%)
Embryos Transferred (Mean)	142 (1.7)	53 (1.6)	20(1.8)	215 (1.7)
Pregnancies/Transfers	28/82(34 %)	11/34 (32.4%)	6/11 (54.5%)	45/126 (35.7%)
Deliveries/Babies	24/27*	8/10	2/2	34/39
Spontaneous Abortions	4 (14.3%)	3 (27.3%)	1 (16.7%)	8 (17.8%)

*Three pregnancies ongoing.

normal embryos or those with a balanced chromosomal complement for transfer in 82 (74.5%) cycles. Of those 102 (11.2%) blastomeres with no results, conversion procedure had failed in 9%, with the remaining containing no nucleus in the biopsied blastomeres. Twenty eight (34%) pregnancies were obtained from 82 transfers, resulting in birth of 27 healthy children with three pregnancies still ongoing. Only 4 of 28 pregnancies were spontaneously aborted, representing a considerable reduction of spontaneous abortion rate prior to PGD (see Chapter 6) (Table 5.15).

In 58 PGD cycles from couples with maternally derived translocations, testing was performed using PB1 and PB2 FISH analysis. Of 667 oocytes from these cycles tested, FISH results were available in 529 (79.3%), allowing preselection of normal or balanced embryos for transfer in 34 (58.6%) cycles, resulting in 11 (32.4%) clinical pregnancies, 8 of which resulted in the delivery of 10 unaffected children and three were spontaneously aborted (see Table 5.15) The confirmatory testing was possible in two of three spontaneously aborted embryos, showing the presence of de novo translocations different from the expected meiotic outcomes [60].

The proportion of abnormal oocytes and embryos detected varied depending on the type of translocations and their origin. The results of testing for maternally or paternally derived translocations in reciprocal or Robertsonian translocations are summarized in Tables 5.16 and 5.17 which do not include 6 cycles for other rearrangements. As seen in Table 5.16, testing of 825 (685 were with results) oocytes and embryos from 82 PGD cycles for maternally derived reciprocal translocations resulted in prediction of unbalanced embryos, leaving only 158 (23.0%) suitable for transfer, of which 81 (11.8%) were determined to be balanced and 77 (11.2%) normal. On the other hand, the testing of 456 embryos from 50 PGD cycles for paternally derived reciprocal translocations resulted in

Table 5.16. Results of PGD for reciprocal translocations

Maternally Derived	
Total Studied	825 from 82 cycles [685 (83%) with results]
Unbalanced	528 (77%)
Balanced	81 (11.8%)
Normal	77 (11.2%)
Suitable for Transfer	158 (23.0%)
Paternally Derived	
Total Studied	456 from 50 cycles [396 (86.8%) with results]
Unbalanced	278 (70.2%)
Balanced	61 (15.4%)
Normal	57 (14.4%)
Suitable for Transfer	118 (29.8%)

Table 5.17. Results of PGD for Robertsonian translocations

Maternally Derived	
Total Studied	184 from 23 cycles [166 (90.2%) with results]
Unbalanced	110 (66.3%)
Balanced	36 (21.7%)
Normal	20 (12.0%)
Suitable for Transfer	53 (33.7%)
Paternally Derived	
Total Studied	156 from 22 cycles [140 (89.7% with results]
Unbalanced	76 (54.3%)
Balanced	40 (28.5%)
Normal	24 (17.1%)
Suitable for Transfer	64 (45.7%)

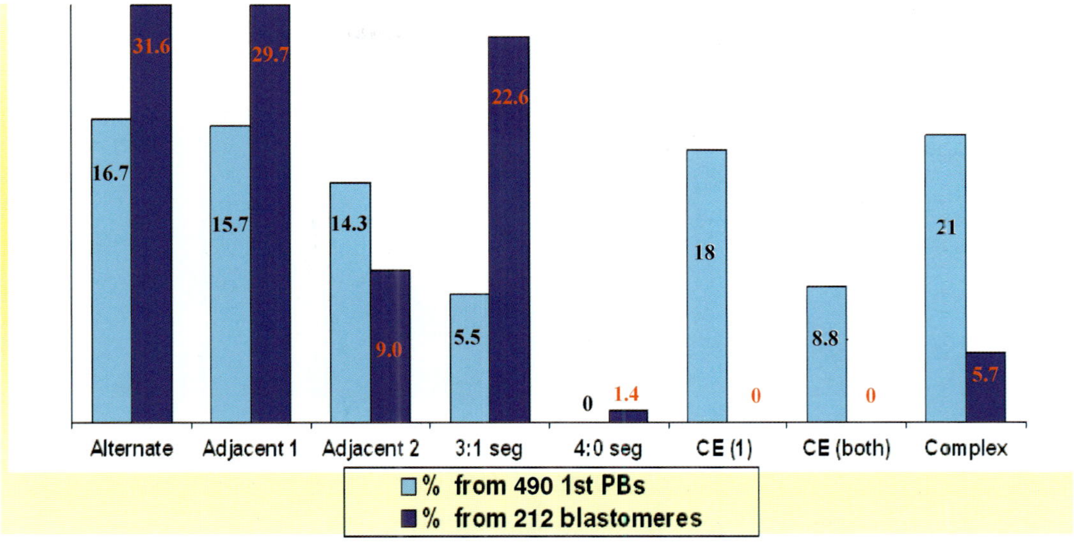

Figure 5.18. Distribution of segregation patterns. Bar graph demonstrating the distribution of observed segregation modes for female reciprocal translocations by PB1 analysis versus inferred segregation patterns by blastomere analysis. CE-chromatid exchange.

prediction of 278 (70.2%) unbalanced embryos, leaving 118 (29.8%) embryos suitable for transfer, of which 61 (15.4%) were determined to be balanced and 57 (14.4%) normal.

As seen in Table 5.17, testing of 184 (166 were with results) oocytes and embryos obtained from 23 PGD cycles for maternally derived Robertsonian translocations resulted in prediction of 110 (66.3%) unbalanced embryos, with the remaining 53 (33.7%) embryos suitable for transfer, which included 36 (21.7%) balanced and 20 (12.0%) normal. Similarly, the testing of 156 (140 with results) embryos obtained from 22 PGD cycles for paternally derived Robertsonian translocations allowed identification of 76 (54.3%) unbalanced embryos, with the remaining 64 (45.7%) suitable for transfer, including 40 (28.6%) balanced and 24 (17.1%) normal ones.

The overall results for the total of 183 PGD cycles for translocation show that normal/balanced embryos were available for transfer in 126 (69%) of the clinical cycles. Fourty-five (35.7%) clinical pregnancies resulted from 126 transfer cycles in which the mean number transferred was 1.7 embryos. Thirty-four of these pregnancies resulted in the delivery of 39 healthy children, while three pregnancies are still ongoing and 8 (17.8%), overall, resulted in spontaneous abortions. The data on the pregnancy outcomes of these patients prior to undertaking PGD showed that 105 (85.4%) of their 123 pregnancies resulted in spontaneous abortions. This demonstrates the tremendous positive impact PGD has on the clinical outcome of pregnancies in couples carrying both reciprocal and Robertsonian translocations (see Chapter 6).

An extremely poor reproductive outcome, with more than three quarters of the pregnancies resulting in spontaneous abortions, may clearly be explained by the poor meiotic outcomes, which vary depending on the type of translocations and their origin. To investigate the meiotic outcome of translocations in relation to the type and origin, a segregation patterns from 130 patients carrying balanced translocations were analyzed which is presented in Figures 5.18 and 5.19. The meiotic outcomes were inferred either from PB (Figure 5.18) or blastomere analysis (Figure 5.19), with meoitic outcome detection rates by each of these methods being comparable, except for chromatid exchanges, detected only by sequencial PB1 and PB2 analysis (18%), and complex errors which were higher in PB analysis (21% vs. 5.7%); 3:1 segregation was found more frequently in blastomere analysis (5.5% vs. 22.6%).

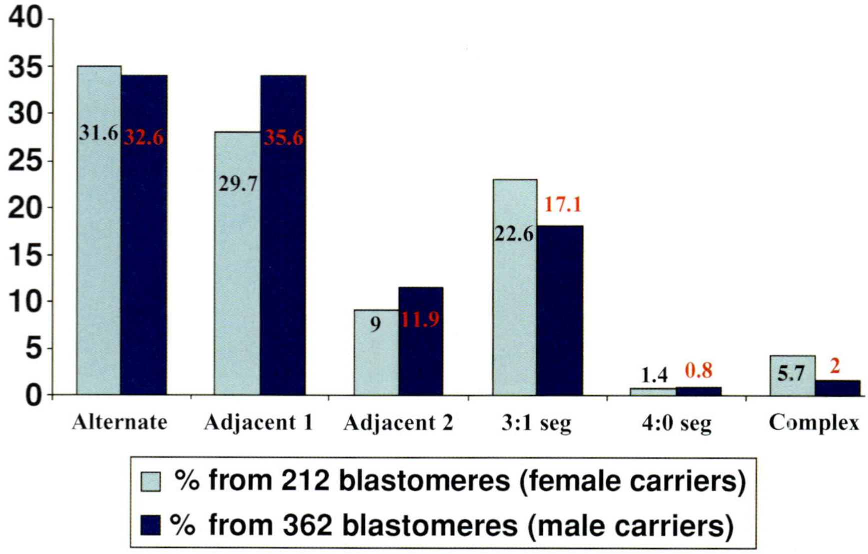

Figure 5.19. Distribution of segregation patterns Bar graph demonstrating the distribution of inferred segregation modes for female reciprocal translocations compared with male reciprocal translocations.

Segregation patterns for paternally and maternally derived translocation showed similar tendencies, predominantly represented by alternate (32.6 and 31.6%, respectively) and adjacent I (35.6 and 29.7%, respectively), and much lower for adjacent II (11.9 and 9%, respectively). These meiotic outcomes may explain the proportion of balanced and unbalanced embryos detected, which were predicted in 77% embryos obtained from maternally derived reciprocal translocations, leaving only 23% suitable for transfer, including 11.8% balanced and 11.2% normal. On the other hand, unbalanced embryos were predicted in 70.2% embryos obtained from paternally derived reciprocal translocations, leaving 29.8% embryos suitable for transfer, including 15.4% balanced and 14.4% normal. Testing of embryos for maternally derived Robertsonian translocations resulted in prediction of 66.3% unbalanced embryos, with the remaining 33.7% embryos suitable for transfer, which included 21.7% balanced and 12.0% normal. Similarly, the testing of embryos for paternally derived Robertsonian translocations allowed identification of 54.3% unbalanced embryos, with the remaining 45.7% suitable for transfer, including 28.6% balanced and 17.1% normal ones. Clinical pregnancies were obtained in 35.7% of transfer cycles, with 26.2% overall delivering healthy children. The data on the meiotic outcome may explain the observed 85.4% spontaneous abortion rate in patients prior to undertaking PGD, which was reduced to 17.8% after PGD, demonstrating the tremendous positive impact of PGD on the clinical outcome of pregnancies in couples carrying translocations (see Chapter 6).

Although the application of the conversion technique to visualize chromosomes in single blastomeres improves the accuracy of the diagnosis by analysis of metaphase chromosomes using a combination of commercially available probes, a high frequency of mosaicism in the cleavage-stage embryos, arising from anaphase lag or nuclear fragmentation, still presents problems for diagnosis. Follow-up analysis of unbalanced embryos, including those in which chromatid malsegregation or recombination was identified by the analysis of PB1, and subsequent testing of PB2 inferred a balanced or normal embryo, revealed a mosaicism rate of 41%. Different cell lines were present, including normal or balanced, which if investigated only by embryo biopsy, may have led to misdiagnosis. Therefore, our PGD strategy for maternally derived translocations is still based on PB1 and PB2 testing, applying the blastomere nucleus conversion technique only if further testing is required.

We have also observed that the embryos with unbalanced chromosome complements have the potential to reach the blastocyst stage of embryo development in extended culture. Of 992 (71.6%)

unbalanced embryos identified in 1387 embryos with FISH results, 250 were cultured for a further period, of which 78 (31.2%) reached the blastocyst stage, confirming our previous results, that some of the detected chromosomal rearrangements, may not be lethal in preimplantation development, being eliminated either during implantation, or postimplantation development [61], explaining an extremely high spontaneous abortion rate in couples carrying translocations. As will be shown in the Chapter 6, our data is also in the agreement with previous reports suggesting as much as a fivefold reduction of spontaneous abortions in PGD cycles for translocations [62, 63].

In addition to avoiding the use of the expensive and time-consuming customized breakpoint-spanning probes [64], the conversion technique used in the present study was highly accurate based on the follow-up testing of embryos predicted to be abnormal and confirmatory prenatal diagnosis of the ongoing pregnancies or testing of the newborn babies. The method also allows balanced and normal embryos to be distinguished, which cannot be achieved by currently available interphase FISH analysis.

Because PGD is practically the only hope for couples with translocations to have an unaffected child without fear of repeated spontaneous abortions, increasing numbers of PGD cycles for this indication have been performed. More than 500 clinical cycles have been performed at the present time, resulting in birth of more than 100 unaffected children [65–71].

Progress was also reported in the development of single-cell comparative genome hybridization (CGH) technique and microarrays to enable PCR-based testing for translocations [72–73]. However, the most important limitation of the present CGH protocol is still the three-day duration of the procedure, which is incompatible with the current laboratory framework for PGD. The other possible approach may be a multiplex fluorescent PCR using polymorphic microsatellite markers for the detection of numerical changes in chromosomes using their allelic fingerprints, which may also be useful for PGD of Robertsonian translocations [74, 75].

The presented data suggest that patients with translocations who are at an extremely high risk for spontaneous abortions should be informed about the availability of PGD for this indication. Awareness of the availability of PGD will permit such couples to establish pregnancies that are unaffected from the onset and offers the opportunity to have children of their own, instead of multiple unsuccessful attempts of prenatal diagnosis and subsequent termination of pregnancy. The application of the method for conversion of interphase nuclei of blastomeres will make it possible to perform PGD for a wide variety of structural rearrangements independent of the availability of specific fluorescent probes.

References

1. Sherman SL, Peterson MB, Freeman S.B. et al. Nondisjunction of chromosome 21 in maternal meiosis I: Evidence for a maternal age-dependent mechanism involving reduced recombination. Hum Mol Genet 1994;3:1529–1535.
2. Hassold T, Merril M, Adkins K, Freemen S, Sherman S. Recombination and maternal age-dependent nondisjunction: Molecular studies of trisomy 16. Am J Hum Genet 1995;57:867–874.
3. Peterson M.B, Mikkelsen M. Nondisjunction in trisomy 21: Origin and mechanisms. Cytogenet Cell Genet 2000;91:199–203.
4. Lamb NE, Freeman S, Savage-Austin A, et al. Susceptible chiasmate configurations of chromosome 21 predispose to nondisjunction in both maternal meiosis I and meiosis II. Nat Genet 1996;14:400–405.
5. Pellestor F, Andreo B, Armal F, Humeau C, Demaille J. Mechanisms of non-disjunction in human female meiosis: The co-existence of two modes of malsegregation evidenced by the karyotyping of 1397 in-vitro unfertilized oocytes. Hum Reprod 2002;17:2134–2145.
6. Dyban A, Fredine M, Severova E, Cieslac J, Wolf G, Kuliev A, Verlinsky Y. Detection of aneuploidy in human oocytes and corresponding first polar bodies using FISH (abstr 97). 7th International Conference on Early Prenatal Diagnosis, May 22–27, Jerusalem, 1994.
7. Verlinsky Y, Cieslak J, Freidin M, et al. Pregnancies following pre-conception diagnosis of common aneuploidies by fluorescent in-situ hybridization. Hum Reprod 1995;10:1923–1927.
8. Munné S, Daily T, Sultan KM, Grifo J, Cohen J. The use of first polar bodies for preimplantation diagnosis of aneuploidy. Hum Reprod 1995;10:1014–1120.
9. Dyban A, Fredine M, Severova E. et al. Detection of aneuploidy in human oocytes and corresponding first polar bodies by FISH. J Assist Reprod Genet 1996;13: 72–77.
10. Pujol A, Boiso I, Benet5 J, et al. Analysis of nine chromosome probes in first polar bodies and metaphase II oocytes for the detection of aneuploidies. Eur J Hum Genet 2003;11:325–336.
11. Verlinsky Y, Cieslak J, Ivakhnenko V, et al. Birth of healthy children after preimplantation diagnosis of common aneuploidies by polar body FISH analysis. Fertil Steril 1996;66:126–129.
12. Verlinsky Y, Cieslak J, Ivakhnenko V, et al. Preimplantation diagnosis of common aneuploidies by the first and second polar body FISH analysis. J Assist Reprod Genet 1998;15:285–289.

13. Verlinsky Y, Cieslak J, Ivakhnenko V, et al. Prepregnancy genetic testing for common age-related aneuploidies by polar body analysis. Genet Testing 1998;1:231–235.
14. Verlinsky Y, Cieslak J, Ivakhnenko V, et al. Prevention of age-related aneuploidies by polar body testing of oocytes. J Assist Reprod Genet 1999;16:165–169.
15. Verlinsky Y, Cieslak J, Ivakhnenko V, et al. Chromosomal abnormalities in the first and second polar body. Mol Cell Endocrinol 2001;183:S47–S49.
16. Verlinsky Y, Cieslak J, Kuliev A. High frequency of meiosis II aneuploidies in IVF patients of advanced maternal age. Reprod Technol 2001;10:11–14.
17. Gianaroli L, Magli MC, Ferraretti AP. The in vivo and in vitro efficiency and eccicacy of PGD for aneuploidy. Mol Cell Endocrinol 2001;183:S13–S18.
18. Munné S. Preimplantation genetic diagnosis of numerical and structural chromosome abnormalities. Reprod BioMed Online 2002;4:183–196.
19. Kuliev A, Verlinsky Y. Current features of preimplantation genetic disgnosis. Reprod BioMed Online 2002;5:296–301.
20. Kahraman S, Bahce M, Samli H, et al. Healthy births and ongoing pregnancies obtained by preimplantation genetic diagnosis in patients with advanced maternal age and recurrent implantation failures. Hum Reprod 2000;15:2003–2007.
21. De Boer KA, Catt JW, Jansen RPC, et al. Moving to blastocyst biopsy for preimplantation genetic diagnosis and single embryo transfer at Sydney IVF. Fertil Steril 2004;82:295–298.
22. Kahraman S, Kumpete Y, Sertyel S., et al. Pronuclear scoring and chromosomal status of embryos in severe male infertility. Hum Reprod 2002;17:3193–3200.
23. Van Blercom J, Davis P, Alexander S. Differential mitochondrial distribution in human pronuclear embryos leads to disproportionate inheritance between blastomeres: relationship to microtubular organization, ATP content and competence. Hum Reprod 2000;15:2621–2633.
24. Ebner T, Yaman C, Mose M, Sommergruber M, Feichtinger O, Tews G. Prognostic value of first polar body morphology on fertilization rate and embryo quality in itracytoplasmic sperm injection. Hum Reprod 2000,15:427–430.
25. Balaban B, Urman B, Isiklar A, Alatas C, Aksoy S, Mercan R. The effect of polar body morphology on embryo quality, implantation and pregnancy rates. Fertil Steril 2001;76 (suppl. 1):S8.
26. Miller KF, Sinoway CE, Fly KL, Falcone T. Fragmentation pf the polar body at the time of ICSI does not predict fertilization or early embryo development but may be associated with improved pregnancy and implantation. Fertil Steril 2001;76 (suppl. 1):S201.
27. Verlinsky Y, Munné S, Cohen J, et al. Over a decade of preimplantation genetic diagnosis experience: A multicenter report. Fertil Steril 2004;82:292–294.
28. Kuliev A, Cieslak J, Illkewitch Y, Verlinsky Y. Chromosomal abnormalities in a series of 6733 human oocytes in preimplantation diagnosis of age-related aneuploidies. Reprod BioMed Online 2003;6:54–59.
29. Angel R: First meiotic division nondisjunction in human oocytes. Am J Hum Genet 1997;65:23–32.
30. Hunt P, LeMaraire R, Embury P, Sheean L, Mroz K. Analysis of chromosome behaviour in intact mammalian oocytes: Monitoring the segregation of a univalent chromosome during female meiosis. Hum Mol Genet 1995;4:2007–2012.
31. Lamb NE, Feingold E, Savage-Austin A, et al. Characterization of susceptible chiasma configurations that increase the risk for maternal nondisjunction of chromosome 21. Hum Mol. Genet 1997;6:1391–1401.
32. Battaglia DE, Goodwin P, Klein NA, Soules MR. Influence of maternal age on meiotic spindle assembly in oocytes from naturally cycling women. Hum Reprod 1996;11:2217–2222.
33. Eichenlaub-Ritter U, Vogt E, Yiu H. Gosden R. Spindles, mitochondria and redox potential in ageing oocytes. Reprod BioMed Online 2002;5:117–124.
34. Nashmyth K, Peters JM, Uhlman F. Splitting the chromosome: Cutting the ties that bind sister chromatids. Science 2000;288:1379–1384.
35. Yuan L, Liu J, Hoja M, Wilbertz J, Nordqvist K, Hoog C. Female germ cell aneuploidy and embryo death in mice lacking the meiosis-specific protein SCP3. Science 2002;296:1115–1118.
36. Fisher JM, Harvey JF, Morton NE, Jacobs PA. Trisomy 18: Studies of the parent and cell division of origin and effect of aberrant recombination on nondisjunction. Am J Hum Genet 1996;56:669–675.
37. Kuliev A, Cieslak J, Zlatopolsky Z, Ilkevitch Y, Kirilova I, Verlinsky Y. Aneuploidy rescue after female meiosis I and follow up analysis of its outcome in resulting preimplantation embryos. Am J Hum Genet 2003;73(suppl.): 189.
38. Kuliev A, Verlinsky Y. Meiotic and mitotic nondisjunctions: lessons from preimplantation genetic diagnosis. Hum Reprod Update 2004;10:401–407.
39. Kim NH, Chung HM, Cha KY, Chung KS. Microtubule and microfilament organization in maturing human oocytes. Hum Reprod 1998;13:2217–2222.
40. Barrit J, Brenner C, Cohen J, Matt D. Mitochondrial DNA rearrangement in human oocytes and embryos. Mol Hum Reprod 1999;5:927–933.
41. Perez G, Flaherty S, Barry M, Matthews C. Preliminary observations of polar body extrusion and pronuclear formation in human oocytes using timeplapse video cinematography. Hum Reprod 1997;12:532–541.
42. Gianaroli L, Magli MC, Ferraretti AP. et al. Pronuclear morphology and chromosomal abnormalities as scoring criteria for embryo selection. Fertil Steril 2003;80:837–844.
43. Munné S, Bahce M, Sandalinas M. Differences in chromosome susceptibility to aneuploidy and survival to first trimester. Reprod BioMed Online 2004;8:81–90.
44. Munné S, Sandalinas M, Escudero T, et al. Some mosaic types increase with maternal age. Reprod BioMed Online 2002;4:223–232.
45. Silber S, Sadowy S, Lehahan K, Kilani Z, Gianaroli L, Munné S. High Rate of chromosome mosaicism but not aneuploidy in embryos from karyotypically normal men requiring TESE. Reprod BioMed Online 2002;4(suppl. 2):20.
46. DeBaun MR, Niemitz EL, Feinberg AP. Association of in vitro fertilization with Beckwith-Wiedemann syndrome and epigenetic alterations of LIT1 and H19. Am J Hum Genet 2003;72:156–160.
47. Gicquel C, Gaston V, Maldenbaum J., et al. In vitro fertilization may increase the risk of Beckwith-Videmann syndrome related to the abnormal imprinting of the KCNQ1OT Gene. Am J Hum Genet 2003;72:1338–1341.

48. Maher ER, Brueton LA, Bowdin SC, et al. Beckwith-Wiedemann syndrome and assisted reproduction technology (ART). J Med Genet 2003;40:62–64.
49. Niemitz EL, Feinberg AP. Epigenetics and assisted reproductive technology: A call for investigation. Am J Hum Genet 2004;74:599–609.
50. Halliday J, Oke K, Breheny S, Algar E, Amor JA. Beckwith-Wiedemann syndrome and IVF: A case-control study. Am J Hum Genet 2004;75:526–528.
51. Lucifero D, Chaillet JR, Trasler M. Potential significance of genomic imprinting defects for reproduction and assisted reproductive technology. Hum Reprod Update 2004;10:3–18.
52. Munné S, Morrison L, Fung J, et al. Spontaneous abortions are reduced after preconception diagnosis of translocations. J Assist Reprod Genet 1998;15:290–296.
53. International Working Group on Preimplantation Genetics. Preimplantation Genetic Diagnosis—Experience of Three Thousand Clinical Cycles. Report of the 11th Annual Meeting International Working Group on Preimlantation Genetics, in conjunction with 10th International Congress of Human Genetics, May 15, Vienna, 2001. Reprod BioMed Online 2001;3:49–53.
54. Verlinsky Y, Kuliev A. (eds.). Preimplantation Diagnosis of Genetic Disorders: A New Technique for Assisted Reproduction. Wiley Liss, New York, 1993.
55. Munné S, Bahce M, Sadowy S, Cohen J. Case report: Chromatid exchange and predivision of chromatids as other sources of abnormal oocytes detected by preimplantation genetic diagnosis of translocations. Prenat Diagn 1998;18:1450–1458.
56. Verlinsky Y, Kuliev A. Atlas of Preimplantation Genetic Diagnosis. Parthenon, New York, 2000.
57. Verlinsky Y, Evsikov S, Karyotyping of human oocytes by chromosomal analysis of the second polar body. Mol Hum Reprod 1999;5:89–95.
58. Verlinsky Y, Evsikov S. A simplified and efficient method for obtaining metaphase chromosomes from individual human blastomeres. Fertil Steril 1999;72:1–6.
59. Willadsen S, Levron J, Munné S, et al. Rapid visualization of metaphase chromosomes in single human blastomeres after fusion with in-vitro matured bovine eggs. Hum Reprod 1999;14:470–474.
60. Scriven PN, Handyside AH, Mackie Ogilvie C. Cromosome translocations: Segregation modes and strategies for preimplantation genetic diagnosis. Prenat Diagn 1998;18:1437–1449.
61. Evsikov S, Cieslak J, Verlinsky Y. Survival of unbalanced translocations to blastocyst stage. Fertil Steril 2000;74:672–676.
62. Munné S, Sandalinas M, Escudero T. Outcome of preimplantation genetic diagnosis of translocations. Fertil Steril 2000;73:1209–1218.
63. Munné S, Cohen J, Sable D. Preimplantation genetic diagnosis for advanced maternal age and other indications. Fertil Steril 2002;78:234–236.
64. Cassel MJ, Munné S, Fung J, Weier HUG. Carrier-specific breakpoint-spanning DNA probes: An approach to preimplantation genetic diagnosis in interphase cells. Hum Repro 1997;12:2019–2027.
65. Verlinsky Y, Cieslak J, Evsikov S, Galat V, Kuliev A. Nuclear transfer for full karyotyping and preimplantation diagnosis of translocations. Reprod BioMed Online 2002;5:302–307.
66. Fisher J, Escudero T, Chen S, et al. Obstetric outcome of 100 cycles of PGD of translocations and other structural abnormalities. Reprod BioMed Online 2002;4(suppl. 2):26.
67. Gianaroli L, Magli MC, Ferraretti AP, et al. Robertsonian and reciprocal translocations. Reprod BioMed Online 2002;4(suppl. 2):26–27.
68. Lim CK, Min JH, Song GJ, et al. Reliability of PGD with FISH analysis in reciprocal or Robertsonian translocation carriers. Reprod BioMed Online 2002;4(suppl. 2):29.
69. Van Assche E, Staessen C, Ogur G, et al. PGD for reciprocal and Robertsonian translocations in 41 treatment cycles. Reprod BioMed Online 2002;4(suppl. 2):27.
70. Munné S, Fisher J, Escudero T, et al. Outcome of >100 cycles of preimplantation genetic diagnosis of translocations. Fifth International Symposium on Preimplantation Genetics, June 5–7, Antalya, Turkey, 2003, p. 31.
71. Cieslak J, Zlatopolsky Z, Galat V, Masciangelo C, Tur-Kaspa I, Kuliev A, Verlinsky Y. Preimplantation diagnosis in 146 cases of chromosomal translocations. Fifth International Symposium on Preimplantation Genetics, June 5–7, Antalya, Turkey, p. 23. Istanbul Memorial Hospital, Istanbul, Turkey 2003.
72. Malmgren H, Sahlén S, Inzunza J. Single cell CGH analysis of human preimplantation embryos from PGD patients with balanced structural chromosome aberrations. Reprod BioMed Online 2002;4(suppl. 2):14–15.
73. Osborne EC, Trounson AO, Cram DS. Diabgnosis of Robertsonian translocations using microarrays. Fifth International Symposium on Preimplantation Genetics, June 5–7, 2003, Antalya, Turkey, p. 31. Istanbul Memorial Hospital, Istanbul, Turkey 2003.
74. Katz MG, Mansfield J, Gras L, Trounson A, Cram DS. Diagnosis of trisomy 21 in preimplantation embryos by single-cell DNA fingerprinting. Reprod BioMed Online 2002;4:43–50.
75. Kuliev A, Verlinsky Y. Current features of preimplantation genetic disgnosis. Reprod BioMed Online 2002;5:296–301.

6

Clinical Outcomes Following Preimplantation Genetic Diagnosis

Data on the clinical outcomes were reported from 1416 PGD embryo transfer cycles performed in our centre [1] and from 1670 PGD transfer cycles collected from 25 different centres by ESHRE PGD Consortium [2]. These transfers resulted in 338 (23.9%) and 309 (18.5%) clinical pregnancies, respectively, and birth of 539 unaffected children, overall (260 and 269, respectively), with multiple pregnancies observed in approximately one third of the cases. The overall congenital malformation rate was 5.4% and 6.6%, respectively, which is not different from population prevalence, of which only 2.25% and 3.9%, respectively, were attributable to the major abnormalities.

Overall, approximately 7000 PGD cycles have presently been performed, with further wider application not only in the United States and Western Europe, but also in the Eastern Mediterranean and Asian countries. Also, an increasing number of PGD cases are being performed for conditions that have never been practiced in traditional prenatal diagnosis, such as preimpantation gender determination for social reasons, performed in India [3] and Jordan [4]. However, the majority of PGD cycles are still performed for chromosomal disorders, with the ratio of PGD cycles for chromosomal and single-gene disorders of approximately 3:1, the latter including late-onset disorders with genetic predisposition, the proportion of which is gradually increasing. Approximately 2000 PGD cycles have currently been performed for single-gene disorders, with outcome data available for hundreds of clinical pregnancies and babies, suggesting the acceptable accuracy of PGD, which may be further improved by the above-mentioned new developments in DNA analysis of single cells (see Chapter 2). As presented in Chapter 3, a relatively high pregnancy rate (approximately 39%) was observed in PGD for single-gene disorders, despite transfer of only 2.2 embryos per cycle on the average. This may be explained by the fact that these are fertile couples of younger reproductive age (under 35 years) compared with the poor-prognosis IVF patients referred for aneuploidy testing.

In PGD for chromosomal disorders, the indications are also expanding, with further obvious interest in PGD for translocations, because of a strong impact of PGD on reducing the spontaneous abortion rate in carriers of balanced translocations. The majority of these cycles have been performed in two centres, including Reprogenetics and ours [5–7], with an increasing number of other centres, which performed a few dozen PGD cycles for translocations, such as the Brussels [8], SISMER [9], London [10], and Seoul [11] centres. The available experience demonstrates a clear advantage of PGD for translocations over traditional prenatal diagnosis, as shown above, attributable to a poor meiotic outcome, particularly in reciprocal translocations (see Section 5.2). As mentioned, the accuracy of PGD for translocation has recently been improved by the introduction of increasing numbers of subtelomeric probes and the technique of blastomere nucleus conversion to metaphase, which will allow a reliable testing for any complex chromosomal rearrangement. The experience of more than 500 cycles accumulated at the present time further confirms the previous observations of more than a fourfold reduction of spontaneous abortion rate in translocation carriers, making PGD a preferred option for chromosomal translocations over traditional prenatal diagnosis [7, 12–15]. Although the proportion of PGD cycles in which there were balanced or normal embryos available for transfer, especially in reciprocal translocations, was not high, if the transfer of normal or balanced embryos was possible, the pregnancy rate was comparable to those PGD cycles performed for Mendelian disorders (see Chapter 5).

There is further strong evidence supporting the clinical usefulness of PGD for assisted reproduction, as more than half of the preimplantation embryos are chromosomally abnormal from the onset and unsuitable for transfer in IVF patients of advanced maternal age. It is evident that the further improvement of IVF efficiency will be unrealistic without the preselection of aneuploidy-free oocytes and embryos. With the current tendency for limiting the number of transferred embryos to only two, and even to only one in blastocyst transfer, in the attempt to avoid the complications due to multiple pregnancies, the preselection of chromosomally normal embryos will soon become a standard IVF practice, as there is no meaning of deliberately transferring untested embryos, more than 50% of which are aneuploid, definitely compromising the outcome of IVF in a traditional setting. In the meantime, the information on the availability of such an option should be provided at least to poor-prognosis IVF patients, so they could have the opportunity of improving their chances to become pregnant and avoid the establishment of pregnancy destined to be lost due to chromosomal aneuploidy. The clinical outcome is presently available in hundreds of clinical pregnancies resulting from PGD for aneuploidies, supporting the initial observations on the positive impact on the implantation and pregnancy rates and the improvement of the pregnancy outcomes. This is not surprising, as only 1 in 10 of recognized pregnancies is chromosomally abnormal, in contrast to 1 in 2 in preimplantation embryos, suggesting that the majority of these chromosomally abnormal embryos seem to be eliminated before implantation. Incidental transfer of these embryos in the absence of chromosomal testing should contribute to implantation and pregnancy failures in IVF patients of advanced reproductive age, or affect pregnancy outcome through spontaneous abortions. A clinical impact of aneuploidy testing, in terms of the improved outcome of pregnancies through the reduction of spontaneous abortions, has been observed not only for IVF patients with advanced reproductive age, but also for other poor-prognosis patients, including those with repeated IVF failures and repeated spontaneous abortions, poor responders and TESE patients. Although randomized controlled studies will be still required to further quantify the clinical impact of preselection of aneuploidy-free zygotes for embryo transfer, the available results suggest the clinical relevance of the preselection of aneuploidy-free oocytes and embryos.

The positive clinical impact of aneuploidy testing has recently been documented also in terms of doubling implantation rate in IVF patients age 40 and older, provided that sufficient number of zygotes for testing are available [16]. Clinical outcomes of 138 PGD cycles were compared with the matched control group of women (average age 39.4 and 39.8 years, average number of two-pronuclei zygotes 8.7 and 8.9, average number of attempts 1.5 and 1.7), showing a higher implantation rate in the PGD group (17.6 in PGD compared with 10.6 in control group), despite transferring two embryos in PGD in contrast to 3.7 on the average in the control group. The differences were even more significant when only those couples were selected who have had less than two failed cycles, more than eight available zygotes for preselection, or both of these parameters, when the implantation rate almost doubled.

In a similar study we analyzed the clinical outcome of 340 PGD cycles in comparison with 546 non-PGD cycles (Table 6.1), involving 218 (64%) and 485 (89%) transfers, respectively, which resulted in 20.2 and 18.6% pregnancy rates, respectively. Despite the average transfer of 1.8 embryos in PGD compared with 2.8 in non-PGD group, the implantation rate was 14.6% in contrast to 8.1% in

Table 6.1. Improvement of implantation rate after PGD by polar body analysis

	PGD By PB FISH Analysis	Non-PGD
Cycles	340	546
Embryo transfers	218 (64%)	485 (89%)
Ongoing pregnancies	44 (20.2%)	90 (18.6%)
Embryos transferred	398	1364
Mean embryos ET	1.8	2.8
Implantation rate	14.6%[a]	8.1%[b]

[a,b] Chi-square, $P < 0.001$.

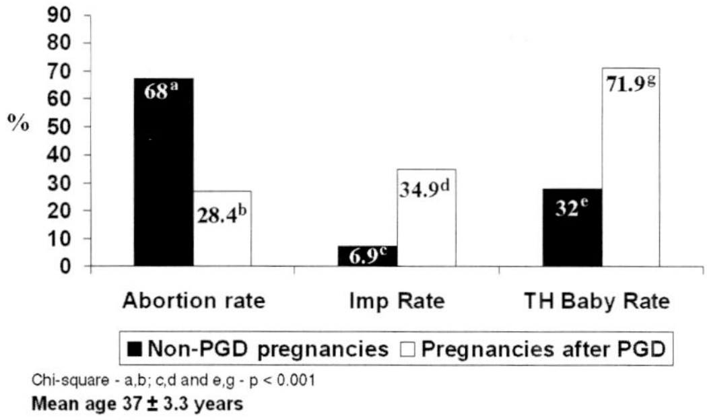

Figure 6.1. Bar graph representing the obstetrical history of 432 patient/pregnancies before and after preimplantation genetic diagnosis for aneuploidy. Retrospective analysis shows a significantly lower abortion rate, a significantly higher implantation rate, and a greater than twofold increase in the take-home baby rate. Of these PGD patients, 82.1% were over age 35.

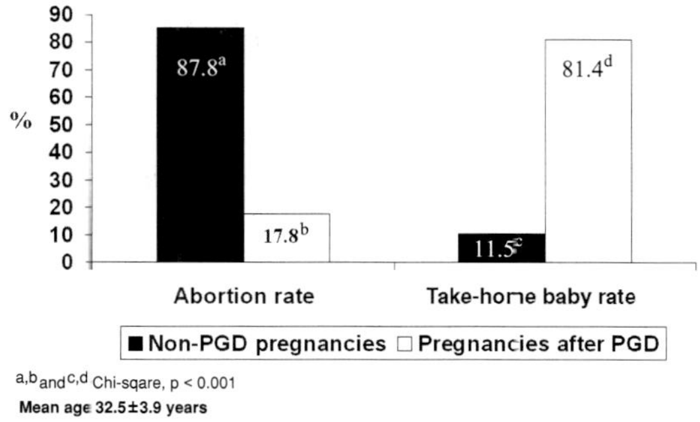

Figure 6.2. Bar graph representing the retrospective analysis of the previous obstetrical history in relation to the 44 patient/pregnancies that occurred following preimplantation genetic diagnosis for chromosome translocations. A significantly lower abortion rate (17.8%) was realized along with a significantly higher take-home baby rate of 81.4%.

non-PGD group, thus supporting the above data on the positive impact of PGD on implantation rate.

The improvement of the reproductive outcome was particularly obvious from the analysis of the reproductive history of PGD patients, i.e., from comparison of the reproductive outcome in the same patients with and without PGD, as previous reproductive experience of the patients is the best control for PGD impact [17]. Analysing the outcome of transfer of euploid embryos in 699 cyles, an implantation rate per pregnant patient of 53.2% was observed, with a total of 171 clinical pregnancies generated, which resulted in birth of 210 children from 193 couples, and 34 spontaneous abortions, giving a take-home baby rate per pregnant patient of 88.6%. Previous reproductive history of these patients showed 3.6% take home baby rate and only 10.9% implantation rate. This makes obvious the clinical usefulness of PGD for aneuploidy for IVF patients with poor reproductive performance.

This is in agreement with our study, involving the analysis of the reproductive outcome of 432 pregnancies from PGD cycles for advanced reproductive age (mean maternal age 37 years) in comparison with their reproductive history prior to undertaking PGD (Figure 6.1). As can be seen from these data, there were significant differences in all parameters tested, including a fivefold increase in implantation rate, almost a threefold reduction of spontaneous abortion rate, and more than a twofold increase of take-home baby rate after PGD in contrast to that prior to PGD.

The impact was even higher in PGD for translocations, shown in Figure 6.2, suggesting as much as approximately a fivefold reduction of spontaneous abortion rate, and approximately a sevenfold increase of take home baby rate after PGD.

In the light of these data, the current IVF practice of selection of embryos for transfer based on morphological criteria may hardly be an acceptable procedure for IVF patients of advanced reproductive age in the future. In addition to an extremely high risk of establishing an affected pregnancy from the onset, this will significantly compromise the very poor chances of these patients to become pregnant, especially with the current tendency of limiting the number of transferred embryos to only two, leaving only a single embryo on the average with a potential chance of reaching the term. Although culturing embryos to day 5 (blastocyst) before transfer may allow to some extent the preselection of developmentally more competent embryos compared with day 3, at least 1 in 5 or even more, aneuploid embryos will still be capable of developing to blastocysts [18–20]. As we demonstrated above, an unexpected high proportion of embryos even with autosomal monosomies reach the blastocyst stage, as well as those with unbalanced translocations. So these abnormal embryos will not be eliminated in the current shift to the blastocyst transfer, and may implant and lead to spontaneous abortions, compromising the outcome of pregnancies resulting also from implanted normal embryos in multiple pregnancies. In fact, multiple pregnancies represent a severe complication of

IVF, as can be seen from the mortality rate associated with triplet pregnancies, including stillbirth, neonatal death, and perinatal death of 32/1000, 79/1000, and 109/1000, respectively [21]. The risk of acquired disorders is also increased following the birth of twins and triplets compared with singleton birth. For example, the estimated rate of cerebral palsy is 8.8/1000 in twins and 16.9/1000 in triplets, compared with 2.7/1000 in natural conceptions. This is even higher in survivors to one year, which is 7.3/1000 in twins and 29/1000 in triplets, in contrast to 1.6/1000 in singletons [22, 23]. As these children are also characterized by low birth weight, a higher prevalence of neurological, psychomotor, and mental abnormalities is also observed. So the avoidance of multiple pregnancies is one of the urgent problems of ART, which may in the future be avoided by preselection and transfer of a single blastocyst with the greatest developmental potential to result in healthy pregnancy. Such testing is presently possible for the nuclear abnormalities and may soon become realistic for cytoplasmic disorders as well.

However, patients should be aware of the limitations of the methods for aneuploidy testing, as at least half of chromosomally abnormal embryos are with mosaicism, which is the major challenge in improving the accuracy of PGD for aneuploidies performed by embryo biopsy. As the overall prevalence of aneuploidies in oocytes and embryos seems to be comparable, suggesting that mosaic embryos may originate from the aneuploid oocytes through the process known as "trisomy rescue," the further improvement of PGD accuracy may in the future require testing of both oocytes and the resulting embryos. This may be achieved by a sequential biopsy of both PB1 and PB2 and single blastomere from the resulting embryo, so that both meiotic and mitotic errors could be excluded. In addition, the information of both the oocyte and the embryo chromosome sets will make it possible to detect the potential uniparental disomy cases, which may be present in one third of the normal disomic embryos, originating from trisomic oocytes. Collecting of this unique information may also be useful in finding a possible explanation at least for some of the cases of imprinting disorders, such as BWS, reported in association with assisted reproductive technology, discussed previously in more detail in section 5.1.4.

In conclusion, PGD is a new approach to detect and avoid transfer of embryos with genetic abnormalities as an alternative to embryo transfer based on morphological criteria in the current IVF practice. Although PGD was originally introduced for preexisting genetic conditions as an alternative option to prenatal diagnosis, it has become of special value for assisted reproduction practices, because genetic factors contribute considerably to infertility problems. With the majority of IVF patients being of advanced reproductive age, PGD provides an obvious tool for preselection and avoidance of transfer of the embryos with age-related aneuploidies, the major contributors to spontaneous abortions and implantation failure.

PGD for chromosomal abnormalities is performed by testing either single blastomeres removed from the eight-cell preimplantation embryos or female gametes through genetic analysis of PB1 and PB2, extruded following maturation and fertilization of the oocytes, using FISH analysis with commercially available chromosome-specific probes. At the present time, PGD for age-related aneuploidies has been applied in approximately 5000 IVF cycles from poor-prognosis patients resulting in improved implantation and pregnancy rates in IVF patients of advanced maternal age. Although further studies may be needed to further quantify the impact of PGD on IVF efficiency, the presently available data leave no doubt of the clinical relevance of eliminating chromosomally abnormal embryos from transfer. The clinical value in applying PGD for improving pregnancy and implantation rates in poor-prognosis IVF patients is obvious from the fact that every second oocyte and embryo from such patients has chromosomal abnormalities. No less than three quarters of these chromosomally abnormal embryos should be eliminated before implantation, taking into consideration that only 10% of recognized pregnancies are chromosomally abnormal. So an incidental transfer of these embryos in the absence of chromosomal testing should contribute to implantation and pregnancy failures in IVF patients of advanced maternal age. In addition, the majority of the chromosomally abnormal embryos that implant will further lead to spontaneous abortions and affect the pregnancy outcome.

While it may be predicted that preimplantation genetic counseling and aneuploidy testing will soon become a standard practice for IVF patients of advanced maternal age, it cannot be excluded that preselection of aneuploidy-free embryos may appear even of higher value for younger IVF patients, because of much higher number of oocytes available for testing. So IVF patients will need to be informed about the availability of PGD so they could be able to use the option of preimplantation genetic counseling. This will definitively contribute to improving the standards of assisted reproduction practices, substituting the presently practiced "blind" selection of embryos for transfer using morphological parameters by the preselection of chromosomally normal embryos with the highest possible potential to result in pregnancy. As a tool for reliable preselection of aneuploidy-free embryos, PGD will potentially contribute not only to prevention of the birth of children with chromosomal disorders, but will also be a useful tool for the improvement of the efficiency and standards of assisted reproduction.

References

1. Kuliev A, Verlinsky Y. Current feature of preimplantation genetic disgnosis. Reprod BioMed Online 2002;5:296–301.
2. ESHRE Preimplantation Genetic Diagnosis (PGD) Consortium. Data Collection II (May 2002). Hum Reprod 2002;15:2673–2683.
3. Malpani A, Malpani A, Modi D. The use of preimplantation genetic diagnosis in sex selection for family balancing in India. Reprod BioMed Online 2001;4:16–20.
4. Kilani Z, Haj Hassan L Sex selection and preimplantation genetic diagnosis at the Farah Hospital. Reprod BioMed Online 2002;4:68–70.
5. Munné S, Morrison L, Fung J, et al. Spontaneous abortions are reduced after preconception diagnosis of translocations. J Assit Reprod Genet 1998;15:290–296.
6. Munné S, Sandalinas M, Escudero T. Outcome of premplantation genetic diagnosis of translocations. Fertil Steril 2000;73:1209–1218.
7. Verlinsky Y, Cieslak J, Evsikov S, Galat V, Kuliev A. Nuclear transfer for full karyotyping and preimplantation diagnosis of translocations. Reprod BioMed Online 2002;5:302–307.
8. Van Assche E, Staessen C, Ogur G, et al. PGD for reciprocal and Robertsonian translocations in 41 treatment cycles. Reprod BioMed Online 2002;4(suppl. 2):27.
9. Gianaroli L, Magli MC, Ferraretti AP, et al. Robertsonian and reciprocal translocations. Reprod BioMed Online 2002;4(suppl. 2):26–27.
10. Braude P. Strategies and outcomes of the first 100 cycles of PGD for serious familial genetic conditions. Reprod BioMed Online 2002;4(suppl. 2):37.
11. Lim CK, Min JH, Song GJ, et al. Reliability of PGD with FISH analysis in reciprocal or Robertsonian translocation carriers. Reprod BioMed Online 2002;4(suppl. 2):29.
12. Munné S. Preimplantation genetic diagnosis of numerical and structural chromosome abnormalities. Reprod BioMed Online 2002;4:183–196.
13. Fisher J, Escudero T, Chen S, et al. Obstetric outcome of 100 cycles of PGD of translocations and other structural abnormalities. Reprod BioMed Online 2002;4(suppl. 2):26.
14. Munné S, Fisher J, Escudero T, et al. Outcome of >100 cycles of preimplantation genetic diagnosis of translocations. Fifth International Symposium on Preimplantation Genetics, June 5–7, Antalya, Turkey, 2003, p. 31.
15. Cieslak J, Zlatopolsky Z, Galat V, Masciangelo C, Tur-Kaspa I, Kuliev A, Verlinsky Y. Preimplantation diagnosis in 146 cases of chromosomal translocations. Fifth International Symposium on Preimplantation Genetics, June 5–7, Antalya, Turkey, 2003, p. 23.
16. Munné S, Sandalinas M, Escudero T, et al. Improved implantation after preimplantation genetic diagnosis of aneuploidy. Reprod BioMed Online 2003;7:91–97.
17. Gianaroli L, Magli MC, Ferraretti AP, et al. The beneficial effect of PGD for aneuploidy support extensive clinical application. Reprod BioMed Online 2005;10:633–640.
18. Magli MC, Jones GM, Gras L, et al. Chromosome mosaicism in day 3 aneuploid embryos that develop to morphologically normal blastocysts in vitro. Hum Reprod 2000;15:1781–1786.
19. Sandalinas M, Sadowy S, Alikani M, et al. Developmental ability of chromosomally abnormal human embryos to develop to the blastocyst stage. Hum Reprod 2001;16:1954–1958.
20. Munné S, Bahce M, Sandalinas M. Differences in chromosome susceptibility to aneuploidy and survival to first trimester. Reprod BioMed Online 2004;8:81–90.
21. Newman RB, Luke B.(eds.). Multiple Pregnancy. Lippincott Williams & Wilkins, Philadelphia, 2000, pp. 192–194.
22. Vohr BR, Wright LL, Dusick, et al. Neurodevelopmental and functional outcomes of extremely low birth weight infants in the National Institute of Child Health and Human Development Neonatal Research Network. Pediatrics 2000;105:1216–1226.
23. Hack M, Willson-Costello D, Friedman H, et al. Neurodevelopmental and predictors of outcomes of children with birth weight of less than 1000g: 1992–1995. Arch Pediat Adoles Med 2000;154:725–731.

7

Preimplantation Genetic Diagnosis and Establishment of Human Embryonic Stem Cell Lines

As described in Chapter 4, PGD provides an exciting possibility for obtaining HLA-matched stem cells for treatment of siblings with bone marrow disorders. This involves preselection of HLA-matched embryos in couples with affected siblings requiring compatible stem cell transplantation, obviating the need for therapeutic cloning, which is highly controversial at the present time. However, in emergency situations, it may be too late to wait until the HLA-matched child is born, so other possible approaches need to be developed. For example, some of the tested embryos fail to reach the developmental stage to be considered for transfer, so they might be used for establishment of embryonic stem (ES) cell lines. Alternatively, partially matched embryos may be also considered for this purpose, provided the patients provide consent.

So PGD, in addition to improving access to HLA-compatible stem cell transplantation also provides a novel source for ES cells [1, 2]. Although ES cells are usually derived from culture of the inner cell mass of the preimplantation blastocyst, as shown in mice and in humans [3, 4], possibility of the establishment of ES cells from morula in mink and cow, and also from embryonic germ cells was also reported [5–7]. The current NIH repository of ES cells contains 78 lines, of which only 11 have met NIH scientific criteria (see NIH website), including the presence of L-alkaline phosphatase (TRA-2-39), Oct-4, telomerase, high-molecular-weight glycoproteins (antibodies TRA-1-60, TRA-1-81), stage-specific embryonic antigens (SSEA-3, SSEA-4), euploid karyotype, and teratoma formation in SCID mice [8].

We developed an original technique for the establishment of ES cell lines from human embryos at the morula stage, which were shown to meet all the NIH criteria. Similar to the ES cells established from human blastocysts, the morula-derived ES cells were obtained from embryos donated by patients who gave an informed consent [9]. For establishing human ES cell lines from the morula-stage embryos, zona pellucida is removed and morula placed under middle-density feeder layer. Within several days, cells outgrow and spread into the feeder layer. The primary cell disaggregation is performed with EDTA, and the loose cells are transferred back to the feeder layer to proliferate. Fast proliferating colonies are isolated and propagated further. The typical human morula-derived stem cells are shown in Figure 7.1.

In contrast to ES cell line derivation from morula, the establishment of the human ES cells from blastocyst involves immunosurgery, requiring the isolation and placement of the inner cell mass (ICM) on a feeder layer. The primary cell disaggregation is performed with EDTA and the rounded cells are transferred back to a fresh feeder. No morphological differences between human ES cells originating from ICM and from morula were observed, except for the morula-derived ES cells being more heterogeneous (see Figure 7.1) as well as the pattern of above marker expression (Table 7.1). The established human ES cell lines

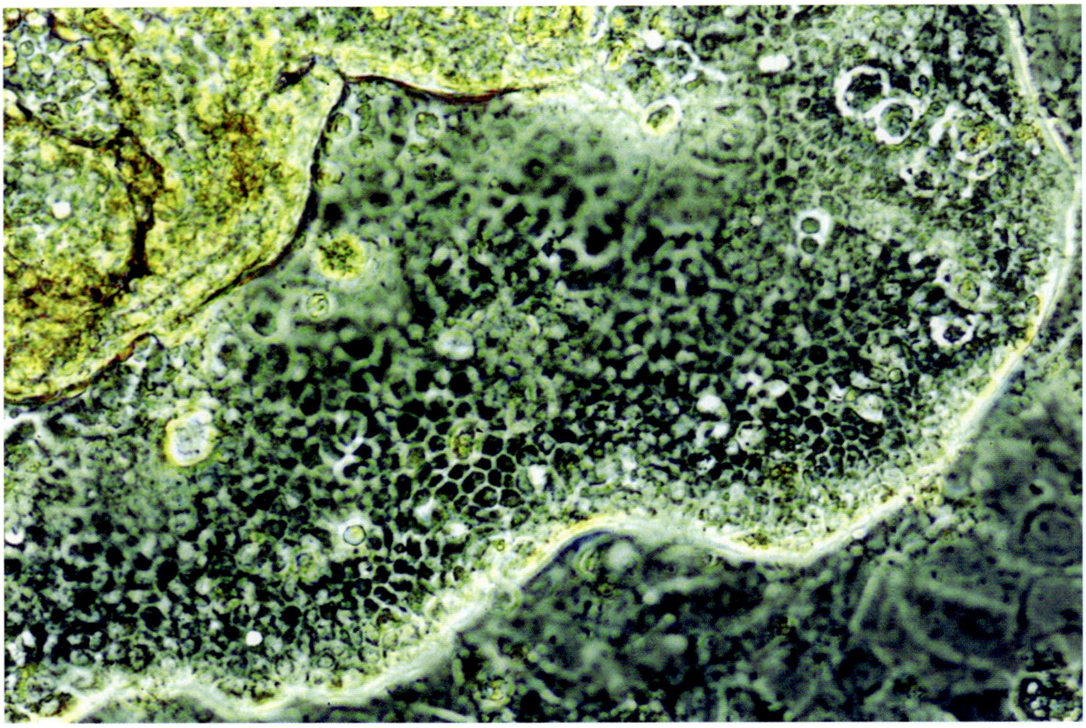

Figure 7.1. Morphology of ES cells derived from morula.

are maintained in vitro from 10–15 passages before freezing in sufficient amounts with control thaw out. The efficiency of obtaining ES cell lines depending on the source is presented in Table 7.2, suggesting no difference.

To induce the differentiation of human ES cells in different cell types, the cells are cultured to form embryonic bodies (Figure 7.2), which are isolated with subsequent disaggregation and plating of clusters of cells. The cultured clusters of cells showed a wide range of cell types belonging to ectoderm, endoderm, and mesoderm, and spontaneously differentiate in vitro into a variety cell types, including neuron-like cells with dendrites and contracting primitive cardiocyte-like cells. The directed differentiation of ES cells was achieved also for ciliated epithelia of the lung, secretary epithelia in the gut-like structures, insulin-producing beta cells, and the production of hematopoietic precursor cells, followed a coculture of embryonic stem cells with mouse bone marrow cell line [10–14]. It was also shown that the neural progenitors differentiate in vitro into astrocytes, oligodendrocytes, and mature neurons, which were able to respond to host signals and were capable of constructing the neuronal and glial lineages, following their transplantation into the developing brain [15]. Controlling differentiation into pure populations of specific neural cells may eventually form the basis of therapy for some neurodegenerative disorders and spinal injuries. These developments provide an

Table 7.1. Markers for human ES-cells derived from different stages (ICM-ES, inner-cell-mass-derived SC; Bd-ES, blastocyst-derived SC; Md-ES, morula-derived ES)

Cell Lines	TRA-2-39 (AP)	Oct-4	SSEA-3	SSEA-4	TRA-1-60	TRA-1-80
ICM-ES	+	+	+	+	+	+
Bd-ES	+	+	+	+	+	+
Md-ES	+	+	+	+	+	+

Table 7.2. Efficiency of establishing human stem cell lines

Embryos	Embryo stage	Cell lines	Sex M/F
46	Morula	8 (17%)	3M/5F
39	Blastocyst	7 (18%)	3M/4F
32	ICM	5 (16%)	3M/2F

obvious potential for the therapeutic use of embryonic stem cells in clinical practice.

The above developments have made it possible to initiate the establishment of a repository of the ES cell lines with different genetic abnormalities. Although initially the major goal of the establishment of human ES cell lines was the development of cell-replacement therapies, it is presently obvious that human ES cell lines will have an important role in the studies of mechanisms of genetic disorders through generating the sources of normal and genetically abnormal cells and tissues. The ability to obtain the human ES cell lines with specific genetic disorders that could produce unlimited quantities of the disease tissue where the disease has a genetic basis makes it realistic to undertake research on the primary disturbances of the cellular processes in the genetically abnormal cells and to identify the molecular mechanisms that might be blocked to prevent the disease progression. Therefore, there is the obvious need for establishment of ES cell lines originating from embryos with genetic and chromosomal abnormalities to provide the basis for understanding of the mechanisms of phenotype realization of genetic defects and for the development of new approaches for their possible treatment.

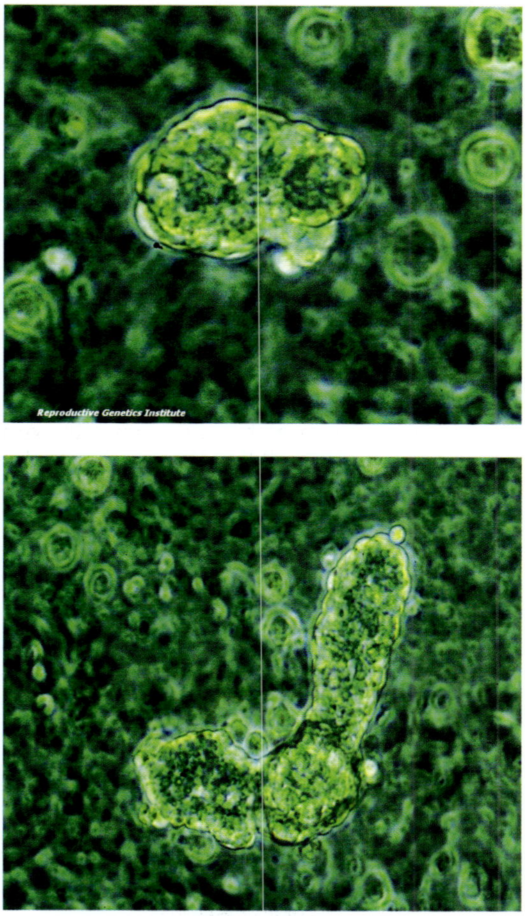

Figure 7.2. Formation of embryonic bodies in the process of differentiation in vitro. Two different morula-derived ES cell lines are shown.

The establishment of ES cell lines with genetic disorders has become possible with introduction of PGD, so that the mutation-free embryos are transferred while those affected provide the valuable source of ES cell lines with genetic abnormalities, using the above techniques for the initiation of ES cells lines from preimplantation embryos. This also provides a unique opportunity of investigating the potential of establishing ESC lines depending on the genotype.

Based on our ongoing PGD work described in previous chapters, we attempted the establishment of ES cell lines from embryos with a variety of single-gene and chromosomal disorders. Our preliminary results show a poor outcome of ES cell lines from the embryos with autosomal aneuploidies, which might be expected to have a selective disadvantage in cell culture. Those ES cell lines that were established were shown to have normal karyotype (Figure 7.3). It was recently reported that in the process of the propagation and maintenance of the human ES cell lines there may be a risk of the de novo chromosomal abnormalities [16]. The human ES cell line in which incidental chromosomal abnormalities were detected was approved by NIH as a source of human ES cells and distributed to more than 100 research institutions around the world. This human ES cell line was obtained from the Wisconsin-based stem cell registry WiCell, and was shown to have normal karyotype, 46,XX, following establishment and maintenance in culture for several months, and also was shown to have the ability to differentiate with the formation of neural and beating cardiac muscle cells. However, karyotyping changes involving the gain

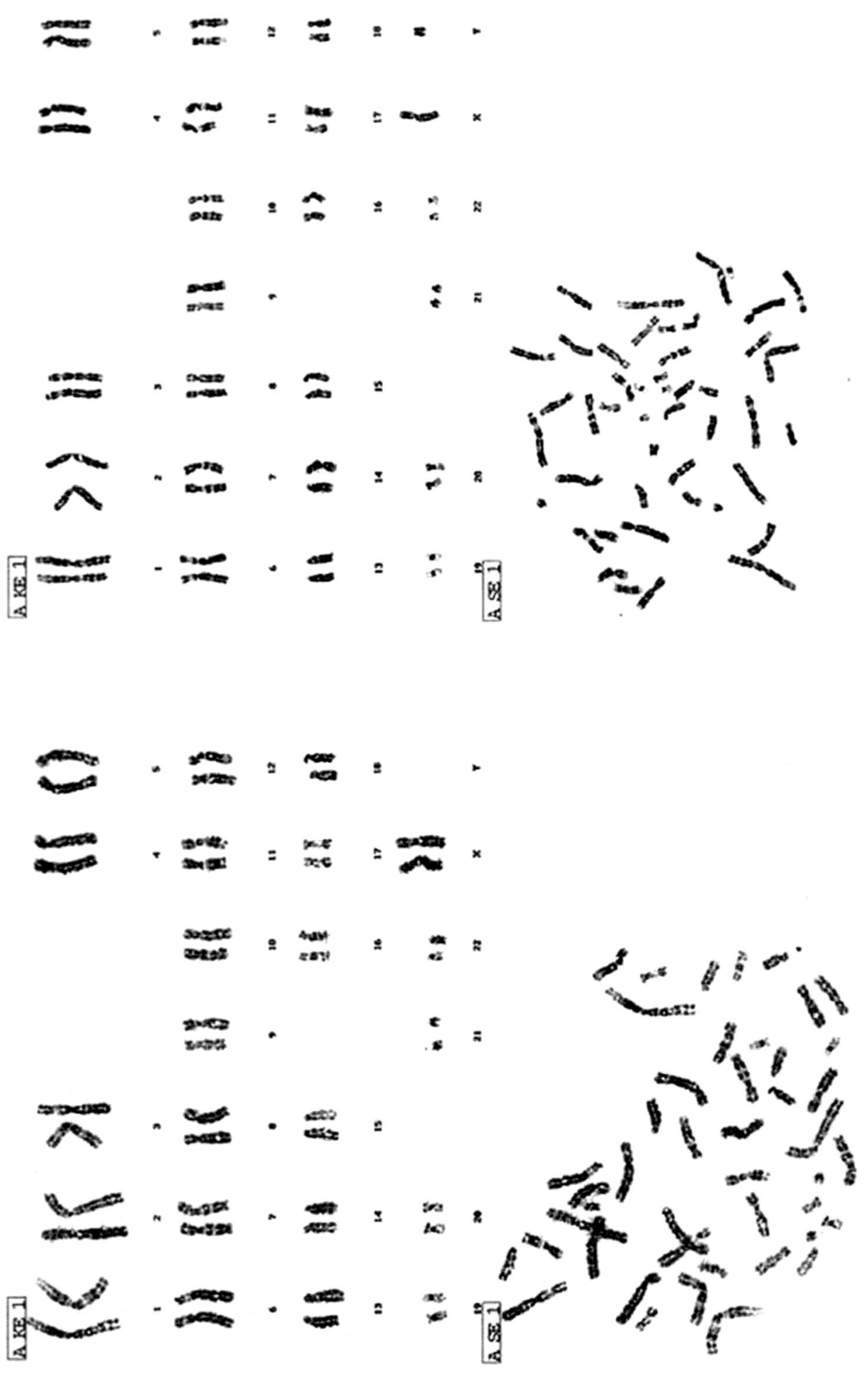

Figure 7.3. Karyotype of two ES cell lines derived from normal embryos. (*Left*) Normal karyotype 46, XX and (*right*) normal karyotype 46, XY.

of the chromosome 17q were observed in three independent human ES cell lines on five independent occasions, together with the occasional gain of chromosome 12, which was suggested to be attributable to a selective advantage of these cells to the propagation of undifferentiated cells. This phenomenon creates concern over the use of ES cells for stem-cell-based therapies, because cytogenetic changes may be only part of the genetic abnormalities acquired in the process of the establishment, maintenance, and differentiation of human ES cell lines.

This also makes initial genetic testing of the embryos used as a source of ES cell lines one of the basic requirements, so PGD is an important source for the ES cell lines, because the embryos obtained from PGD are well tested, with the genotype of the potential ES cell line known from the onset. Our current experience shows that the efficiently of the derivation of ES cell lines from the embryos with single-gene disorders is not affected, as may be demonstrated by the establishment of the first ES cell line repository with different genetic abnormalities, which presently contains more than 100 human ES cell lines. This repository contains human ES cell lines obtained from embryos with single-gene disorders, such as thalassemia (HBB), neurofibromatosis type 1 (NF1), Marfan syndrome (FBN1), myotonic dystrophy (DMPK), Becker muscular dystrophy (BMD), fragile-X syndrome (FMR1), which are described below.

7.1 Marfan Syndrome (FBN1)

Marfan syndrome is an autosomal dominant disease caused by mutation of the gene located in chromosome 15 (15q21.1). The disease leads to skeletal abnormalities, including scoliosis, chest wall deformity, tall stature, and abnormal joint mobility. Ectopia lentis occurs in up to 80% of patients and almost always is bilateral. The leading cause of premature death is progressive dilation of the aortic root and ascending aorta causing aortic incompetence and dissection.

As can be seen from the pedigree presented in Figure 7.4, the couple presented for PGD because the female partner was a carrier of G7712A mutation in exon 62. PGD was performed using PB1 and PB2 analysis, which predicted eight mutant, and eight mutation-free oocytes. Three embryos resulting from these oocytes were transferred, and three of those that reached blastocyst stage were frozen. One of the mutant embryos, resulting from oocyte #5, was donated for research and used for establishment of ES cell line, which was confirmed to contain G7712A mutation in exon 62 (hESC-154).

7.2 Thalassemia (HBB)

The list of mutations for which we performed PGD was presented in Chapter 3. Thalassemias are among the most common single-gene disorders, requiring lifelong blood transfusion and iron chelation therapy, with the only radical treatment being HLA-compatible bone marrow transplantation, as describe in Chapters 3 and 4. As shown in those chapters, thalassemia is one of the major indications of PGD combined with preimplantation HLA typing, so this provided the possibility of the establishment of ES cell lines, one of which is shown in Figure 7.5.

As seen from the pedigree, the parents were carriers of different mutations, the father carrying Cd39 and the mother IVS I-110 mutation, both being the most common mutations in the Mediterranean region. They had one previous child affected with severe beta-thalassemia, requiring HLA-compatible bone marrow transplantation. So the couple requested PGD and HLA typing, which was performed by blastomere biopsy and multiplex PCR analysis, involving simultaneous testing of both mutation and HLA markers, described previously in Chapter 4. Of 10 embryos tested, three were affected (embryos #1, 6, and 7), one contained only one chromsome 11 with Cd 39 mutation, while the remaining six were carriers of one of the mutations. Only one of these embryos was a full HLA match and transferred. Of the six affected embryos four were donated for research, including embryo #1, which was also an HLA non-match, resulting in the establishment of ES cell line hESC-164. The follow-up testing showed that the cells are compound heterozygous affected for Cd39 and IVSI-110 mutations.

7.3 Fragile-X Syndrome (FRM1)

As described in Chapter 3, FRM1 is the most common genetic disorder, with a prevalence of 1 in every 2000 children, and is characterized by moderate to severe mental retardation, microorchidism, large ears, prominent jaws, and high-pitched,

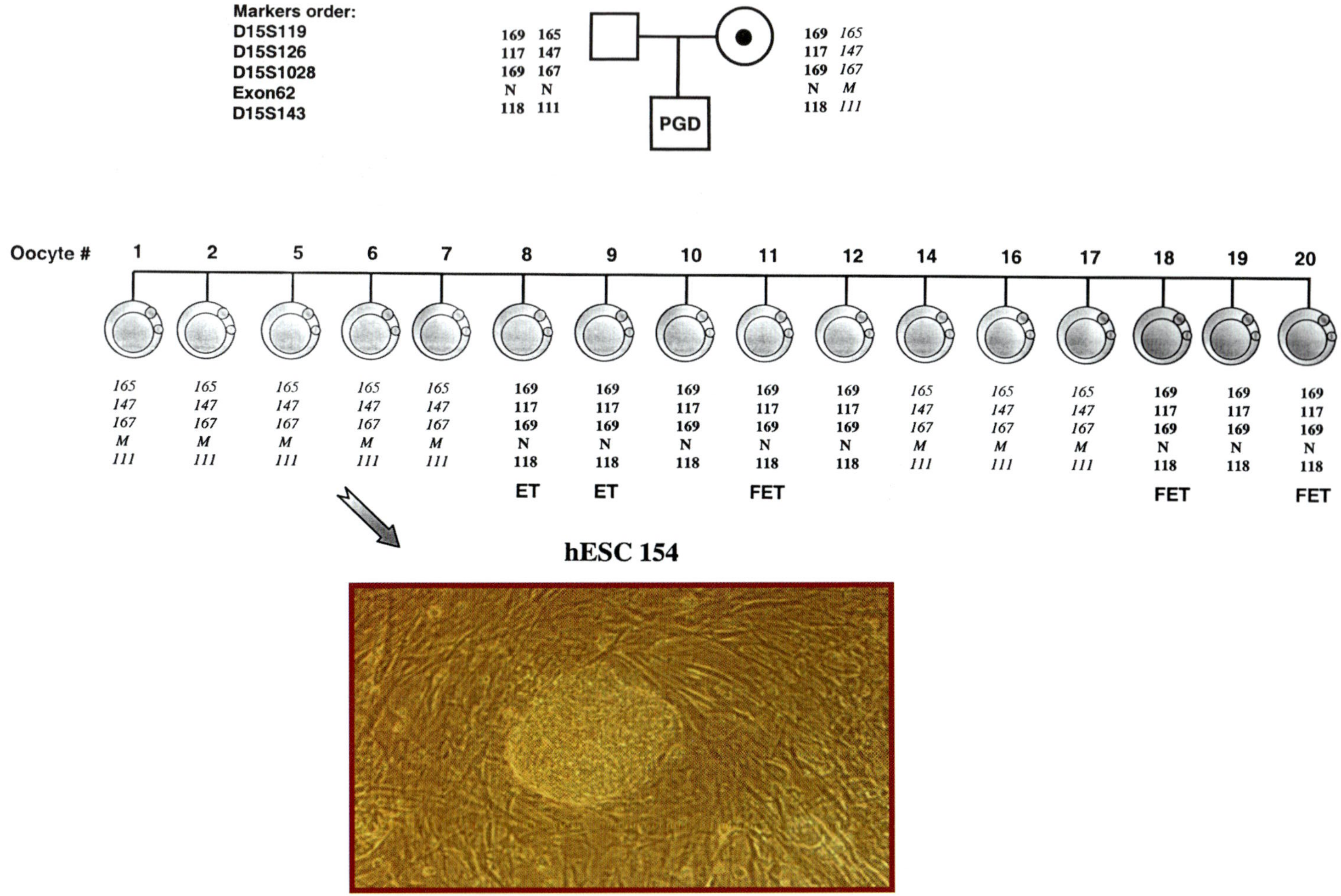

Figure 7.4. PGD for Marfan syndrome with the establishment of embryonic stem cell line. (*Top*) Family pedigree with paternal and maternal haplotypes (marker order is given on left) showing the ongoing unaffected pregnancy as a result of PGD. (*Middle*) Results of sequential PB1 and PB2 analysis of 16 oocytes, six of which were predicted to be mutation-free, and the other six affected. Two embryos were transferred back to the patient (ET), three were frozen and used for frozen embryo transfer, while one of the affected embryos (embryo #5) was donated for research and used for establishment of ES cell line (*bottom*) with confirmed affected genotype.

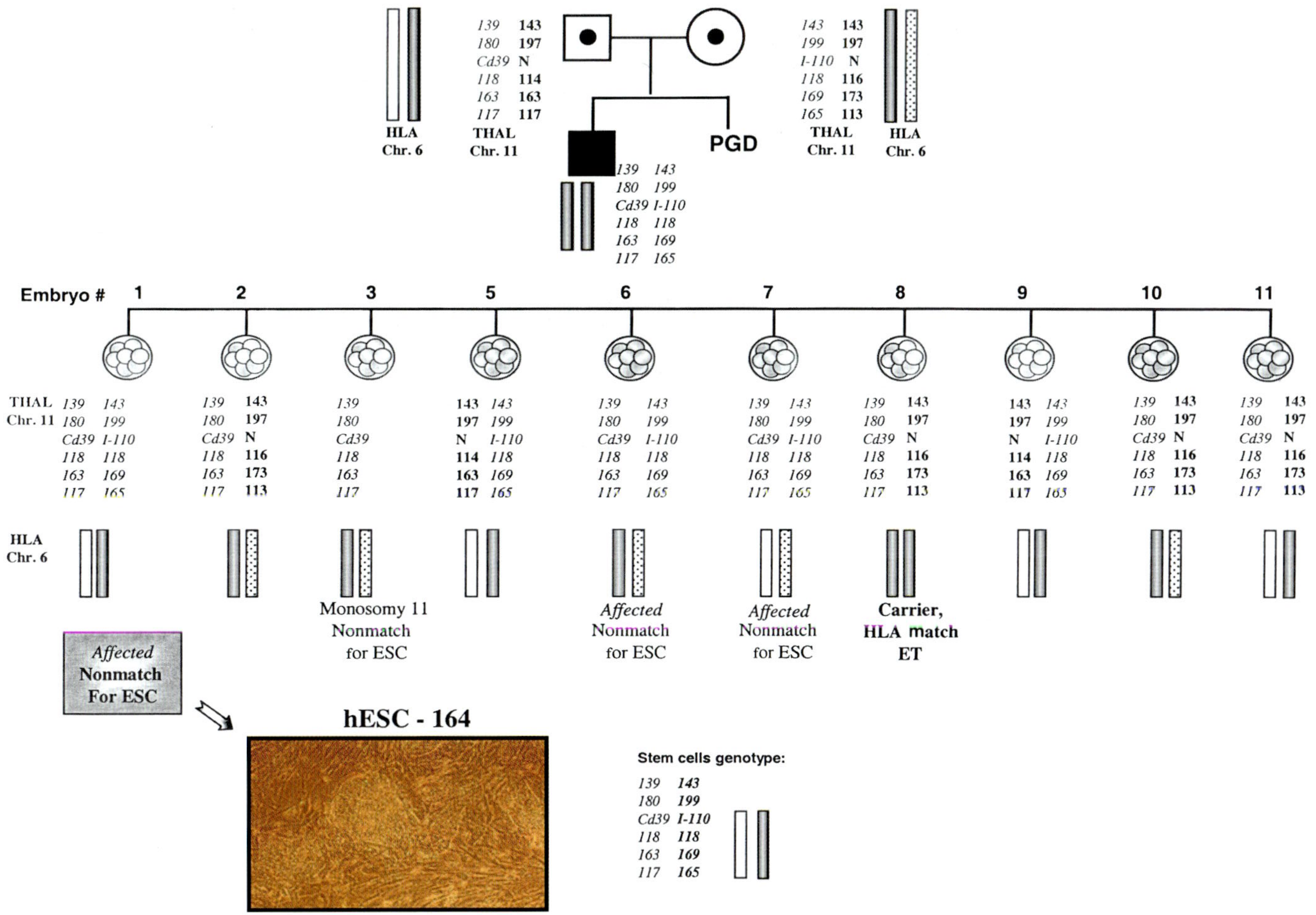

Figure 7.5. PGD for thalassemia and HLA typing, with establishment of ES cell line. (*Top*) Family pedigree with paternal and maternal haplotypes: father is carrier of Cd39 mutation; mother, IVSI-110; HLA haplotypes as darker and lighter bars, darker corresponding to haplotypes of sibling with thalassemia. (*Middle*) Results of blastomere biopsy analysis of 10 embryos, six of which were predicted to be unaffected, one being also HLA-matched to the affected siblings, which was transferred (embryo #8). Of the remaining four embryos, one was monosomic for chromosome 11 of paternal origin with mutation Cd39, and three were affected, one of which (embryo #1) was donated for research and used for establishment of ES cell line (*bottom*) with confirmed double heterozygous affected genotype with Cd39/IVSI-110.

jocular speech, resulting from an expansion of a CGG repeat. As seen from the pedigree (Figure 7.6), the couple presented for PGD with three previous unsuccessful pregnancies, one of which resulted in birth of affected child with fully expanded allele. PGD was performed by PB1 and PB2 testing for maternal mutation, using simultaneous amplification of seven linked polymorphic markers listed in Figure 7.6, which showed that 10 of 16 tested oocytes were free of mutation, which were either transferred or frozen for future use by the couple. Of six embryos resulting from oocytes containing the mutant gene, embryo #13 was the source of embryonic stem cell line, confirmed to contain the fully expanded allele of FRM1 gene.

7.4 Myotonic Dystrophy (DMPK)

Similar to FRM1, DMPK is an autosomal dominant dynamic mutation in chromosome 19 (19q13.2–13.3), resulting from the expansion of triplet repeat, with a prevalence of 1 in 8000 live births. This is a severe neurodegenerative disorder chalacterized by myotonia, muscle wasting in the distal extremities, cataract, hypogonadism, defective endocrine functions, and cardiac arrythmias. As mentioned in Chapter 3, DM was one of the most frequent indications for PGD, which also provided the possibility for obtaining affected embryos for the establishment of the ES cell lines with DMPK. One of the PGD cycles for the maternally derived mutation is presented in Figure 7.7, showing the results of blastomere biopsy of six embryos, which resulted in the transfer of two mutation-free embryos (embryos #5 and 7). Two of the remaining four embryos were with aneuploidies, monosomy 16 in embryo #8 and monosomy 22 in embryo #10, and two affected embryos were donated for research and resulted in the establishment of two embryonic stem cell lines, hESC-148 and hESC-153.

7.5 Neufibromatosis Type 1 (NF1)

PGD for NF1 was described in Chapter 4, resulting in the birth of healthy unaffected twins in the case presented in Fig. 7.8. Three of the embryos with mutation Trp → Ter (TGG → TGA) in exon 29 of the NF1 gene were donated for research and used

Table 7.3. List of ES cell lines with genetic disorders in RGI Repository*

Cell line ID	Genetic disorder
SC-125	Female, fragile-X syndrome (FMR1), carrier
SC-128	Fanconi anemia-A (FANCA), 14 bp deletion, carrier
SC-137	Neurofibromatosis, type 1 (NF1)
SC-138	Neurofibromatosis, type 1 (NF1)
SC-139	Neurofibromatosis, type 1 (NF1)
SC-140	Neurofibromatosis, type 1 (NF1)
SC-141	Neurofibromatosis, type 1 (NF1)
SC-148	Myotonic dystrophy (DMPK)
SC-153	Myotonic dystrophy (DMPK)
SC-154	Marfan syndrome (FBN1)
SC-158	Beta-globin mutation IVS 1-1, carrier
SC-164	Beta-thalassemia IVSI-110/cd39
SC-170	Becker muscular dystrophy (BMD)

*The list presently includes also one EC cell line with adrenoleukodystrophy, 3 with Huntington disease; one with muscular dystrophy of Duchenne type and 2 with chromosomal disorders, including trisomy 14 and trisomy 13 due to translocation t(4;13).

for the establishment of three ES cell lines with NF1, shown in Figure 7.8.

7.6 Becker Muscular Dystrophy (BMD)

Becker muscular dystrophy is an X-linked disorder located in Xq21.2, resembling Duchenne muscular dystrophy (DMD), which is the most common form of muscular dystrophy. The clinical features of BMD, however, are more benign and realized later than DMD, which typically presents in boys age 3–7. As shown in the pedigree of the couple presenting for PGD (Figure 7.9), there were three affected boys in the family of the mother, who is a carrier of the deletion in intron 45 of the gene. Of eight embryos tested for deletion using seven polymorphic markers, only one was mutant (embryo #1), so donated for the establishment of embryonic stem cell line. Of the remaining seven normal embryos, four reached the blastocyst stage, of which two were transferred (embryos #2 and 4) and two were frozen for subsequent transfer (embryos #6 and #8). The established ES cell line, hESC-170, was tested by the same set of markers used for PGD and was confirmed to contain the mutant gene.

Twenty ES cell lines overall, listed in Table 7.3, currently represent the world's first repository of ES cell lines with genetic disorders (17), which are presently used for research

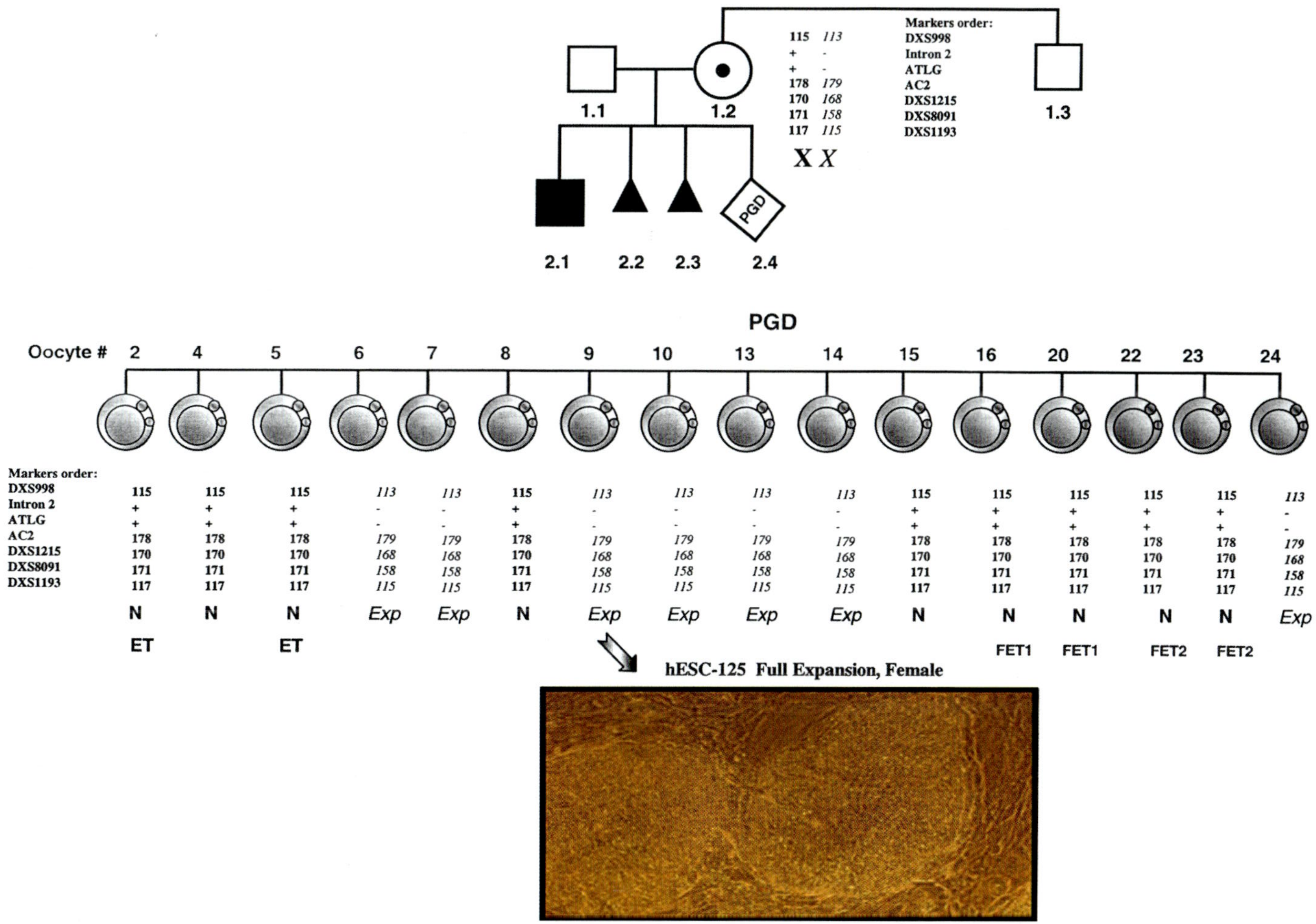

Figure 7.6. PGD for fragile X syndrome (FMR1) with establishment of ES cell line. (*Top*) Family pedigree, showing maternal (1.2) haplotypes (non-bolded with expansion), and reproductive history with one affected son (2.1), two spontaneous abortions, and ongoing unaffected pregnancy following PGD. (*Middle*) Results of polar body analysis of 16 oocytes, nine of which were predicted to be unaffected, and seven with the expansion. Two of the embryos deriving from mutation-free oocytes (oocytes #2 and 5) were transfered and four frozen. Of seven affected embryos originating from the mutant oocytes, one (embryo deriving from oocyte #9) was donated for research and used for establishment of ES cell line (*bottom*). The ES cell line was confirmed to have female genotype with fully expanded allele.

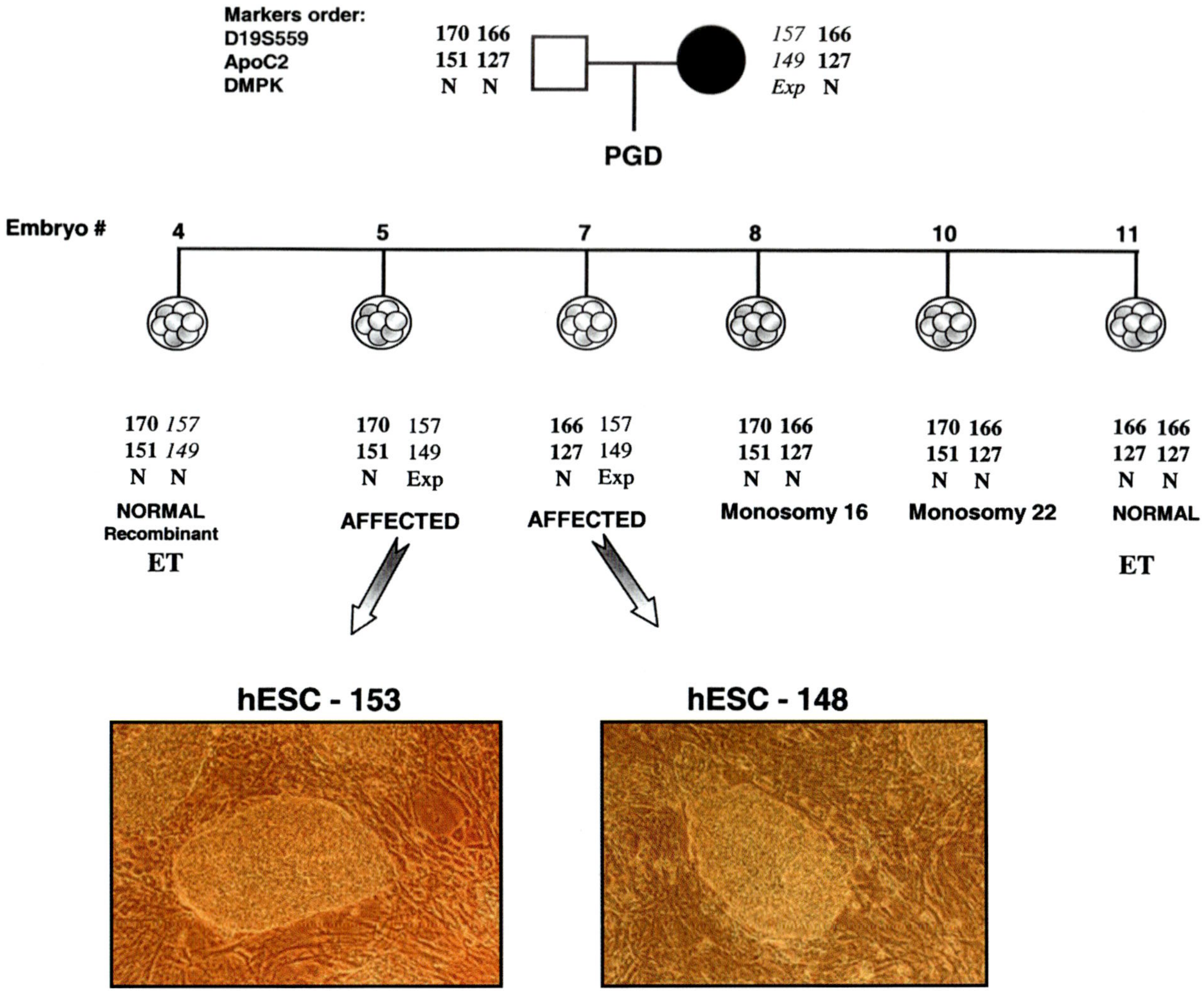

Figure 7.7. PGD for myotonic dystrophy and aneuploidy with establishment of two mutant ES cell lines. (*Top*) Paternal and maternal haplotypes (marker order is given on left). (*Middle*) Results of blastomere analysis of six embryos, two of which were predicted to be mutation-free and transferred (embryos #4 and 11), two were mutation-free but chromosomally abnormal (embryo #8 with monosomy 16 and #10 with monosimy 22), and the remaining two were with expansion and donated for research (embryos #5 and 7). These embryos were used for establishment of ES cell lines (*bottom*) with confirmed affected genotypes.

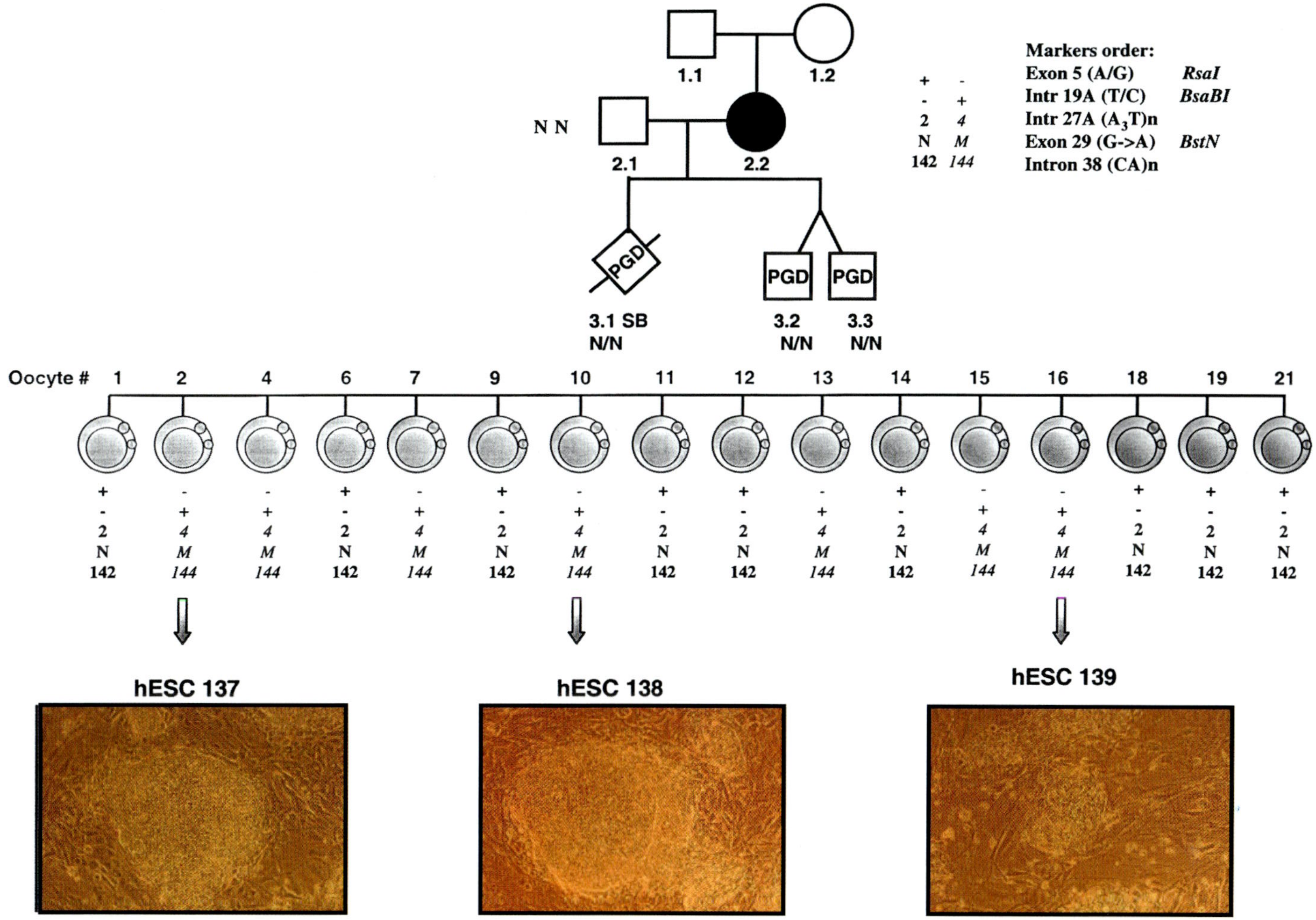

Figure 7.8. PGD for NF1 with establishment of three ES cell lines (initial PGD cycles for this couple were described in Section 3.6.2). (*Top*) Family pedigree showing affected mother (2.2) with de novo NF1 mutation. Two pregnancies were obtained from the transfer of mutation-free embryos following PGD, one of which resulted in a stillbirth (3.1) and the other in a healthy twin pregnancy (3.2 and 3.3), all three babies being confirmed to be unaffected. (*Middle*) Results of polar body analysis of 16 oocytes, seven of which were predicted to be mutant, and three of the embryos deriving from these mutant embryos were donated for research (embryos #2, 10, and 16). These embryos were used for establishment of ES cell lines (*bottom*), which were confirmed to be affected.

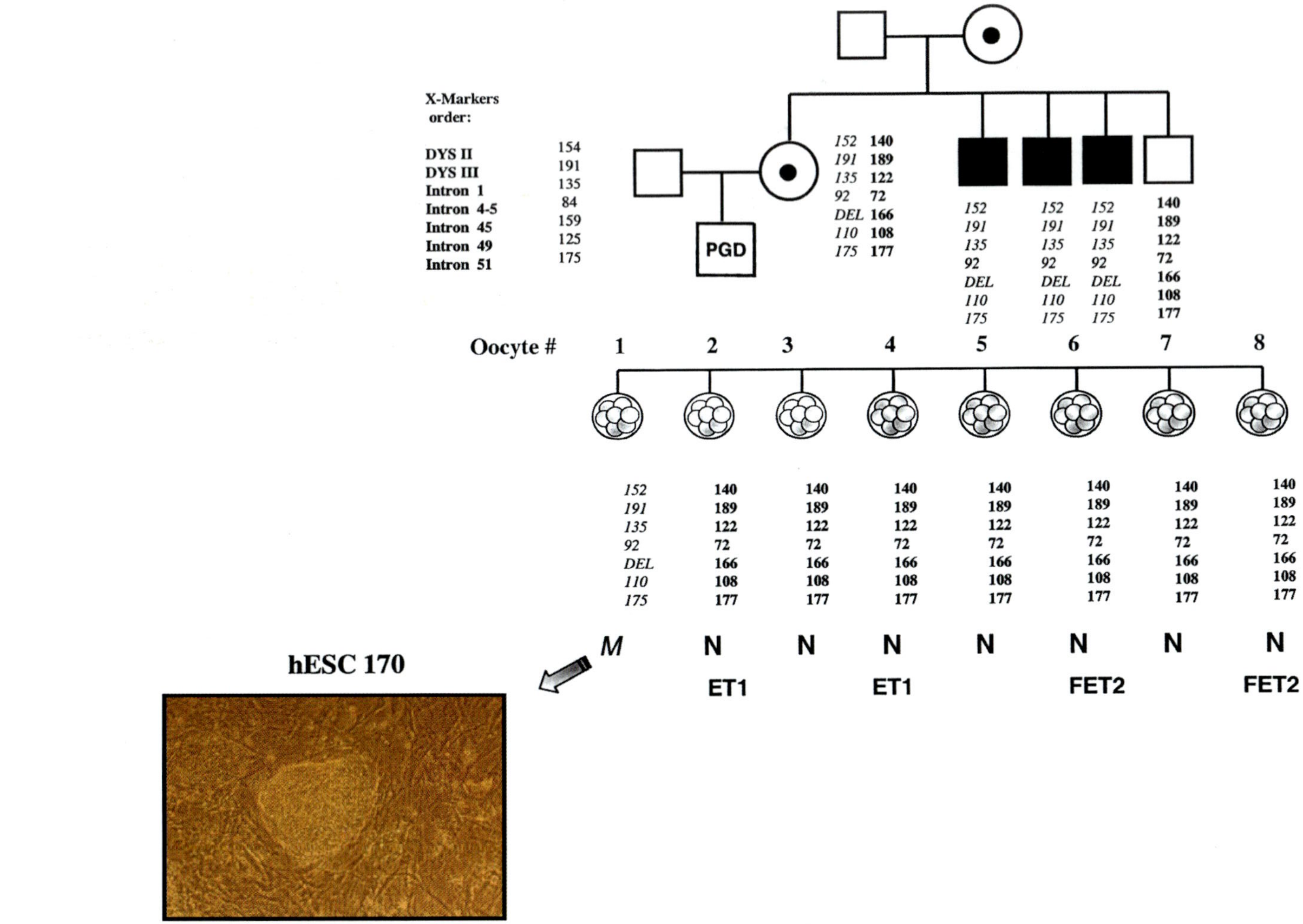

Figure 7.9. PGD for BMD with establishment of ES cell line. (*Top*) Family pedigree showing a carrier mother who had three affected bothers with the deletion. PGD allowed having an ongoing unaffected pregnancy. (*Middle*) Results of blastomere analysis of eight embryos, identifying only one mutant embryo (embryo #1), which was the source for the establishment of ES cell line (*bottom*), and was confirmed to be affected.

purposes and are also available on request (Reproductive Genetics Institute, Inc., Chicago, IL).

References

1. Pickering S, Braude P, Patel M, Burns J, Bolton V, Minger S. Preimplantation genetic diagnosis as a novel source of embryos for stem cell research. Reprod BioMed Online 2003;7:353–364.
2. Kuliev A, Verlinsky Y. Thirteen years' experience of preimplantation diagnosis. Report of the Fifth International Symposium on Preimplantation Genetics. Reprod BioMed Online 2004;8:229–235.
3. Martin GR. Isolation of a pruripotent cell line from early mouse embryos cultured in medium conditioned by teratocarcinoma stem cells. Proc Natl Acad Sci USA 1981;78:7634–7638.
4. Tompson JA, Itskovitz-Eldor J, Shapiro SS, et al. Embryonic stem cell lines derived from human blastocysts. Science 1998;282:1145–1147.
5. Soukoyan MA, Vatolin SY, Golubitsa AN, et al. Embryonic stem cells derived from morulae, inner cell mass and blastocyst of mink:comparison of their pluripotencies. Mol Reprod Dev 1993;36:148–158.1993.
6. Stice SL, Strelchenko NS, Keefer CL, Matthews L. Pluripotent bovine embryonic stem cell lines direct embryonic development following nuclear transfer. Biol Reprod 1996;54:100–110.
7. Shamblott MJ, Axelman J, Littlefield JM, et al. Human embryonic germ cell derivatives express a broad range of developmentally distinct markers and proliferate extensively in vitro. Proc Natl Acad Sci USA 2001;98:113–118.
8. National Institutes of Health Guidelines for Research Using Human Pluriopotent Stem Cells. NIH Stem Cells Information Archives, August 25, 2000 (www.nih.gov/news/stemcell/stemcellguidelines.htm).
9. Strelchenko N, Verlinsky O, Kukharenko V, Verlinsky Y. Morula-derived human embryonic stem cells. Reprod BioMed Online 2004;9:623–629.
10. Trounson A, Pera M. Human embryonic stem cells. Fertil Steril, 2001;76:660–661.
11. Assady S, Maor G, Amit M, et al. Insulin production by human embryonic stem cells. Diabetes 2001;50:1691–1696.
12. Kaufman DS, Hanson ET, Lewis RL, Auerbach R, Thomson JA. Hematopoietic colony-forming cells derived from human embryonic stem cells. Proc Natl Acad Sci USA 2001;98:10716–10721.
13. Clark AT, Bodnar MS, Fox M, et al. Spontaneous differentiation of germ cells from human embryonic stem cells in vitro. Hum Mol Genet 2004;13:727–739.
14. Bielby RC, Bocaccini AR, Polak JM, Buttery LDK. Osteoblasts derived from embryonic stem cells undergo in vivo mineralization on a PDLLA scaffold (abstr 227). Abstracts of 2nd Annual ISSCR Meeting, Boston, 2004.
15. Reubinoff B. Human ES cells: A potential platform for cell and gene therapy. Fifth International Symposium on Preimplantation Genetics, June 5–7, Antalya, Turkey, p. 34. Istanbul Memorial Hospital, Istanbul, Turkey, 2003.
16. Draper JD, Smith K, Gokhale P, et al. Recurrent gain of chromosomes 17q and 12 in cultures human embryonic stem cells. Nat Biotech 2004;22:53–54.
17. Verlinsky Y, Strelchenko N, Kukharenko V, et al. Human embryonic stem cells with genetic disorders. Reprod BioMed Online 2005;10:105–110.

8

Ethical, Social, and Legal Issues

Preventive approaches to congenital disorders always raise ethical problems, because traditionally the emphasis should be on treatment rather than on the avoidance of birth of children with congenital disorders, which is unfortunately unavailable for most genetic disorders at the present time. On the other hand, the most ethically acceptable preventive approaches are indeed those that involve the primary preventive measures, which are better tolerated by society, than the secondary preventive measures involving pregnancy termination. As described in Chapter 1, one of the best examples of the most efficient primary preventive measures may be a population-based fortification of the major foodstuffs by folic-acid-containing multivitamins, which has been demonstrated to result in significant reduction of neural tube defects and congenital malformations overall. Still, such programs have been introduced only in a few populations, so the lack of similar preventive measures in most of communities may be the reason that thousands of children with some of the congenital disorders continue to be born who otherwise might have been born healthy, which therefore may represent an important legal, social, and ethical issue. What is of special importance is that these primary preventive measures are ethically acceptable in any population, because they provide the actual gain in infants free of congenital malformations rather than the avoidance of birth of affected children.

The same is true for preimplantation genetic diagnosis (PGD), which is also a primary preventive measure, although applied on a family level, allowing genetically disadvantages couples to produce unaffected children of their own, who might not otherwise be born at all because of fear of these couples to reproduce and face prenatal diagnosis and termination of pregnancy [1, 2]. So any legal restrictions of these patients' choices may only force them to achieve their goal by traveling to the other countries where the regulations regarding PGD are more liberal. The most recent review on the status of PGD in different countries [3–8] showed that international legal practices range from explicit legalization (e.g., the Netherlands, United Kingdom, France, Spain) to more or less "lawless control" as in Belgium and the United States, to legal prohibition through restrictive laws as in Italy, Germany, Austria, and Switzerland. However, even in these countries, there is a tendency to ease such legal restrictions. For example, there is no interdiction of PGD in Austria, neither through the law on reproductive medicine, nor through the law on genetic engineering, unless the polar body or blastomere biopsy would be misinterpreted as "interference in the germ cell lineage," which would be prohibited [7]. In France, PGD is under the control of CNMBRDP, which is a governmental commission also controlling IVF [4, 5]. According to the regulation, an agreement is required to perform the embryo biopsy and genetic/FISH testing, the evolution of which and indications being the subject for a follow-up by the representatives on a regular basis. Only a few centers in France are allowed to perform PGD for the initial five years, subject to renewal afterwards; the regulations are under the "ethical" law, requiring a forthcoming reexamination. Similarly, PGD in the United Kingdom is regulated by HFEA, which is also a governmental organization, which provides the license for performing PGD and also must approve any new condition for which it is performed. For example, PGD for chromosomal aneuploidies, practiced for more than 10 years in many other countries, has only recently been allowed by HFEA. HFEA also initially refused to allow preimplantation HLA typing without PGD, and has changed its position only recently. Finally, PGD and many aspects of IVF have recently been forbidden in Italy by act of Parliament, according to which only three oocytes are allowed to be aspirated for fertilization in vitro, clearly following the opinions of the hierarchy of the Roman Catholic Church [9]. Of course, this makes PGD impossible in Italy, despite the fact that this country has been among the most actively involved in the development and application of PGD for genetic and chromosomal disorders [2].

As described in Chapter 2, with the introduction of the methods of preconception diagnosis by polar body diagnosis, PGD has become ethically acceptable even in countries with restrictive laws, such as in Germany, where no manipulations are allowed after conception. Polar body testing may also resolve the issue of PGD in Austria and Switzerland [7], and may in the future make PGD an acceptable procedure even in those countries where no preventative measures have ever been allowed on religious grounds. No doubt preconception diagnosis will also make PGD even more attractive in Muslim countries, where blastomere biopsy is currently acceptable and a more preferred option over prenatal diagnosis. Preconcention diagnosis may no longer be restricted to maternally derived genetic abnormalities because the recent progress in sperm duplication may in the future also allow sperm testing for paternally derived abnormalities prior to fertilization (see Chapter 2).

The available experience shows that PGD has become a routine procedure in an increasing number of countries, such the United States and Belgium, where no strict governmental regulations for PGD exist [3, 6, 7]. It is, therefore, not surprising that the largest experiences in PGD for genetic and chromosomal disorders have been accumulated in these countries. The guidelines and standards for appropriate PGD practice have recently been developed by the Preimplantation Genetic Diagnosis International Society (PGDIS) and may be followed to achieve the required standards of PGD [10]. There are also regulations developed by national scientific societies, such as in Japan, where no active PGD program is currently available, but there are regulations for PGD developed by the National Society of Obstetrics and Gynecology and by the Japan Society of Human Genetics, which provide guidelines for genetic diagnosis [8].

The ethical issues of PGD have been recently evolving together with the development of methods and with the expansion of PGD indications. Initially, when PGD was applied only to preexisting conditions, with the only goal of avoiding the risk of birth of children with genetic disorders present at birth, PGD allowed avoiding traditional prenatal diagnosis and termination of pregnancy, so making prevention of genetic disorders more ethically acceptable. Some of couples may have had the experience of repeated pregnancy terminations before having a normal child, while others could not accept prenatal diagnosis and termination of pregnancy at all. So PGD initially was an important alternative, so that at-risk couples had the choice of either going through prenatal diagnosis and termination of pregnancy, or controlling their pregnancy outcome by testing the oocytes or embryos before implantation to ensure that the pregnancy is unaffected from the onset. Accordingly, not informing genetically disadvantages couples about PGD availability, which may have affected their possible choices, may present an important ethical and legal issue. It is especially important for those conditions, such as translocations, carriers of which have an absolutely miserable pregnancy outcome. As mentioned in Chapter 5, analysis of meiotic outcome in carriers of translocation leaves little chance for prenatal diagnosis to be useful in identifying a balanced or normal fetus, as the carriers of translocations have more than an 80% prospect of loosing their pregnancy by spontaneous abortions. So PGD for such couples may clearly be the only hope, providing a realistic option of having unaffected children of their own.

As shown in Chapter 3, PGD will have increasing practical implications with the current progress in improvement of the quality of life, life expectancy, and the possibility to reproduce by genetically affected patients. For example, the life expectancy in cystic fibrosis (CF) may presently not be different from normal individuals, who may be able to procreate and have their own children. Similarly, with the success in stem cell transplantation, children with thalassemia may be radically cured, and these so called "ex-thalassemics" require PGD to avoid 50% risk of producing their own thalassemic children. As seen from Chapter 3, PGD has already been applied for homozygous or double heterozygous affected individuals with CF, thalassemias, and phenylketonuria (PKU), who were able to have their own unaffected children following PGD. On the other hand, this may still create the feeling that some extreme variations of the genotype are rejected by society, so the couples may face a complex decision of transferring back the embryos with different genotypes. For example, some couples may elect to transfer the embryos carrying the affected genes, such as for deafness or achondroplasia, so using PGD to conceive a disabled child, which sets a poor precedent for later patients facing complex familial decisions [11].

The important breakthrough from the ethical and social point of view was the introduction of PGD for diseases with genetic predispositions, especially when it has become possible to avoid the transfer of the embryos carrying the genes predisposing to common disorders of adult life. Although there is no difference in the application of PGD for early- or late-onset disorders with genetic predisposition from the application of PGD to chromosomal disorders and autosomal recessive metabolic disorders with the onset at birth or early childhood, the discomfort of PGD for disorders with genetic predisposition can be explained by the fact that this has been controversial or even unacceptable in the practice of prenatal diagnosis. The same diagnosis is of course possible by chorionic villus sampling (CVS) or amniocentesis with the only difference that if the fetus carries the gene predisposing to late-onset diseases with genetic predisposition, such as Alzheimer disease (AD) or other late-onset diseases with genetic predispositions described in Chapter 3, the couple would have to make an important decision of pregnancy termination, which could hardly be justified on the basis of genetic predisposition alone, taking into consideration that the clinical manifestation of the disease might not be realized at all in some proportion of cases. Alternatively, PGD technology allows genetic testing of human eggs and embryos before pregnancy, therefore making it totally realistic to establish only potentially normal pregnancies without a disease with early- or late-onset disorder with genetic predisposition. Thus, the prospective at risk-parents have to be informed about the availability of the PGD technology to allow them to make the decision themselves about their reproductive options. With no current prospect for treatment of most late-onset diseases with genetic predisposition, such as AD, which may arise despite presymptomatic diagnosis and follow-up, prevention of inherited predisposition to late-onset disease may be the only possible option for couples at risk, because the carriers of mutations causing the above group of diseases not only have up to 100% lifetime risk of developing disease, but also pass this genetic predisposition to their children. The extremely difficult life experience of families affected by any catastrophic early- or late-onset inherited disorder of seeing suffering from the disease and being anxious that they themselves will be soon affected, makes them responsible to ensure that future generation will not be faced with the same difficulties [12, 13].

As for helping couples with their fully responsible decision to use the option of PGD to avoid the inheritance of the causative gene by their progeny, such as the gene for AD, although one of the partners may not be around to see this child grow up, of course societal discussions on the issue will be of great use [14]. First of all, the situation when only one parent supports the child to grow up and takes the responsibility for his or her future is not rare. On the other hand, this is not much different from that in parents who may get cancer or killed in car accident, which are the main killers in Western countries. At least using PGD is better than having children without testing, because these children will have a 50% chance of having AD or other dominantly inherited predisposition to severe late-onset disorders with genetic predisposition. The possibility that there may be soon approaches to prevent the clinical manifestation of these disorders in carriers of the mutant gene should not be excluded either.

PGD for common late-onset disorders provides a novel nontraditional option for patients who may wish to avoid the transmission of the predisposing mutant gene to their potential children. This may appear for some patients the only reason for undertaking pregnancy, as the pregnancy may be established free from an inherited predisposition from the very onset. Because, as mentioned, such diseases never present at birth or early childhood and even later may not be expressed at all, the application of PGD is still controversial. However, with no current prospect for treatment of many of them, PGD may be offered as the only relief for such at-risk couples.

Therefore, prospective parents should be informed about this emerging new technology, so they could make their reproductive choice based on all available options. This seems to be ethically more acceptable than a denial of the information on the availability of PGD. Presented results of PGD for early- or late-onset disorders in Chapter 3 demonstrate the extended practical implications of PGD, providing prospective couples at genetic risk with wider reproductive options for having unaffected children of their own.

One of the most recent ethical issues of PGD has been preimplantation HLA typing, because PGD for this indication is done for the benefit of a potential recipient rather than for the embryo itself, particularly when there is no need for testing of causative gene [11, 15, 16]. This may lead to feelings of moral outrage in some, while others may justify the action by saving a child's life from a severe disease. It is of interest that the majority of Americans are supportive of using PGD to ensure that an infant will provide an HLA match to donate stem cells or even tissue to an older sibling [17].

However, the attitudes may be different depending on whether genetic testing in the embryo is done or not. If preimplantation HLA typing is performed in combination with PGD, with the primary purpose being testing for the causative gene, such as in the case of Fanconi anemia (FA) [18], it appeared morally more acceptable than preimplantation HLA typing as a sole purpose. The example of the latter situation may be leukemia or sporadic Diamond-Blackfan anaemia (DBA) in elder children that may be cured by HLA-matched stem cell transplantation [19], which, however, does not present any benefit to the embryos tested. For example, as mentioned, such parents have been denied permission for preimplantation HLA typing in the United Kingdom until recently. The moral dilemma also stands on the need of parents to have another child.

Some issues associated with preimpalnattion HLA typing are related to the actual indications for preimplantation HLA testing, which seem to be similar to the indications for stem cell transplantation, because preimplantation HLA typing has the objective of improving access to an HLA-identical stem cell transplant, which is the key in achieving an accceptable engraftment and survival in stem cell therapy. No doubt the indications will be modified with progress in treatment of bone marrow disorders. For example, with the current success in cure rate by chemotherapy, acute lymphoid leukemia (ALL) and acute myeloid leukemia (AML) may no longer be indications [20], but the option of stem cell therapy should still be available for patients taking into consideration a sizeable proportion of patients for whom chemotherapy was not effective and who may still require compatible stem cell transplantation, especially if the parents plan to have another child anyway. So all the conditions for which bone marrow or cord blood stem cell transplantation is required are also indications for preimplantation HLA typing.

The other important issue is the applicability of preimplantation HLA typing for sporadic conditions, such as DBA, for which there are also inherited forms. Accordingly, the inherited forms might require PGD for the mutations involved, to exclude the risk of transplantation of compatible stem cells that might contain exactly the same mutation as the sibling [21]. For example, there are already two known mutations causing DBA, one in the gene encoding ribosomal protein S19 on chromosome 9, and another gene mapped to chromosome 8 (see Chapter 4). It cannot be excluded that additional mutations will be found for some of the other sporadic forms, also requiring PGD [22]. Still without such information, couples have to make the decision about the need for undertaking transplantation because of serious iron overload in the patients urgently requiring compatible stem cell transplantation from the family member. Therefore, all known mutations causing the disease should be excluded by detailed mutation testing in parents and affected children, and this should also be confirmed by ongoing follow-up studies of the HLA-matched children born after preimplantation HLA typing.

The other controversial issue involves the written consent form, which is signed by parents for the embryo, similar to the situation when parents sign the consent form for umbilical cord blood stem cell collection and storage. It may be also argued that parents do not actually need a baby, and have it merely as a means to save its older sibling, so it would then become a commodity in some peoples' eyes, although parents usually claim that another child is needed for their family anyway, a decision that is solely the parents' right.

Although at the present time only umbilical cord blood stem cells are being collected from the "designer babies" at birth, presenting no harm for the baby, it is argued that the same approach may be used for organ donation. While with the progress in differentiation of cord blood stem cells into the other types of cells [23] such possibilities cannot be entirely excluded, it should be mentioned that preimplantation HLA typing also allows avoiding many ethical issues of reproductive and therapeutic cloning, as it provides a more ethically acceptable option of selecting an HLA-matched progeny, rather that obtaining custom-made embryonic stem cells following somatic nuclear transfer and cloning.

The other nonmedical use of PGD is for sex selection, which has recently been considered acceptable for social reasons, provided that it is applied for selection of sex of second or subsequent children [24, 25]. Although this was not widely used, in some countries, such as India or Jordan, there was need for PGD use for sex balancing, and this seems to be well justified [26, 27].

Finally, PGD raises many ethical issues, which are not unique to its clinical practice and instead are the same as in assisted conception [11]. One of the major criticisms concerns the selection of the embryos according to certain genetic parameters and destruction of others. In fact, the selection of a few embryos for transfer from approximately a dozen available after hyperstimulation is a routine practice of IVF, the remaining embryos being either frozen or discarded. Such embryo selection is usually done routinely based on morphological criteria, which has the goal of identifying the embryos with highest developmental potential. PGD, on the other hand, allows improvement of the embryo selection, by applying genetic tests, which has shown that perfectly morphologically normal embryos may be chromosomally abnormal, and so destined to be lost during preimplantation and postimplantation development. As described in Chapter 5, approximately half of oocytes and embryos obtained from women of advanced reproductive age are chromosomally abnormal, suggesting that it might no longer be an acceptable practice to select embryos on morphological grounds. In other words, the advent of PGD is a natural evolution of assisted reproduction, allowing replacement of an almost "blind" selection of embryos on morphological grounds by chromosomal testing, to ensure the transfer of chromosomally normal embryos, improving significantly the chances of IVF patients to become pregnant. It may be hoped that the genetic testing of the oocytes and embryos may also be further extended for cytoplasmic abnormalities, which together with testing of nuclear abnormalities will in the future allow the identification of a single viable embryo for transfer, which will insure the highest possible efficiency of IVF, allowing a singleton unaffected pregnancy and birth of a healthy baby.

References

1. Vastag B. Merits of embryo screening debated. JAMA 2004;291:927–927.
2. Verlinsky Y, Munné S, Cohen J., et al. Over a decade of preimplantation genetic diagnosis experience: A multicenter report. Fertil Steril 2004;82:292–294.
3. Gleicher N. Editorial. J Assist Reprod Genet 2004;21:3.
4. Plachot M, Cohen J. Regulations for PGD in France. J Assist Reprod Genet 2004;21:5–6.
5. Menezo VJR, Frydman R, Frydman N. PGD in France. J Assist Reprod Genet 2004;21:7–9.
6. Hill B, Li M. What regulation for PGD? J Assist Reprod Genet 2004;21:11–13.
7. Feichtinger W. Preimplantation Diagnosis: A European clinician's point of view. J Assist Reprod Genet 2004;21:15–17.
8. Takeshita N, Kubo H. Regulating preimplantation genetic diagnosis: How to control PGD. J Assist Reprod Genet 2004;21:19–25.
9. Benagiano P, Gianaroli L. The new Italian IVF legislation. Reprod BioMed Online 2004;9:117–125.
10. PGDIS. Guidelines for good PGD practice. Reprod BioMed Online 2004;9:430–434.
11. Edwards RG. Ethics of PGD: Thoughts on the consequences of HLA typing in embryos. Reprod BioMed Online 2004;9:222–224.
12. Simpson JL. Celebrating preimplantation genetic diagnosis of p53 mutations in Li-Fraumeni syndrome. Reprod BioMed Online 2001;3:2–3.
13. Cram D. Preimplantation genetic diagnosis for familial cancer. Reprod BioMed Online 2001;3:3–4.
14. Towner D, Loewy RS. Ethics of preimplantation diagnosis for a woman destined to develop early-onset Alzheimer disease. JAMA 2002;287:1038–1040.
15. Damewood MD. Ethical implications of a new application of preimplantation diagnosis. JAMA 2001;285:3143–3144.
16. Fost NC. Conception for donation. JAMA 2004;291:2125–2126.
17. American Society of Reproductive Medicine (ASRM). Most Americans find genetic testing of embryos acceptable in conceiving a child who will donate tissue to save an older sibling. ASRM Bull May 6, 2004.
18. Verlinsky Y, Rechitsky S, Schoolcraft W, Strom C, Kuliev A. Preimplantation diagnosis for Fanconi anemia combined with HLA matching. JAMA 2001;285:3130–3133.
19. Verlinsky Y, Rechitsky S, Sharapova T, Morris R, Tharanissi M, Kuliev A. Preimplantation HLA typing. JAMA 2004;291:2079–2085.
20. Fasth A, Wahlstrom J. Preimplantation testing to produce an HLA-matched donor infant. JAMA 2004;292:803.
21. Wagner JE, Kahn JP, Wolf SM, Lipton JM. Preimplantation testing to produce an HLA-matched donor infant. JAMA 2004;292:803–804.
22. Orfali KA, Ohene-Abuakwa Y, Ball SE. Diamond Blackfan anemia in the UK: Clinical and genetic heterogeneity. Br J Haematol 2004;125:243–252.
23. Weimann JM, Johansson CB, Trejo A, Blau HM. Stable reprogrammed heterokaryons form spontaneously in Purkinje neurons after bone marrow transplantation. Nat Cell Biol 2003;5:959–966.
24. American Society of Reproductive Medicine (ASRM). Ethics Committee: Preconception gender selection

for nonmedical reasons. Fertil Steril 2001;75:861–864.

25. Robertson JA. Extending preimplantation genetic diagnosis: The ethical debate. Ethical issues in new uses of preimplantation genetic diagnosis. Hum Reprod 2003;18:465–471.
26. Malpani A, Malpani A, Modi D. The use of preimplantation genetic diagnosis in sex selection for family balancing in India. Reprod BioMed Online 2001;4:16–20.
27. Kilani Z, Haj Hassan L. Sex selection and preimplantation genetic diagnosis at the Farah Hospital. Reprod BioMed Online 2002;4:68–70.

Index

INDEX